Resistive and Reactive Circuits

OTHER BOOKS BY ALBERT P. MALVINO

Transistor Circuit Approximations
Electronic Instrumentation Fundamentals
Digital Principles and Applications (with D. Leach)
Electronic Principles
Experiments for Electronic Principles (with G. Johnson)

Resistive and Reactive Circuits

Albert Paul Malvino, Ph.D.

McGraw-Hill Book Company

NEW YORK	KUALA LUMPUR	PANAMA
ST. LOUIS	LONDON	RIO DE JANEIRO
SAN FRANCISCO	MEXICO	SINGAPORE
DÜSSELDORF	MONTREAL	SYDNEY
JOHANNESBURG	NEW DELHI	TORONTO

Library of Congress Cataloging in Publication Data

Malvino, Albert Paul.
  Resistive and reactive circuits.

1. Electric circuits. I. Title.
TK454.M24   621.3815'3   73-18271
ISBN 0-07-039856-9

*The editors for this book were Alan W. Lowe and
Cynthia Newby, the designer was Marsha Cohen, and
its production was supervised by James E. Lee and
Laurence Charnow. It was set in Palatino by
Progressive Typographers.
It was printed by The Murray Printing Company
and bound by Rand McNally & Company.*

To my first son or fourth daughter
and to my dear wife who is now convinced
I know nothing about mathematics.

With doubt and dismay you are smitten
You think there's no chance for you son?
Why the best books haven't been written
The best race hasn't been run!

Barton

Contents

Preface

The typical dc-ac textbook does a poor job of preparing the technician for modern electronic courses; too much time is wasted on magnetic circuits, intricate ac calculations, power-oriented topics, three-phase systems, etc. The topics that really count, such as Thevenin's theorem, superposition theorem, resistive ac circuits, etc., get a thin treatment, with almost no follow-up in succeeding chapters.

To select the right topics for this dc-ac textbook, I used *Electronic Principles* as my guide, because it covers the whole field of analog electronics including bipolars, FETs, ICs, op amps, etc. My reasoning was this: if a dc-ac textbook prepares a technician to study a modern book like *Electronic Principles*, then the technician has learned all the vital dc-ac topics needed for everyday electronics.

Part 1 of this book is about resistive circuits with dc or ac sources, the special case so prominent in today's electronics because of direct-coupled circuits. On completion of this first half of the book, the technician will be ready to begin a basic electronics course.

Part 2 of this book is about reactive circuits. Chaps. 9 through 12 cover transients and ac theory without trigonometry, complex numbers, and the classic phasor approach. Because of this, the reader can cover the first 12 chapters entirely with algebra. These first 12 chapters may be adequate for many programs that wish to cover practical dc-ac theory without complex numbers and phasors.

The final part of the book, Chaps. 13 through 18, requires a knowledge of trigonometry. Here you will find the extensive use of complex numbers and phasors that typifies in-depth ac analysis. We discuss complex numbers and phasors in detail; however, trigonometry is assumed as a prerequisite for Chaps. 13 through 18.

I think you will enjoy the format of this book. It includes two or three key questions at the beginning of longer topics; these questions are designed to focus your attention on what's important, to point you in the right direction before turning you loose. To make sure you understand the key ideas, multiple-choice tests are included throughout each chapter; the answers to these tests are at the very end of the chapter.

Many thanks to Lloyd Temes of Staten Island Community College, New York. He reviewed the manuscript and offered many excellent suggestions.

Albert Paul Malvino

PART 1. RESISTIVE CIRCUITS

1. Introduction

A *physical quantity* is anything you can detect with your senses or with measuring instruments. Electricity is easier to understand after learning basic ideas about physical quantities.

1-1. UNITS AND PREFIXES

In this section, your job is to learn

What are units of measure?
How are prefixes used?

Units of measure

Saying something has a length of 7 means nothing; but saying it has a length of 7 ft gives the exact size. To discuss any physical quantity, therefore, we must always specify a *number* and a *unit of measure,* an amount of the physical quantity used as a reference.

British units of measure are familiar in English-speaking countries. Length is measured in inches, feet, yards, miles, and so on. Time is measured in seconds, minutes, hours, days, etc. As a rule, British units are rarely used in electrical work, especially not in electronics.

Metric units of measure are standard in electronics. Table 1-1 lists some physical quantities, their metric units and abbreviations, and the equivalent amounts in British

TABLE 1-1. *METRIC UNITS*

QUANTITY	UNIT	ABBREVIATION	BRITISH EQUIVALENT
Length	meter	m	39.4 in
Force	newton	N	0.225 lb
Time	second	s	1 s
Temperature	degree Celsius	°C	1.8°F

units. As shown, length is measured in *meters* (m), where

$$1 \text{ m} = 39.4 \text{ in}$$

Force is measured in *newtons* (N), where

$$1 \text{ N} = 0.225 \text{ lb}$$

Time is measured in *seconds* (s), and temperature in degrees *Celsius* (°C).

Metric units are simpler than British units, because each physical quantity has only a single unit of measure. In British units, for instance, length is measured in inches, feet, yards, miles, etc. But in metric units, length is measured in meters only. This single unit of measure for each physical quantity simplifies calculations; answers for length always come out in meters; answers for time always come out in seconds; answers for force always come out in newtons; and so on.

Later chapters introduce other physical quantities and their metric units. For convenience, all units of measure found in this book are listed at the front of the book. You can refer to these from time to time as needed.

Milliunits

Milli (abbreviated m) stands for one-thousandth, 0.001, or 10^{-3}. When used as a prefix, it converts a unit into a *milliunit*, an amount that is one-thousandth of the unit. For instance, a millisecond is one-thousandth of a second. It's written as 1 ms, and it equals

$$1 \text{ ms} = 0.001 \text{ s} = 10^{-3} \text{ s}$$

To convert any amount into milliunits, move the decimal point three places to the right and replace 10^{-3} by m. For example, given

$$0.007 \text{ s}$$

move the decimal point three places to the right to get

$$7(10^{-3}) \text{ s}$$

Then replace (10^{-3}) by m to get

$$7 \text{ ms}$$

The final answer is read as seven milliseconds.

Here are more examples of millisecond conversion:

$$0.025 \text{ s} = 25(10^{-3}) \text{ s} = 25 \text{ ms}$$

$$0.175 \text{ s} = 175(10^{-3}) \text{ s} = 175 \text{ ms}$$

$$0.0004 \text{ s} = 0.4(10^{-3}) \text{ s} = 0.4 \text{ ms}$$

Microunits

Micro (abbreviated μ) stands for one-millionth, 0.000001, or 10^{-6}. When used as a prefix, it converts a unit into a *microunit,* an amount that is one-millionth of the unit. To convert any amount into microunits, move the decimal point six places to the right and replace 10^{-6} by μ.

As an example, given

$$0.000005 \text{ s}$$

move the decimal point six places to the right to get

$$5(10^{-6}) \text{ s}$$

Then replace 10^{-6} by μ to get

$$5 \ \mu\text{s}$$

This final answer is read as five microseconds.

Here are more examples:

$$0.0000003 \text{ s} = 0.3(10^{-6}) \text{ s} = 0.3 \ \mu\text{s}$$

$$0.00006 \text{ s} = 60(10^{-6}) \text{ s} = 60 \ \mu\text{s}$$

$$0.000175 \text{ s} = 175(10^{-6}) \text{ s} = 175 \ \mu\text{s}$$

Incidentally, μ (*mu*) is the twelfth letter of the Greek alphabet. Greek letters are used a great deal in electronics. For your convenience, the front of the book lists all the Greek letters used in this book. Refer to these as needed.

Prefixes

Table 1-2 shows all the prefixes used in electronics. We use these prefixes to indicate multiples or submultiples of a unit of measure. Learn each prefix, its abbreviation, and its equivalent power of 10.

EXAMPLE 1-1.
An *ohm* (abbreviated Ω, Greek letter *omega*) is a unit of measure discussed in the next chapter. Express 7000 Ω in kilohms.

TABLE 1-2. PREFIXES

PREFIX	ABBREVIATION	POWER OF 10
giga	G	10^9
mega	M	10^6
kilo	k	10^3
milli	m	10^{-3}
micro	μ	10^{-6}
nano	n	10^{-9}
pico	p	10^{-12}

SOLUTION.

$$7000 \ \Omega = 7(10^3) \ \Omega = 7 \ k\Omega$$

We moved the decimal point three places to the left and replaced 10^3 by k.

EXAMPLE 1-2.
Express 2,200,000 Ω in megohms.

SOLUTION.

$$2,200,000 \ \Omega = 2.2(10^6)\Omega = 2.2 \ M\Omega$$

Here we move the decimal point six places to the left and replace 10^6 by M.

EXAMPLE 1-3.
Convert 6.8 kΩ to ohms. Also, 2 MΩ to ohms.

SOLUTION.

$$6.8 \ k\Omega = 6.8(10^3) \ \Omega = 6800 \ \Omega$$

and $$2 \ M\Omega = 2(10^6) \ \Omega = 2,000,000 \ \Omega$$

Test 1-1   (answers at end of chapter)

1. Which of these is not a metric unit?
    (*a*) meter   (*b*) foot   (*c*) second   (*d*) newton .... ( )
2. Which of these does not belong?
    (*a*) inch   (*b*) foot   (*c*) yard   (*d*) meter .... ( )
3. Which of these is closest in meaning to unit of measure?
    (*a*) meter   (*b*) second   (*c*) reference   (*d*) prefix .... ( )
4. A force of 4 N is closest to
    (*a*) 0.5 lb   (*b*) 1 lb   (*c*) 2 lb   (*d*) 4 lb .... ( )

5. 5 μs equals
      (*a*) 0.005 s   (*b*) 0.000005 s   (*c*) 0.00005 s   (*d*) 0.05 s .... (   )
6. 2 ms is closest to
      (*a*) 0.04 s   (*b*) 0.0001 s   (*c*) 0.005 s   (*d*) 0.001 s .... (   )
7. 47 kΩ is closest to
      (*a*) 500 Ω (*b*) 5000 Ω (*c*) 50,000 Ω (*d*) 5 MΩ .... (   )

1-2. WHERE FORMULAS COME FROM

A *formula* is a compact mathematical summary of how quantities are related. You will see many formulas in this book. Unless you know where each one comes from, you will be confused and discouraged as they accumulate.

Fortunately, there are only three ways formulas come into existence. Knowing what they are makes all the difference when studying electricity. In fact, knowing the origin of a formula tells you why the formula is true. In what follows, therefore, be sure to learn the answers to

What is a defining formula?
An experimental formula?
A derived formula?

Defining formula

In Gaelic (the language of ancient Ireland), *fear* means man, *bean* stands for woman, and *paiste* is child. To understand Gaelic, you would have to memorize

$$fear = \text{man} \qquad\qquad (1\text{-}1a)$$

$$bean = \text{woman} \qquad\qquad (1\text{-}1b)$$

$$paiste = \text{child} \qquad\qquad (1\text{-}1c)$$

and so on.

In a similar way, to understand electricity you have to memorize the meaning of new words such as *current, voltage,* and *resistance.* But a hazy idea of these new words is not good enough. Your idea of *current* must be the same as my idea of current; your concept of *voltage* must duplicate my concept of voltage; your definition of *resistance* must be identical to my definition of resistance. The only way to get this precision is with *defining formulas.*

A defining formula is an equation between a quantity being defined or explained and other quantities already known. As an example, average speed is defined as the distance traveled divided by the time of travel. In symbols,

$$s = \frac{d}{t}$$

where s = average speed

$\quad d$ = distance

$\quad t$ = time

If a car travels 120 mi in 2 h, it has an average speed of

$$s = \frac{120 \text{ mi}}{2 \text{ h}} = 60 \text{ mi/h}$$

Where do defining formulas come from? They're invented; made up out of the blue. For instance, if we were trying to solve traffic problems, we might define *traffic* as the number of cars passing a point divided by the time during which they pass the point. In symbols, the defining formula may look like

$$\tau = \frac{n}{t} \qquad\qquad (1\text{-}2)$$

where τ = traffic

$\quad n$ = number of cars passing a point

$\quad t$ = time during which cars pass

(τ, *tau*, is the nineteenth letter of the Greek alphabet; see front of book.) With this defining formula, 40 cars passing a point in 2 min means the traffic is

$$\tau = \frac{n}{t} = \frac{40 \text{ cars}}{2 \text{ min}} = 20 \text{ cars/min}$$

Since we made up Eq. (1-2), there's no way to derive or prove it with other formulas. The same is true of all defining formulas; you can't prove them because someone made them up. Proof of a defining formula is meaningless, because defining formulas are like word equations, Eqs. (1-1a) through (1-1c). These word equations are inventions, made up at some point in time. Likewise, defining formulas were made up at some point in time.

So when you encounter a defining formula, don't look for proof; it doesn't exist. Your job is to *memorize* defining formulas much the same as you would memorize the words of a foreign language.

Experimental formula

An *experimental formula* is different; it summarizes a relation that has existed in nature since the beginning of time. Here's how an experimental formula is discovered. Someone notices a quantity seems to be related to other quantities in a way unknown up to that moment. If clever enough, the person can write a formula summarizing the relation. Then, by repeated experiments he may be able to prove the formula. If so, he has *discovered* an experimental formula.

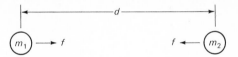

Figure 1-1. Gravitational
attraction of masses.

An example is Newton's law of gravitational attraction:

$$f = \frac{m_1 m_2}{Kd^2}$$

This says the force of attraction between two masses equals the product of the masses, divided by a constant times the square of the distance (Fig. 1-1).

Newton arrived at this formula by reason and guesswork; he did not derive it from other formulas. In fact, he was uncertain about the formula for 15 years until he was able to confirm it by astronomical observations. The only proof of Newton's law of gravitational attraction is that it agrees with results of experiments and observations.

Another example of an experimental formula is the way a light beam bounces off a mirror. Let θ stand for angle (θ is *theta*, the eighth letter of the Greek alphabet). Then,

$$\theta_i = \theta_r$$

where θ_i is the angle of incidence and θ_r is the angle of reflection of a ray of light on a polished surface (Fig. 1-2). This equal-angle formula is proved by repeated experiments or observations; it cannot be derived from other formulas.

To summarize, an experimental formula represents the discovery of a relation that already exists, whereas a defining formula is the invention of a new relation. Keep this difference in mind throughout the book.

Derived formula

If you add or subtract the same number from both sides of an equation, the equation is still valid. There are many other operations, such as multiplication, division, factoring, substitution, etc., that preserve the equality of both sides of an equation. For this reason, we can get all kinds of new formulas by mathematical juggling.

A *derived formula* is one you create with mathematics. This means you start with a set of formulas, and by different mathematical operations (factoring, substitution, etc.) you arrive at a new formula not in the original set of formulas. When this happens, you have come up with a derived formula.

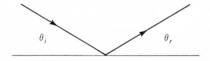

Figure 1-2. Angle of incidence
equals angle of reflection.

As a simple example, the defining formula for average speed is

$$s = \frac{d}{t}$$

Multiplying both sides by t gives

$$d = st$$

This is a derived formula; it says distance traveled equals the product of average speed and time. A car with an average speed of 65 mi/h will during a 2-h trip travel a distance of

$$d = st = 65 \times 2 = 130 \text{ mi}$$

As another example, suppose we have these two equations:

$$x = 3t$$

$$y = 6t$$

By rearranging the first equation,

$$t = \frac{x}{3}$$

Substituting this into the second equation gives

$$y = 6\frac{x}{3}$$

or

$$y = 2x$$

This derived formula says y equals two times the value of x, a relation implied in the original set of equations.

In summary, a derived formula is a relation buried in the original set of formulas; you need mathematics to dig it out. The information in a derived formula is already implied in the starting equations; you merely rearrange the equations until you can see the new information.

Summary

In studying electricity, you have to know what kind of formula you're dealing with; otherwise, you won't know what's going on. As an aid, the first time we use a formula we will indicate whether it is a defining, experimental, or derived formula.

To build your understanding of electricity on solid logical ground, always remember these distinctions:

1. Defining formulas are invented; they can't be proved experimentally or mathematically. The defined quantity appears for the first time in a formula; before this, no

other formula has contained the quantity being defined. (About 30 percent of the formulas in this book are defining formulas.)

2. Experimental formulas represent discoveries of existing relations; they are proved only by experiment or observation. The experimental formulas in this book have been proved so often that they are sometimes called *laws*. (Less than 10 percent of the formulas in this book are experimental formulas.)

3. Derived formulas are the only ones we can prove mathematically, that is, by rearranging other formulas. In deriving formulas, we can use any valid mathematical operation that preserves the equality of both sides of an equation. (More than 60 percent of the formulas in this book are derived formulas.)

EXAMPLE 1-4.

Liquid pressure p is defined as

$$p = \frac{f}{A}$$

where f = force of liquid on an area
A = area on which force is exerted

Derive a formula for force in terms of pressure and area.

SOLUTION.

Easy. Multiply both sides by A to get

$$f = pA$$

This derived formula says the force equals the pressure times the area.

EXAMPLE 1-5.

Suppose we are given this formula:

$$R = \frac{V}{I}$$

Derive a formula for V, and a formula for I.

SOLUTION.

Multiplying both sides by I gives

$$V = RI$$

This derived formula says V equals the product of R and I. Next, divide both sides by R to get

$$I = \frac{V}{R}$$

This derived formula says I equals the ratio of V to R.

math can prove a derived formula

Test 1-2

1. When Galileo dropped weights off the leaning tower of Pisa, he was looking for what kind of formula?
   (*a*) defining   (*b*) experimental   (*c*) derived   .... ( )
2. Which of the following formulas is unprovable?
   (*a*) defining   (*b*) experimental   (*c*) derived   .... ( )
3. Observation is to experimental formula as mathematics is to which of these formulas?
   (*a*) defining   (*b*) experimental   (*c*) derived   .... ( )
4. Which of these comes closest in meaning to defined?
   (*a*) unnatural   (*b*) unknown   (*c*) observed   (*d*) invented   .... ( )
5. These words can be rearranged to form a sentence: PROVE MATH CAN DERIVED A FORMULA. Is the sentence
   (*a*) true   (*b*) false   .... ( )

1-3.   EARLY DISCOVERIES

About 600 B.C. the Greeks discovered an unusual property of amber (a semiprecious stone). After being polished with fur, the amber attracted dust, lint, and other objects. This phenomenon became known as *electricity* (derived from *elektron,* Greek for amber). The rubbed amber was described as electrified or *charged.*

When two pieces of charged amber are near each other, they *repel* as shown in Fig. 1-3*a*. Early experimenters also found that glass could be charged by rubbing, and that two pieces of charged glass *repel* (Fig. 1-3*b*). Furthermore, they found that a piece of charged amber *attracts* a piece of charged glass (Fig. 1-3*c*). Based on these experiments, early workers concluded the charge on amber represented one kind of charge, and the charge on glass another kind.

Other charged objects acted either like charged amber or like charged glass. As a result, the first major conclusion about electricity was this:

1. *There are only two kinds of charge.*
2. *Like charges repel; unlike charges attract.*

1-4.   THE FLUID THEORY

Franklin (1750) made an outstanding contribution with his *fluid theory.* He visualized electricity as an invisible fluid; when a body has more than a normal amount of this fluid, it has a positive charge; if it has less than a normal share, it has a negative charge. As part of the fluid theory, Franklin assumed the charge on glass was positive and the charge on amber was negative.

The fluid theory was easy to visualize, and agreed with all experiments before the beginning of the twentieth century. Especially important, when a wire was connected

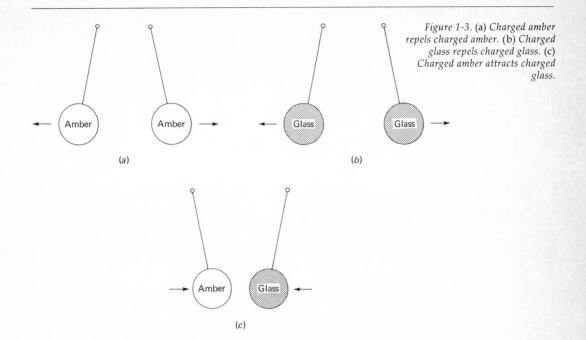

Figure 1-3. (a) Charged amber repels charged amber. (b) Charged glass repels charged glass. (c) Charged amber attracts charged glass.

between a positively charged body and a negatively charged body (see Fig. 1-4), the force of attraction *disappeared* after a few seconds. This implied the bodies had lost their charges. In terms of the widely held fluid theory, charges had *flowed* from positive (excess) to negative (deficiency). Flow of charge from positive to negative is called *conventional flow.*

From about 1750 to 1900 an immense number of concepts and formulas came into existence based on conventional flow.

1-5. COULOMB'S LAW

Coulomb (1785) discovered a formula for the force between charged bodies (see Fig. 1-5). By ingenious experiments, he proved that force is directly proportional to the

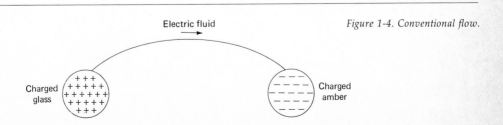

Figure 1-4. Conventional flow.

product of the charges and inversely proportional to the square of the distance between charges.[1] In other words, he proved this experimental formula:

$$f = \frac{Q_1 Q_2}{Kd^2} \qquad (1\text{-}3)$$

where f = force on each charge
Q_1 = charge on first body
Q_2 = charge on second body
K = a number that depends on the units of measure
d = distance between charges

This famous formula is called *Coulomb's law.*

As an example of how it works, suppose the two charges of Fig. 1-5 repel each other with a force of 10 N. If you halve the distance between these charges, the force will increase to 40 N. Or with the original distance, if you double one of the charges, the force will increase to 20 N.

The main things to remember about Coulomb's law are

1. Increasing either charge increases the force.
2. Bringing the charges closer together increases the force.

1-6.   ELECTRIC FIELD

To turn the pages of this book, something has to make physical contact with the pages. In general, to move anything, one body has to touch another. But there are three major exceptions: gravity, electricity, and magnetism. Somehow, masses, charges, and magnets interact without touching to produce gravitational, electrical, and magnetic forces. This interaction without touching is called *action at a distance.*

Our minds prefer to think in terms of physical contact. This section is about the *electric field,* an attempt to fill the empty space around a charge with something we can visualize. In particular, find the answers to

What is an electric field?
What is a constant field?

[1] Two quantities are directly proportional if doubling one doubles the other; they're inversely proportional if doubling one halves the other.

Figure 1-5. Coulomb force between charges.

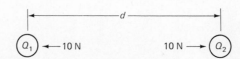

Basic idea

Figure 1-6*a* shows two positive charges, one stationary and the other movable. Because of the repulsion, the movable charge is pushed to the right. Likewise, in Fig. 1-6*b* the stationary charge pushes the movable charge upward. In general, a stationary positive charge pushes a movable positive charge radially outward (like spokes of a wheel). To indicate the direction of force, we can visualize the arrows shown in Fig. 1-6*c*.

In Fig. 1-6*c* the space around the positive charge is called a *field.* Because of the force, we say a *field of force* surrounds the positive charge. To distinguish this field from others, we call it an *electric field of force,* or simply an electric field. (Other kinds of fields are gravitational and magnetic fields.)

Negative charges also have electric fields. It's a custom always to draw arrows in the direction of force felt by a *positive charge.* This is why the field around the negative charge of Fig. 1-6*d* points inward; this is the direction of force felt by a positive charge.

The use of arrows for the electric field is a visual aid. Because the electric field is invisible, we draw or visualize lines of force around the charge. By looking at the lines of force, we can tell at a glance how other charges are affected. The field arrows of Fig. 1-6*c* indicate positive charges are pushed outward, and negative charges are pulled inward. On the other hand, the field arrows of Fig. 1-6*d* indicate that positive charges are pulled inward, and negative charges are pushed outward.

Figure 1-6. (a) *Movable charge pushed right.* (b) *Movable charge pushed upward.* (c) *Electric field around positive charge.* (d) *Electric field around negative charge.*

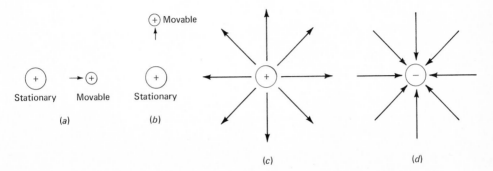

Constant-force field

Of special importance is the *constant electric field,* one whose force is everywhere constant or the same in value. Figure 1-7 shows a way to produce a constant electric field. The glass plate is positively charged, and the amber plate negatively charged. A positive charge Q between the plates is subject to a downward push from the upper plate and a downward pull from the lower plate.

Either experimentally or with advanced mathematics, it's possible to prove the same amount of force is exerted on a charge anywhere between the plates. If a movable charge is subject to a force of 2 N at point A, this charge will also be subject to a force of 2 N at point B and at point C. Because the force is constant in value, this special kind of field is called a constant electric field.

A simple explanation of constant force is this. Midway between the plates, the downward push equals the downward pull. When the positive charge is closer to the upper plate, the downward push is greater but the downward pull is less, and the sum of these two adds to the same total force as before. Likewise, when the positive charge is nearer the lower plate, there's more pull from the lower plate than push from the upper plate; the sum adds up to the same total force as before.

There are other ways to set up a constant electric field. When a wire is connected across a battery, the battery produces a constant electric field inside the wire because one battery terminal has a positive charge and the other terminal a negative charge. Any movable charges inside the wire feel a constant force. Because of this, the field can force charges to flow through the wire. More is said about this later.

Test 1-3

1. The earth attracts other bodies with mass. This is an example of
   (*a*) gravitational field   (*b*) magnetic field   (*c*) electric field   .... ( )
2. The electric field around a charge extends how far?
   (*a*) 1 mi   (*b*) 1 in   (*c*) infinity   (*d*) none of these   .... ( )

Figure 1-7. Constant electric field.

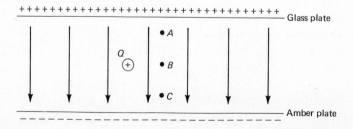

3. The direction of the electric field is always
   (*a*) away from the negative charge   (*b*) toward the positive charge
   (*c*) tangential to a radial line   (*d*) away from the positive charge .... ( )
4. In a constant electric field, which of these remains the same?
   (*a*) charge   (*b*) speed   (*c*) force   (*d*) gravity .... ( )
5. The field around a positive charge is always
   (*a*) constant   (*b*) stronger away from the charge   (*c*) outward
   (*d*) inward .... ( )
6. Electric field is to charge as gravitational field is to
   (*a*) mass   (*b*) charge   (*c*) magnet   (*d*) volume .... ( )

1-7. ELECTRON THEORY

In 1897, Thomson discovered a new atomic particle: the *electron*. The charge of an elec-tron is the same as the charge on rubbed amber, because the two charges repel. Since Franklin defined amber charge as negative, the electron charge automatically is neg-ative.

The planetary concept of matter is well known. Matter is made up of atoms. Each atom is a positively charged nucleus surrounded by orbiting electrons. The outward push of centrifugal force on each electron is exactly balanced by the inward pull of the nucleus. In this way, electrons travel in stable orbits, similar to the way planets move around the sun.

As you read this section, be on the lookout for answers to these questions

What is a conductor?
What is an insulator?

Conductors

A neutral atom of copper has 29 protons in the nucleus and 29 orbiting electrons. As shown in Fig. 1-8*a*, 2 electrons are in the first orbit, 8 in the second, 18 in the third, and 1 in the outer orbit.

The nucleus and the inner-orbit electrons are called the *core*. Because of this, we can visualize a copper atom as shown in Figure. 1-8*b*. The core has a net charge of +1 (29 protons and 28 electrons). This is why Fig. 1-8*c* shows the copper atom as a +1 charge orbited by a −1 charge.

The outer orbit of a copper atom is so large that the attraction between the core and the outer-orbit electron is very weak. Because of this, it's easy to dislodge the outer-orbit electron from the atom, leaving a positively charged core as shown in Fig. 1-8*d*.

Figure 1-9*a* represents a piece of copper wire; the circled plus signs are cores and the minus signs are outer-orbit electrons. The outer-orbit electrons are so weakly at-

Figure 1-8. (a) Copper atom.
(b) Copper core and outer-orbit
electron. (c) Positively charged
core and negatively charged
electron. (d) Charged atom.

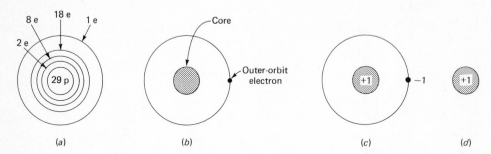

tracted by the cores that they can easily move from one core to another. This is the reason they are often called *free electrons.*

If we place positive charges on the left end of the wire and negative charges on the right end (Fig. 1-9b), a constant electric field will be inside the wire, pointing from left to right as shown. (A battery is one way to supply the positive and negative charges at each end of the wire.) The constant electric field will try to push cores right and free electrons left. The cores cannot move because they are held in position by other atoms; however, free electrons move easily to the left.

As the electrons move to the left, they leave positive cores at the right end of the wire. Free electrons can then enter the right end of the wire and go into orbit around the positive cores. At the same time, free electrons leave the left end of the wire. In this way, we can get a flow of free electrons through the wire. On the average, the wire remains electrically neutral; that is, the total number of free electrons in the wire approximately equals the total number of cores.

All metals and a few nonmetals are *conductors,* materials with enough free electrons to allow significant flow of charges. The best conductor is silver, then copper, then gold. These three metals have a single loosely held electron in the outer orbit.

Figure 1-9. (a) Piece of copper
wire. (b) Constant electric field
through copper wire.

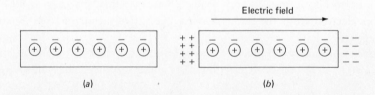

Insulators

At the other extreme are *insulators,* materials with almost no free electrons. When you apply a constant electric field, you get almost zero flow. Examples are bakelite, glass, rubber, plastics, etc.

Insulators are important. They prevent the flow of charges in undesired paths. For instance, copper wire is usually covered with an insulating layer such as varnish or plastic to keep free electrons inside the wire.

Test 1-4

1. A copper core attracts a free electron with
   (*a*) no force   (*b*) weak force   (*c*) great force .... ( )
2. A free electron tends to move in the direction of
   (*a*) the electric field   (*b*) opposite the electric field
   (*c*) toward the negative end of a conductor
   (*d*) away from the positive end of a conductor .... ( )
3. Conductor is to free electrons as pipe is to
   (*a*) atoms   (*b*) molecules   (*c*) pressure   (*d*) water .... ( )
4. Which of the following is the opposite of conductor?
   (*a*) broken pipe   (*b*) train   (*c*) wire   (*d*) insulator .... ( )
5. The following can be rearranged into a sentence: INSULATORS NO FREE ELEC-TRONS HAVE. Is the sentence
   (*a*) true   (*b*) false .... ( )

Insulators have no free Electrons

1-8. CONVENTIONAL VERSUS ELECTRON FLOW

The fluid theory says positive charges flow in the direction of the electric field, from the positive end of a conductor to the negative end (Fig. 1-10*a*). The discovery of electrons created a problem about direction of flow; since electrons have negative charges, they move exactly opposite conventional flow (Fig. 1-10*b*).

Figure 1-10. (a) Conventional flow is from positive to negative. (b) Electron flow is from negative to positive.

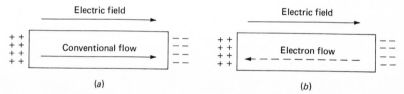

We are in a dilemma. On the one hand, many people refuse to change the defining, experimental, and derived formulas based on conventional flow; they argue that above the atomic level it makes no difference whether you imagine positive charges flowing in the direction of the electric field or negative charges flowing the opposite way. On the other hand, some people feel conventional flow is a complete myth; they argue all conventional concepts and formulas must be changed no matter how difficult the transition from conventional to electron flow.

You can escape the dilemma by learning the answers to

What is action flow?
What is reaction flow?

Action flow

Here's a simplified explanation of how charges flow in a wire. Free electrons are the only particles that move. In Fig. 1-11*a*, the electric field pushes all free electrons to the left. Each time a free electron leaves the wire, a positive core is left behind. Figure 1-11*b* shows a positive core at *A*, just after a free electron has left the wire. The free electron at *B* is subject to the attraction of core *A* and the push of the electric field; therefore, this free electron can move from *B* to *A*. When this happens, a positive core is left at *B* (see Fig. 1-11*c*).

Next, the free electron at *C* is subjected to the push of the electric field and the pull of core *B*; therefore, this free electron can go into orbit around core *B*. This leaves a positive core at *C* (Fig. 1-11*d*).

Reaction flow

Cores don't move, but *positive charge is transferred in the direction of the electric field.* Look at Figs. 1-11*b* through *d*; you can see that positive charge is being transferred in a direction opposite electron flow. The transferred positive charge eventually reaches the right end of the wire. When this happens, a free electron can enter the wire and go into orbit around the positive core. In this way, the wire remains neutral.

The flow of negative and positive charges in a wire is reminiscent of the action-reaction principle in physics. Each action (an electron moving from one atom to another) produces an equal and opposite reaction (transfer of positive charge). In other words, an electron flow in a wire produces an equal and opposite conventional flow. Because electron flow and conventional flow *coexist* in a wire, we are free to calculate the value of either; the other has the same magnitude but opposite direction.

This book emphasizes whatever kind of flow suits the discussion, usually electron flow for atomic-level explanations and conventional flow for circuit analysis. In marking diagrams a solid arrow is used for conventional flow (Fig. 1-10*a*) and a dashed arrow for electron flow (Fig. 1-10*b*).

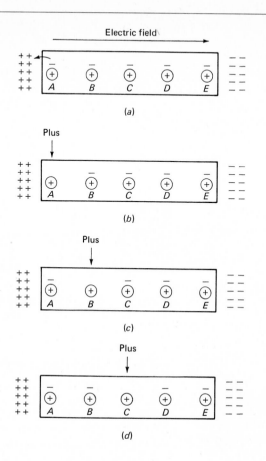

Figure 1-11. *Action flow is to left, and reaction flow is to right.*

Test 1-5

1. Action flow involves the motion of
   (a) atoms (b) molecules (c) electric field (d) electrons .... ( )
2. Reaction flow is to conventional flow as action flow is to
   (a) positive charge (b) electron flow (c) positive flow
   (d) none of these .... ( )
3. Which of these belongs least?
   (a) action flow (b) conventional flow (c) proton flow
   (d) electron flow .... ( )
4. When a free electron leaves a copper atom, the atom
   (a) moves in the opposite direction (b) becomes neutral
   (c) becomes positively charged (d) none of these .... ( )

SUMMARY OF FORMULAS

EXPERIMENTAL

$$f = \frac{Q_1 Q_2}{K d^2} \tag{1-3}$$

Problems

1-1. Which physical quantity does each of the following represent: 7 N, 25°C, 36 s, 5 mm, 2 ms.

1-2. Express 0.000007 s in microseconds.

1-3. How many kilometers does 2000 m equal?

1-4. Write each of the following as a power of 10:
   a. 5 ms
   b. 100 μs
   c. 2 MΩ
   d. 30 kΩ

1-5. Convert each of these as indicated:
   a. 5 mm to micrometers
   b. 8000 s to milliseconds
   c. 5 ms to microseconds
   d. 25 ps to seconds

1-6. Convert each of the following:
   a. $5(10^{-3})$ s to milliseconds
   b. 0.05 μs to picoseconds
   c. 0.5 ps to nanoseconds
   d. 0.0003 μs to nanoseconds

1-7. A volt (abbreviated V) is a unit of measure discussed in the next chapter. Convert each of these as indicated:
   a. 0.07 V to millivolts
   b. 4 mV to microvolts
   c. $2(10^{-5})$ V to millivolts
   d. 5000 mV to volts

1-8. Convert each of these:
   a. 60,000 Ω to kilohms
   b. 5,000,000 Ω to megohms
   c. 220 kΩ to ohms
   d. 7 MΩ to ohms

1-9. An ampere (abbreviated A) is a unit of measure discussed in the next chapter. Convert each of the following as indicated:

    a. 6 A to microamperes

    b. $3(10^{-2})$ A to milliamperes

    c. 80,000 mA to amperes

    d. 0.03 μA to nanoamperes

1-10. Given the following formula

$$I = \frac{Q}{t}$$

derive a formula for Q and a formula for t.

1-11. Suppose you are given

$$W = fd$$

What is a formula for f? For d?

1-12. Given

$$V = \frac{W}{Q}$$

derive a formula for W. Derive a formula for Q.

1-13. Suppose you have these two formulas:

$$P = VI$$

$$V = RI$$

Derive a new formula for P in terms of R and I. Derive a formula for P in terms of V and R.

1-14. Two charges repel each other with a force of 10 N. If the distance between the charges is cut in half, what is the force of repulsion?

1-15. Suppose a pair of unlike charges attracts with a force of 3 N. If both charges are doubled, what is the new force of attraction?

1-16. Two like charges repel with a force of 18 N. What is the value of force if

    a. One of the charges is tripled in value.

    b. The distance between the original charges is tripled.

1-17. If an atom has 41 protons in its nucleus and 39 orbiting electrons, which way does its electric field point?

ANSWERS TO TESTS

1-1. *b, d, c, b, b, d, c*

1-2. *b, a, c, d, a*

1-3. *a, c, d, c, c, a*

1-4. *b, b, d, d, b*

1-5. *d, b, c, c*

2. Three Electrical Quantities

Current, voltage, and *resistance* are the three basic electrical quantities. You have to understand them before you can understand how electric circuits work.

2-1. CURRENT

As you read this section, get the answers to

> *What is charge?*
> *How is current defined?*

Charge

Charge is an amount of electricity. The proton is the smallest positive charge in existence; the electron, the smallest negative charge. These two fundamental charges have the same magnitude, but are opposite in sign. Detecting the motion of a single proton or electron is difficult; this is why we need a larger unit of measure.

The *coulomb* (abbreviated C) is the metric unit of measure for charge. A coulomb is the charge on $6.24(10^{18})$ protons or electrons. So the defining formula for charge in coulombs is

$$Q = \frac{n}{6.24(10^{18})}$$

where Q = charge in coulombs
n = number of protons or electrons

As an example, the charge on $24(10^{18})$ electrons is

$$Q = \frac{24(10^{18})}{6.24(10^{18})} = 3.85 \text{ C}$$

The use of coulombs as the unit of measure for charge is standard. Almost everybody uses this basic unit of measure. For this reason, whenever you calculate the value of Q, the answer is always in coulombs; this is the only unit that's possible in the metric system.

Definition of current

In Fig. 2-1 the electric field forces electrons to pass through the cross section. The stronger the electric field, the greater the number of electrons that pass through this cross section in each second. When the field doesn't vary with time, the flow of electrons is steady, that is, the same number of electrons pass through the cross section during each second.

For the case of steady flow, *current* is defined as the number of coulombs that have passed through the cross section, divided by the time during which they have flowed. As a defining formula,

$$I = \frac{Q}{t} \tag{2-1}$$

where I = current
Q = charge
t = time

Current was originally known as *intensity* of flow; this is why the letter I appears in the defining formula.

Here are examples of calculating current. If 12 C pass through the cross section in 4 s, the current equals

$$I = \frac{Q}{t} = \frac{12 \text{ C}}{4 \text{ s}} = 3 \text{ C/s}$$

(Read the answer as three coulombs per second.) Or if 0.8 C flow through the cross section in 0.05 s, the current is

$$I = \frac{Q}{t} = \frac{0.8 \text{ C}}{0.05 \text{ s}} = 16 \text{ C/s}$$

Figure 2-1. Free electrons flowing through wire.

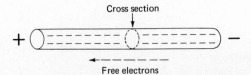

Cross section

Free electrons

Current is the same as *rate of flow*, because its numerical value tells us the number of coulombs that pass through the cross section during each second. In the first of the preceding examples, 3 C flow through the cross section during each second. In the second example, 16 C pass through the cross section during each second. The rate of flow is greater in the second case than in the first.

The ampere

The *ampere* (abbreviated A) is the metric unit of measure for current. It's defined as

$$1 \text{ ampere} = 1 \text{ coulomb per second}$$

or

$$1 \text{ A} = 1 \text{ C/s}$$

An ampere is the value of current when a single coulomb passes through the cross section during each second.

In the foregoing examples, we calculated currents of 3 C/s and 16 C/s. These can be written as

$$3 \text{ C/s} = 3 \times 1 \text{ C/s} = 3 \times 1 \text{ A} = 3 \text{ A}$$

and

$$16 \text{ C/s} = 16 \times 1 \text{ C/s} = 16 \times 1 \text{ A} = 16 \text{ A}$$

The use of ampere as the unit of measure for current is standard. Because of this, you always specify current in amperes rather than coulombs per second. Whenever Eq. (2-1) is used, Q is expressed in coulombs and t in seconds. For this reason, I is always in amperes. Therefore, whenever you calculate current, the answer is automatically in amperes. For instance, if 5 C flow through the cross section in 2 s, the current is

$$I = \frac{5}{2} = 2.5 \text{ A}$$

Chapter 7 discusses an *ammeter*, an instrument that measures current. When connected in an electric circuit, it indicates the number of coulombs passing through it during each second.

EXAMPLE 2-1.
$9(10^{18})$ electrons pass through a cross section in 5 s. How much charge is this in coulombs? What is the current?

SOLUTION.
With the defining formula for charge in coulombs, we get

$$\frac{9(10^{18})}{5}$$

$$Q = \frac{n}{6.24(10^{18})} = \frac{9(10^{18})}{6.24(10^{18})} = 1.44 \text{ C}$$

and Eq. (2-1) gives

$$I = \frac{Q}{t} = \frac{1.44}{5} = 0.288 \text{ A} = 288 \text{ mA}$$

(The answer is read either as 0.288 amperes or as 288 milliamperes.)

EXAMPLE 2-2.
10 C pass through a battery in 2 h. What does the current equal?

SOLUTION.
In electricity the basic unit of time is always the *second*. Therefore, whenever you calculate current, you always have to express the time in seconds. Since 1 h equals 3600 s,

$$I = \frac{Q}{t} = \frac{10}{2 \times 3600} = 0.00139 \text{ A} = 1.39 \text{ mA}$$

EXAMPLE 2-3.
The current in a transistor equals 5 mA.   How much charge passes through the transistor in 3 s?

SOLUTION.
Start with

$$I = \frac{Q}{t}$$

and multiply both sides by t to get

$$Q = It$$

This derived formula says charge equals current times time. Since $I = 5$ mA and $t = 3$ s,

$$Q = It = 0.005 \times 3 = 0.015 \text{ C}$$

Test 2-1   (answers at end of chapter)

1. Which of these comes closest in meaning to current?
    (*a*) charge   (*b*) force on charge   (*c*) flow   (*d*) rate of flow .... ( )
2. Which of these does not belong?
    (*a*) charge   (*b*) current   (*c*) rate of flow   (*d*) intensity of flow   ...... ( )
3. Electron is to current as bullet is to
    (*a*) target   (*b*) rate of fire   (*c*) speed   (*d*) gun .... ( )

4. Ampere means
    (*a*) rate of flow   (*b*) current   (*c*) coulomb per second
    (*d*) charges in motion .... ( )
5. $I = Q/t$ is defined for a flow that is
    (*a*) steady   (*b*) transient   (*c*) in a conductor   (*d*) in motion .... ( )

2-2. VOLTAGE

The current in a conductor depends on the force pushing free electrons; the greater the force, the larger the current. It is difficult to build instruments that measure this force; however, it is easy to build instruments that measure *voltage*. As will be proved, the force on the electrons is directly proportional to voltage. Therefore, measuring voltage is equivalent to measuring the force on the electrons.

To understand voltage, learn the answers to

> *What is work?*
> *What is potential energy?*
> *How is voltage defined?*

Work

In physics, work is done when a force moves through a distance. If the force is constant, the work done is defined as the product of force and distance. As a defining formula,

$$W = fd \tag{2-2}$$

where W = work
      f = force
      d = distance

As an example, if 3 N push a body through a distance of 4 m, the work done equals

$$W = fd = 3 \text{ N} \times 4 \text{ m} = 12 \text{ N·m}$$

(Read the answer as twelve newton meters.)

In the metric system, the *joule* (abbreviated J) is the unit of measure for work. It's defined as

$$1 \text{ joule} = 1 \text{ newton meter}$$

or

$$1 \text{ J} = 1 \text{ N·m}$$

A joule is the amount of work done when a force of 1 N moves through a distance of 1 m.

In the foregoing example, we calculated a work of 12 N·m. This can be written as

$$12 \text{ N·m} = 12 \times 1 \text{ N·m} = 12 \times 1 \text{ J} = 12 \text{ J}$$

Whenever force is in newtons and distance in meters, the work is in joules.

In electricity, force is always expressed in newtons, and distance in meters; there-fore, whenever you use Eq. (2-2), the answer is automatically in joules. If 5 N move through a distance of 8 m, the work equals

$$W = 5 \times 8 = 40 \text{ J}$$

Examples of work

Here are some subtle examples of work. Figure 2-2*a* shows a force of 3 N pushing down on a positive charge. Since the electric field is constant between the plates, the force is constant even though the charge is moved. To lift the charge from *B* to *A*, we have to exert a force of 3 N through a distance of 4 m. Therefore, the work done in moving the charge from *B* to *A* equals

$$W = fd = 3 \times 4 = 12 \text{ J}$$

As another example, Fig. 2-2*b* shows a conductor with a length of 8 m. If each coulomb of free electrons in the wire is subject to a force of 2 N, the work done in moving each coulomb through the wire equals

$$W = fd = 2 \times 8 = 16 \text{ J}$$

Potential energy

A body has *energy* if it can do work on another body. This is why energy is defined as the work a body can do. The amount of energy a body has equals the work

Figure 2-2. (a) Work done lifting positive charge against opposing electric field. (b) Work done moving charges through a wire.

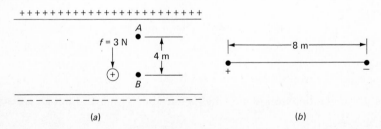

(a)   (b)

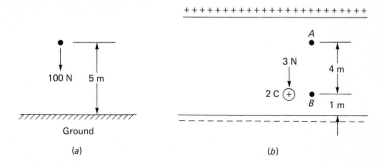

Figure 2-3. (a) *Potential energy of a mass.* (b) *Potential energy of a charge.*

it can do. A body with 12 J of energy can do 12 J of work; one with 5 J of energy can do only 5 J of work.

When a body loses some of its energy, it has done work on another body. The change in energy equals the work done. For instance, if a body initially has 20 J of energy and later has only 6 J of energy, it has done 14 J of work.

Potential energy is the energy a body has because of its position in a field. Figure 2-3a shows a body weighing 100 N. If released, this body can push down with a force of 100 N and move through a distance of 5 m before reaching the ground. Therefore, it can do 500 J of work on another body in its path. Because of this, the elevated body has a potential energy of 500 J with respect to the ground.

As another example, Fig. 2-3b shows a positive charge of 2 C. The electric field pushes down on this charge with a constant force of 3 N. If released, the charge can exert a downward force of 3 N through a distance of 1 m before reaching the lower plate. Therefore, the 2-C charge has a potential energy of

$$W = fd = 3 \times 1 = 3 \text{ J}$$

with respect to the lower plate.

Suppose the charge of Fig. 2-3b were located at point A instead of B. Then, it could exert a downward force of 3 N through a distance of 5 m before reaching the lower plate. In this case, its potential energy with respect to the lower plate would be

$$W = fd = 3 \times 5 = 15 \text{ J}$$

The *difference in potential energy* between two points equals the work the charge can do when it moves from one point to the other. In Fig. 2-3b, if the charge were at A and it moved to B, it would exert a force of 3 N through a distance of 4 m; therefore, the difference in potential energy between A and B is

$$W = fd = 3 \times 4 = 12 \text{ J}$$

Definition of voltage

The *voltage* between two points is defined as the difference in potential energy divided by the charge. In symbols, the defining formula is

$$V = \frac{W}{Q} \qquad (2\text{-}3)$$

where V = voltage
W = work or difference in potential energy
Q = charge

We've already calculated a difference in potential energy of 12 J between A and B in Fig. 2-3b. Therefore, the voltage between these two points equals

$$V = \frac{W}{Q} = \frac{12 \text{ J}}{2 \text{ C}} = 6 \text{ J/C}$$

(Read the answer as six joules per coulomb.)

The numerical value of voltage tells us the number of joules of work that each coulomb of charge can do when it moves from one point to another. In Fig. 2-3b, each coulomb moving from point A to point B can do 6 J of work.

The volt

The *volt* (abbreviated V) is the metric unit of measure for voltage. It's defined as

1 volt = 1 joule per coulomb

or

$$1 \text{ V} = 1 \text{ J/C}$$

A volt is the amount of voltage that exists between two points when each coulomb can do a joule of work as it moves between the two points.

In Fig. 2-3b, we calculated a voltage of 6 J/C. This may be written as

$$6 \text{ J/C} = 6 \times 1 \text{ J/C} = 6 \times 1 \text{ V} = 6 \text{ V}$$

The volt is always used as the unit of measure for voltage. Since we always use joules for work and coulombs for charge, Eq. (2-3) always gives the answer in volts. If 4 C can do 60 joules of work moving between two points, the voltage between the points is

$$V = \frac{W}{Q} = \frac{60}{4} = 15 \text{ V}$$

Electric field and voltage

Electric field and voltage always *coexist;* if one is present, so too is the other. When a conductor has an electric field through it, there's a voltage between the ends of the conductor. Conversely, if you measure a voltage between the ends of a conductor, you will know there's an electric field through it.

Batteries

A *battery* stores *chemical energy* associated with the bonds holding molecules together. When being used, the battery changes this chemical energy into electric potential energy by forcing an excess of free electrons to appear at one of the battery terminals and a deficiency of free electrons at the other terminal. The charge difference means an electric field exists, which implies a voltage. Because of its voltage, a battery is an *energy source.*

The voltage of a battery equals the work each coulomb of charge can do, the difference in potential energy per coulomb between the battery terminals. Each coulomb of free electrons in a 1.5-V flashlight battery can do 1.5 J of work when it travels from the negative to the positive terminal. Similarly, each coulomb of electrons in a 12-V car battery can do 12 J of work when it moves from the negative to the positive terminal.

Figure 2-4a shows the *schematic symbol* for a battery. The upper terminal has a deficiency of free electrons, and therefore is positive; the lower terminal has an excess of free electrons, and is negative. As a memory aid, the short horizontal line at the bottom of the symbol looks like a minus sign; this represents the negative end.

When we connect a conductor across a battery, free electrons flow from the negative terminal through the conductor to the positive terminal as shown in Fig. 2-4b. In terms of electric field, the battery sets up an electric field through the conductor; this field forces free electrons in the conductor to move toward the positive terminal.

Force is proportional to voltage

The greater the voltage between two points, the greater the force on movable charges. We can prove this as follows. In Fig. 2-5, the voltage between A and B is

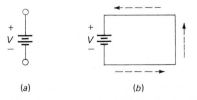

(a)   (b)

Figure 2-4. (a) Schematic symbol for battery. (b) Battery produces electron flow in wire.

Figure 2-5. Force on charge is proportional to voltage.

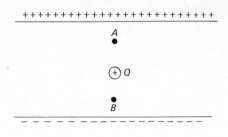

Figure 2-5. Force on charge is proportional to voltage.

$$V = \frac{W}{Q}$$

which can be rewritten as

$$V = \frac{fd}{Q} \qquad (2\text{-}4)$$

where f is the force exerted on charge Q, and d is the distance between A and B.

We will not use Eq. (2-4) in any calculations. The only importance is this: it says voltage and force are directly proportional. In other words, doubling the voltage between two points automatically doubles the force felt by a movable charge between the points. Because of this direct proportion between voltage and force, voltage is sometimes called *electromotive force* (emf).

In a later chapter, we discuss the *voltmeter*, an instrument for measuring the voltage between two points in an electric circuit. When measuring voltage, we are indirectly measuring the force on charges. If a conductor has 6 V across it (Fig. 2-6a) and another has 3 V (Fig. 2-6b), free electrons inside the first conductor are subject to twice as much force as those inside the second conductor.

Polarity and magnitude

To indicate the *polarity* (direction) of voltage, we use plus and minus signs. Figure 2-6a is an example; it shows the polarity of voltage, because point A has a plus sign and point B has a minus sign.

The *magnitude* of voltage is the value of voltage without giving the polarity. Figure 2-6c is an example of this; it shows a value of 6 V, without indicating which end is positive and which is negative.

Figure 2-6. Voltage has polarity and magnitude.

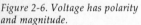

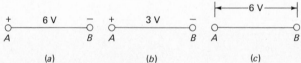

(a)   (b)   (c)

EXAMPLE 2-4.
When a charge of 7 C passes through a conductor, the charge loses a total of 21 J of potential energy. What is the voltage between the ends of the conductor?

SOLUTION.
This is equivalent to saying the difference of potential energy is 21 J and the charge is 7 C. With Eq. (2-3),

$$V = \frac{W}{Q} = \frac{21}{7} = 3 \text{ V}$$

EXAMPLE 2-5.
A charge of 5 C loses 75 J of potential energy passing through a transistor. What is the voltage across the transistor?

SOLUTION.
Easy. With Eq. (2-3),

$$V = \frac{W}{Q} = \frac{75}{5} = 15 \text{ V}$$

EXAMPLE 2-6.
The voltage across a transistor is increased from 10 to 30 V. What happens to the force on the charges inside the transistor?

SOLUTION.
Force is directly proportional to voltage. Therefore, the charges inside the transistor are subject to three times as much force when the voltage is increased from 10 to 30 V.

EXAMPLE 2-7.
A 12-V car battery has a current of 200 A for 10 s while starting an engine. How much work does the battery do?

SOLUTION.
First, get the total number of coulombs that flow during the 10-s interval. Since

$$I = \frac{Q}{t}$$

multiply both sides by t to get

$$Q = It$$

With $I = 200$ A and $t = 10$ s,

$$Q = It = 200 \times 10 = 2000 \text{ C}$$

Second, calculate the work. Since

$$V = \frac{W}{Q}$$

multiply both sides by Q to get

$$W = QV$$

With $Q = 2000$ C and $V = 12$ V,

$$W = QV = 2000 \times 12 = 24{,}000 \text{ J}$$

How much work is this in British units? One joule is approximately 0.75 ft·lb; therefore 24,000 J is approximately 18,000 ft·lb. This is the amount of work done when lifting 1 ton through a distance of 9 ft. We see, therefore, that a car battery stores a remarkable amount of energy.

Test 2-2

1. A body with potential energy must
   (*a*) have motion   (*b*) have weight   (*c*) be in a field
   (*d*) be doing work .... ( )
2. Voltage cannot exist without
   (*a*) electric field (*b*) current   (*c*) work   (*d*) none .... ( )
3. Current is to ampere as voltage is to
   (*a*) coulomb   (*b*) volt   (*c*) work   (*d*) potential energy .... ( )
4. In this series, what comes next?
              AMPERE, COULOMB, SECOND; VOLT, JOULE, _____
   (*a*) ohm   (*b*) current   (*c*) second   (*d*) coulomb .... ( )
5. The following words can be rearranged to form a sentence: ENERGY STORES BAT-
   TERY A CHEMICAL. The sentence is
   (*a*) true   (*b*) false .... ( )
6. Voltage and force are
   (*a*) the same   (*b*) inversely proportional   (*c*) directly proportional
   (*d*) unrelated .... ( )
7. Which does not belong?
   (*a*) polarity   (*b*) voltage   (*c*) magnitude   (*d*) current .... ( )

2-3. RESISTANCE

Resistance is another basic electrical quantity. The first things to learn about it are

> *How is resistance defined?*
> *What is an ohm?*

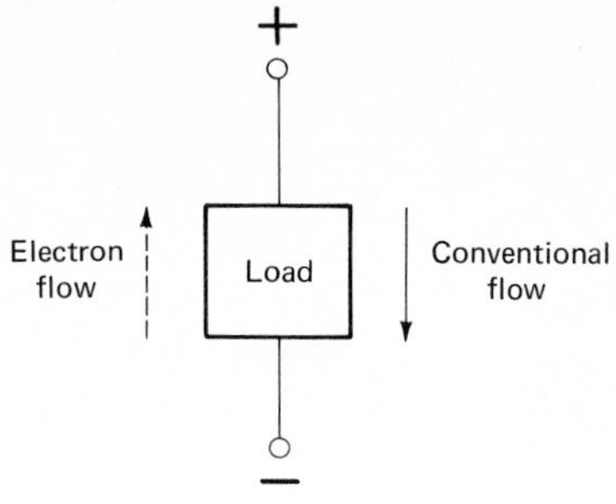

Figure 2-7. Flow directions in a
load.

Loads

A *load* is a device that normally is connected across a source of energy like a battery. Examples of loads are light bulbs, electric motors, electric heaters, etc. In a load free electrons always flow from the negative to the positive end, as shown in Fig. 2-7. Conversely, conventional flow is always from the positive to the negative end.

Definition of resistance

All loads have an important property called *resistance,* the opposition to flow. If charges can move easily through a load, it has low resistance; if not, it has high resistance. The resistance of a load is defined as the voltage across the load divided by the current through the load. As a defining formula,

$$R = \frac{V}{I} \tag{2-5}$$

where R = resistance
$\quad\quad V$ = voltage
$\quad\quad I$ = current

Here are examples. If the voltage across a load is 12 V and the current through it is 2 A, the resistance equals

$$R = \frac{V}{I} = \frac{12 \text{ V}}{2 \text{ A}} = 6 \text{ V/A}$$

(Read this as six volts per ampere.) Or if the voltage is 18 V and the current is 6 A, the resistance is

$$R = \frac{V}{I} = \frac{18 \text{ V}}{6 \text{ A}} = 3 \text{ V/A}$$

The ohm

The *ohm* (abbreviated Ω, Greek letter *omega*) is the metric unit of measure for resistance. It's defined as

$$1 \text{ ohm} = 1 \text{ volt per ampere}$$

or

$$1 \,\Omega = 1 \text{ V/A}$$

An ohm is the value of resistance when a load has a volt across it and an ampere through it.

In the preceding examples, we calculated resistances of 6 V/A and 3 V/A. These may be rewritten as

$$6 \text{ V/A} = 6 \times 1 \text{ V/A} = 6 \times 1 \,\Omega = 6 \,\Omega$$

and

$$3 \text{ V/A} = 3 \times 1 \text{ V/A} = 3 \times 1 \,\Omega = 3 \,\Omega$$

Since the use of ohms is standard in electrical work, you always specify resistance in ohms rather than in volts per ampere.

Whenever Eq. (2-5) is used, voltage is always in volts, and current is always in amperes; therefore, any calculations for resistance automatically result in ohms. For this reason, we can streamline calculations by leaving out the volts and amperes, and expressing the answer in ohms. For instance, a load with 5 V and 2 A has a resistance of

$$R = \frac{5}{2} = 2.5 \,\Omega$$

Chapter 7 discusses the *ohmmeter*, an instrument that measures resistance. When you connect a load across an ohmmeter, you get the resistance in ohms.

EXAMPLE 2-8.
A light bulb has 120 V across it and 2 A through it. What is its resistance?

SOLUTION.

$$R = \frac{V}{I} = \frac{120}{2} = 60 \,\Omega$$

EXAMPLE 2-9.
An electric clothes dryer has 230 V across it and 10 A through it. What is its resistance?

SOLUTION.

$$R = \frac{V}{I} = \frac{230}{10} = 23 \,\Omega$$

EXAMPLE 2-10.

Example 2-7 was about a 12-V battery producing 200 A while starting an engine. What is the resistance of the load on the battery?

SOLUTION.

$$R = \frac{V}{I} = \frac{12}{200} = 0.06 \ \Omega$$

EXAMPLE 2-11.

A transistor radio uses a 9-V battery for its energy source. If the current equals 10 mA, what is the resistance of the load on the battery?

SOLUTION.

$$R = \frac{V}{I} = \frac{9}{0.01} = 900 \ \Omega$$

Test 2-3

1. $R = V/I$ is which kind of formula?
    (a) defining   (b) experimental   (c) derived .... ( )
2. Which is closest in meaning to resistance?
    (a) high voltage   (b) low current   (c) ease of flow
    (d) opposition to flow .... ( )
3. Given

    RESISTANCE, VOLTAGE, _____, VOLT

    What belongs in the blank space?
    (a) ampere   (b) coulomb   (c) ohm   (d) work .... ( )

2-4. OHM'S LAW

Ohm made an important discovery (1846). He found that voltage and current were directly proportional for a given load. For instance, a typical set of voltage-current values for the load of Fig. 2-8a may look like

V	I
3 V	1 A
6 V	2 A
9 V	3 A

The important thing to notice is the direct proportion between V and I. Each increase in V is accompanied by the same *percent change* in I; doubling V doubles I; increasing V by 50 percent increases I by 50 percent, and so on.

In the following discussion, dig out the answers to

What is Ohm's law?
What is a linear resistance?
What are three ways to write Ohm's law?

Graphical meaning of Ohm's law

Figure 2-8*b* shows the graph of the preceding voltage-current values with *V* as the independent variable and *I* as the dependent variable. This linear graph emphasizes the *direction proportion* between *V* and *I*. Any change in *V* produces the same percent change in *I*.

When Ohm graphed the voltage-current values of different loads, he got different lines (see Fig. 2-8*c*). From this, he concluded voltage and current are directly proportional in a given load.

Whenever two quantities are directly proportional, their ratio is a constant. In Fig. 2-8*b*, *V* and *I* are directly proportional; therefore,

$$\frac{V}{I} = \text{a constant}$$

Since resistance is defined as the ratio *V/I*, Ohm concluded *the resistance of a load is constant*. A 100-Ω load has the same resistance no matter what the voltage and current.

Figure 2-8*b* implies the idea of constant resistance. At the first point, $V = 3$ V and $I = 1$ A; therefore,

$$R = \frac{V}{I} = \frac{3}{1} = 3 \ \Omega$$

At the second point, $V = 6$ V and $I = 2$ A; so

$$R = \frac{V}{I} = \frac{6}{2} = 3 \ \Omega$$

Figure 2-8. Illustrating Ohm's law.

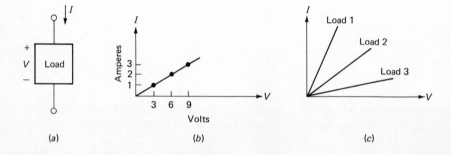

(a) (b) (c)

And at the third point,

$$R = \frac{V}{I} = \frac{9}{3} = 3 \ \Omega$$

In summary, Ohm's law says the resistance of a load is constant. In other words, its V/I ratio does not change even though the voltage and current change.

Exceptions to Ohm's law

Ohm's law is not universal; there are exceptions. For example, if the current in a load is excessively large, the resistance will change. For very large currents, the temperature of the conductor increases; this changes the resistance of the conductor. For this reason, Ohm's law holds only when the *self-heating* of a conductor is negligibly small. We discuss temperature effects later in the chapter.

Even at the same temperature there are some loads that don't obey Ohm's law, that is, their V/I ratio is not constant. From here on, any load that obeys Ohm's law is called a *linear resistance;* the word "linear" reminds us the graph of I versus V is a straight line.

Three ways to write Ohm's law

When R is constant, we can write Ohm's law as

$$R = \frac{V}{I} \qquad\qquad (2\text{-}6a)$$

Multiplying both sides by I gives an alternative form of Ohm's law:

$$V = RI \qquad\qquad (2\text{-}6b)$$

This says the voltage across a load equals the resistance times the current. Dividing both sides of Eq. (2-6b) by R gives another form of Ohm's law:

$$I = \frac{V}{R} \qquad\qquad (2\text{-}6c)$$

This means the current through a load equals the voltage divided by the resistance.

EXAMPLE 2-12.
Figure 2-9a shows the schematic symbol for a resistance. Calculate the resistance.

SOLUTION.
Since $V = 10$ V and $I = 2$ mA,

$$R = \frac{V}{I} = \frac{10}{0.002} = 5000 \ \Omega = 5 \ k\Omega$$

Figure 2-9. Applying Ohm's law.

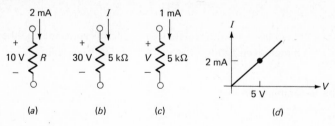

(a)   (b)   (c)   (d)

EXAMPLE 2-13.
Figure 2-9*b* shows the resistance calculated in the preceding example. What is the new value of current?

SOLUTION.
The voltage has been increased to 30 V. In this case, the current equals

$$I = \frac{V}{R} = \frac{30}{5000} = 0.006 \text{ A} = 6 \text{ mA}$$

EXAMPLE 2-14.
Figure 2-9*c* shows the resistance of the preceding example, but this time the current is 1 mA. What does the voltage equal?

SOLUTION.
With Eq. (2-6*b*),

$$V = RI = 5000(0.001) = 5 \text{ V}$$

EXAMPLE 2-15.
A *curve tracer* is an electronic instrument that shows the graph of *I* versus *V*. Suppose we connect a load to a curve tracer and see the display of Fig. 2-9*d*. What is the resistance of the load?

SOLUTION.
The given coordinates are $V = 5$ V and $I = 2$ mA. Therefore,

$$R = \frac{V}{I} = \frac{5}{0.002} = 2500 \ \Omega = 2.5 \text{ k}\Omega$$

Incidentally, if you see a display like Fig. 2-9*d*, you will know right away the load is a linear resistance. Nonlinear resistances have nonlinear graphs.

2-8-14-16
chapter 2

Test 2-4

1. When two quantities are directly proportional, their ratio is
   (*a*) directly proportional   (*b*) inversely proportional   (*c*) a constant
   (*d*) a variable .... ( )
2. Ohm's law means the ratio of voltage to current equals
   (*a*) the resistance   (*b*) a constant   (*c*) a variable
   (*d*) a certain number of ohms .... ( )
3. The following words can be rearranged to form a sentence: OHM'S LAW OBEY
   ALL LOADS. The sentence is
   (*a*) true   (*b*) false .... ( )
4. Which of the following belongs least in the group?
   (*a*) linear   (*b*) Ohm's law   (*c*) constant resistance   (*d*) nonlinear ... ( )
5. Ohm's law is an example of which kind of formula?
   (*a*) defining   (*b*) experimental   (*c*) derived .... ( )
6. Which of these is not a form of Ohm's law?
   (*a*) $I = V/R$ (*b*) $V = RI$ (*c*) $R = V/I$ (*d*) $Q = It$ .... ( )

2-5. MORE ABOUT RESISTANCE

Besides what you learned in the preceding section, you must be able to answer these
questions

What causes resistance?
How do length and area affect resistance?
What does resistivity mean?

Nature of resistance

The voltage across a load forces free electrons to move toward the positive end. But
these free electrons do not move in perfectly straight lines; rather, each electron follows
a zigzag path similar to Fig. 2-10.

*Figure 2-10. Irregular path of
electron moving through
conductor.*

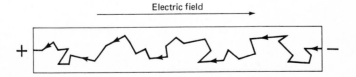

Why? To begin with, the electric field in Fig. 2-10 pushes each electron to the left. In the process, the electron gains speed. Sooner or later, this high-speed electron collides with an atom in its path. When this happens, the electron bounces off the atom and loses some of its speed; in turn, the atom vibrates slightly faster. After each collision, the electron again accelerates, only to collide with another atom in its path.

Resistance depends on these two atomic-level effects:

1. The number of free electrons in a load.
2. The number of collisions each electron has.

The more free electrons in a load, the smaller the resistance of the load. On the other hand, the more collisions each electron has getting through the load, the greater the resistance of the load.

Resistance formula

Starting with the defining formula

$$R = \frac{V}{I}$$

plus formulas for the number of electrons and collisions, it is possible to derive this formula for resistance:

$$R = \rho \frac{l}{A} \tag{2-7}$$

where ρ = a number that depends on atomic structure
  l = length
  A = area of cross section

This equation says resistance is directly proportional to the length of the load and inversely proportional to the cross-sectional area. Given the ρ, l, and A of a conductor, we can calculate its resistance.

Equation (2-7) makes sense. Resistance is directly proportional to length, because more collisions occur in a longer conductor. For instance, in Figs. 2-11a and b, each

Figure 2-11. Resistance depends on length and cross-sectional area.

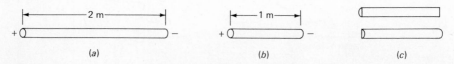

(a)   (b)   (c)

electron enters the right end, travels through the conductor, and leaves the left end. But each electron in Fig. 2-11a has approximately twice as many collisions as each electron in Fig. 2-11b. For this reason, the conductor on the left has double the resistance of the conductor in the center.

Similarly, the smaller the cross section, the fewer the number of free electrons that can pass through it in a given time. As an example, if we cut a conductor in half as shown in Fig. 2-11c, half as many free electrons can pass through the smaller cross section per second. This means each half of Fig. 2-11c has double the resistance of the original conductor.

Resistivity

In Eq. (2-7), ρ is called the *resistivity*; its value depends on the atomic structure of the load. For metric units of measure, the resistivity of copper is

$$\rho_{\text{cop}} = 1.72(10^{-8})$$

With this value, we can calculate the resistance of copper conductors. For instance, if a piece of copper wire is 10 m long and has a cross-section area of $5(10^{-7})$ m², it has a resistance of

$$R = \rho \frac{l}{A} = 1.72(10^{-8}) \frac{10}{5(10^{-7})}$$
$$= 0.344 \ \Omega$$

Table 2-1 shows resistivities for other materials. As we see, silver has the lowest resistivity; copper is next, and then gold. Tungsten, commonly used in electric light bulbs, has about three times the resistivity of copper. Carbon, often used in making resistances, has a resistivity about 800 times that of copper. The last three materials are insulators; notice the tremendous difference between the resistivity of copper and bakelite.

TABLE 2-1.   RESISTIVITY OF COMMON MATERIALS

MATERIAL	ρ^*
Silver	$1.59(10^{-8})$
Copper	$1.72(10^{-8})$
Gold	$2.44(10^{-8})$
Aluminum	$2.83(10^{-8})$
Tungsten	$5.5(10^{-8})$
Carbon	$1.4(10^{-5})$
Bakelite	10^{10}
Glass	10^{12}
Mica	10^{15}

* These values of ρ are for length in meters and area in square meters.

Resistance is directly proportional to ρ

In Eq. (2-7), R is directly proportional to ρ. Therefore, when two conductors have the same length and cross-sectional area, the ratio of their resistances equals the ratio of their resistivities. In symbols,

$$\frac{R_1}{R_2} = \frac{\rho_1}{\rho_2}$$

Equivalently,

$$R_1 = \frac{\rho_1}{\rho_2} R_2 \tag{2-8}$$

This is a useful formula. With two conductors of the same length and area, we can calculate the resistance of either, given the resistance of the other (see Example 2-18).

EXAMPLE 2-16.
Most radio and TV receivers include a *chassis*, a metal frame on which electronic components are mounted. Figure 2-12*a* shows part of an aluminum chassis. Calculate the resistance between the left and right ends of this conductor.

SOLUTION.
The length of the conductor is 1 m. The area of the cross section is

$$A = 0.5 \times 0.002 = 0.001 \text{ m}^2$$

Using the resistivity of aluminum given in Table 2-1,

$$R = \rho \frac{l}{A} = 2.83(10^{-8}) \frac{1}{0.001}$$

$$= 2.83(10^{-5}) \ \Omega = 28.3 \ \mu\Omega$$

This extremely small resistance means the chassis is a good conducting path.

EXAMPLE 2-17.
Figure 2-12*b* shows a *thick-film* resistance made by bonding resistive material to an insulator. Figure 2-12*c* shows the cross section. If $\rho = 0.06$, $l = 0.1$ m, and $A = 3(10^{-6})$ m^2, what is the resistance from point A to point B?

SOLUTION.
With Eq. (2-7),

$$R = \rho \frac{l}{A} = 0.06 \frac{0.1}{3(10^{-6})} = 2000 \ \Omega = 2 \text{ k}\Omega$$

By changing the length and area, a large range of resistance values can be made with this thick-film approach.

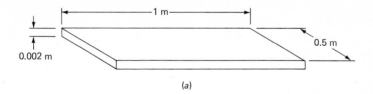

Figure 2-12. (a) *Chassis.* (b) and
(c) *Thick-film resistance.*

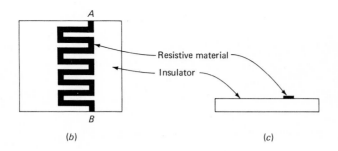

EXAMPLE 2-18.
A piece of copper wire has a resistance of 0.2 Ω. What resistance does a piece of silver wire of the same length and area have? A piece of gold wire?

SOLUTION.
With Eq. (2-8) and the resistivities of Table 2-1, silver wire results in

$$R = \frac{\rho_{\text{sil}}}{\rho_{\text{cop}}} R_{\text{cop}} = \frac{1.59}{1.72} \, 0.2 = 0.185 \ \Omega$$

And gold wire gives

$$R = \frac{\rho_{\text{gold}}}{\rho_{\text{cop}}} R_{\text{cop}} = \frac{2.44}{1.72} \, 0.2 = 0.284 \ \Omega$$

Test 2-5

1. Resistance increases if the number of
    (*a*) atoms increases   (*b*) free electrons increases
    (*c*) collisions increases   (*d*) none of these .... ( )
2. If the diameter is the same, the maximum number of collisions occurs in a length of
    (*a*) 2 in   (*b*) 4 in   (*c*) 1 yd   (*d*) 1 m .... ( )
3. Which of these belongs least?
    (*a*) voltage   (*b*) length   (*c*) area   (*d*) resistance .... ( )

4. Resistance always increases if
   (*a*) length and area increase   (*b*) length decreases and area increases
   (*c*) length stays the same and area increases
   (*d*) length increases and area stays the same   .... ( )
5. Resistivity is
   (*a*) directly proportional to length   (*b*) directly proportional to area
   (*c*) same as V/I (*d*) dependent on material .... ( )

2-6. AMERICAN WIRE GAGE

American wire gage (AWG) sizes are the standard diameters of commercially available wire. Table 2-2 shows AWG sizes commonly used in electronic equipment. First, notice an increase in the AWG number means a smaller diameter. Second, the resistance for 1000 ft of copper wire is given; this value of resistance is at 20°C (68°F), the temperature of a comfortably warm room.

TABLE 2-2. *AWG SIZES AND RESISTANCES*

AWG NUMBER	DIAMETER, IN	OHMS PER 1000 FT*
16	0.051	4.02
18	0.040	6.38
20	0.032	10.2
22	0.025	16.1
24	0.020	25.7
26	0.016	40.8
28	0.013	64.9
30	0.010	103
32	0.008	164

* Copper wire at 20°C (68°F).

AWG-22 wire is especially important because it is used a great deal for wiring radio, TV, and other typical electronic equipment. As shown in Table 2-2, the resistance of AWG-22 copper wire is approximately 16 Ω per 1000 ft.

Since resistance is directly proportional to length, we can calculate the resistance of other lengths of copper wire by using

$$R = \frac{l}{1000} R_{\text{table}} \qquad (2\text{-}9)$$

where l = length, feet
R_{table} = resistance in Table 2-2

EXAMPLE 2-19.
What is the resistance of 4 ft of AWG-22 copper wire? Of 3 in?

SOLUTION.
With Eq. (2-9), 4 ft of AWG-22 copper wire has a resistance of

$$R = \frac{4}{1000} \, 16.1 = 0.064 \; \Omega$$

And 3 in of AWG-22 copper wire has a resistance of

$$R = \frac{0.25}{1000} \, 16.1 = 0.004 \; \Omega$$

EXAMPLE 2-20.
What is the resistance of 350 ft of AWG-16 aluminum wire?

SOLUTION.
Do this in two steps. First, work out the resistance of 350 ft of copper wire. Second, convert this to the resistance of aluminum wire of the same length.
    With Eq. (2-9), 350 ft of AWG-16 copper wire has a resistance of

$$R = \frac{l}{1000} \, R_{\text{table}} = \frac{350}{1000} \, 4.02 = 1.41 \; \Omega$$

With Eq. (2-8), 350 ft of AWG-16 aluminum wire has a resistance of

$$R = \frac{\rho_{\text{alum}}}{\rho_{\text{cop}}} \, R_{\text{cop}} = \frac{2.83}{1.72} \, 1.41 = 2.32 \; \Omega$$

2-7. RESISTORS

A *resistor* is a load whose resistance is the most important property. (Two other properties discussed later are capacitance and inductance.) In this section, you are to find the answers to

> *What are three types of resistors?*
> *What is the color code?*
> *What are the standard resistor values?*

Types of resistors

The simplest way to make a resistor is to wrap wire on a core to get a desired resistance value. For example, if you wrap 1000 ft of AWG-22 copper wire on a cylindrical core, you get a resistance of 16.1 Ω. Using other lengths and AWG numbers, you can get a wide range of resistance values. Resistors of this type are called *wirewound* resistors.
    By mixing powdered carbon and a powdered insulator, we can produce different resistance values. The powdered mixture can then be solidified with a suitable

bonding material. Resistors of this type are called *carbon-composition* resistors. They are commercially available in values from less than 10 Ω to more than 10 MΩ.

By spraying a thin layer of metal on a glass rod, we get a *metal-film* resistor. Resistors of this type are highly accurate and stable in value, typically within 1 percent of the value printed on the resistor body.

The color code

Resistors with axial leads like Fig. 2-13a are usually marked with colored stripes to indicate the value of resistance. The two stripes nearest one end give the first and second digits, the third stripe is the number of zeros, and the fourth stripe is the *tolerance* (maximum percent error). For instance, a 1500-Ω resistor with a tolerance of ±5 percent is related to its stripes as follows:

First stripe	Second stripe	Third stripe	Fourth stripe
↓	↓	↓	↓
1	5	00	±5 percent

Table 2-3 shows the *color code* for resistors; each color is assigned a specific value. After you have memorized the code, you automatically know that black stands for 0, brown for 1, red for 2, and so on. Also notice that gold represents a tolerance of ±5 percent, silver is ±10 percent, and no fourth stripe means ±20 percent.

Here are examples of how to read resistance values. Figure 2-13b shows a resistor with these color stripes: brown, green, red, silver. With Table 2-3, this decodes as follows:

Brown	Green	Red	Silver
↓	↓	↓	↓
1	5	00	±10 percent

Figure 2-13. Color-coded resistors.

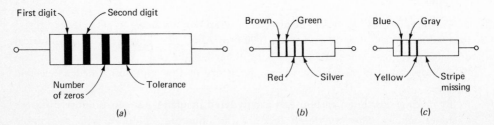

| (a) | (b) | (c) |

TABLE 2-3. *COLOR CODE*

COLOR	VALUE	TOLERANCE, PERCENT
Black	0	
Brown	1	
Red	2	
Orange	3	
Yellow	4	
Green	5	
Blue	6	
Violet	7	
Gray	8	
White	9	
Gold		±5
Silver		±10
None		±20

So, it's a 1500-Ω resistor with a tolerance of ±10 percent. Similarly, the resistor of Fig. 2-13c is coded with blue, gray, and yellow stripes; the fourth stripe is missing. It decodes as

Blue	Gray	Yellow	None
↓	↓	↓	↓
6	8	0000	±20 percent

Therefore, it has a value of 680,000 Ω with a ±20 percent tolerance.

Standard values

Resistors are manufactued in *standard values,* a set of preferred values used in typical electronics work. Table 2-4 shows the first two digits for standard values available in each tolerance. For instance, in 20 percent tolerance you can get

10 Ω, 15 Ω, 22 Ω, 33 Ω, 47 Ω, 68 Ω

100 Ω, 150 Ω, 220 Ω, 330 Ω, 470 Ω, 680 Ω

1 kΩ, 1.5 kΩ, 2.2 kΩ, 3.3 kΩ, 4.7 kΩ, 6.8 kΩ

and so on up to 10 MΩ. Also notice that the better the tolerance, the more resistance values to choose from.

Test 2-6

1. In the resistor color code, white stands for
    (*a*) 2   (*b*) 5   (*c*) 7   (*d*) 9 .... (   )

TABLE 2-4.   STANDARD RESISTOR VALUES

20% TOLERANCE	10% TOLERANCE	5% TOLERANCE
10	10	10
		11
	12	12
		13
15	15	15
		16
	18	18
		20
22	22	22
		24
	27	27
		30
33	33	33
		36
	39	39
		43
47	47	47
		51
	56	56
		62
68	68	68
		75
	82	82
		91

2. A red stripe on a resistor can mean
    (*a*) two zeros   (*b*) four zeros   (*c*) six zeros   (*d*) nine zeros .... ( )
3. Which of these belongs least?
    (*a*) the fourth stripe   (*b*) number of zeros   (*c*) tolerance
    (*d*) maximum per error .... ( )
4. Which of the following values cannot be color-coded?
    (*a*) 240 Ω   (*b*) 5.6 kΩ   (*c*) 8200 Ω   (*d*) 333,000 Ω .... ( )
5. If the third stripe is black, the resistance value must fall in the range of
    (*a*) 10 to 91 Ω   (*b*) 100 to 910 Ω   (*c*) 1 kΩ to 9.1 kΩ
    (*d*) 10 kΩ to 91 kΩ .... ( )

2-8. TEMPERATURE COEFFICIENT

Resistance changes with temperature. Briefly, here is the reason. The atoms of any material are held together by chemical bonds (ionic or covalent). You can visualize these atoms and bonds as shown in Fig. 2-14. At absolute zero temperature (−273°C), the atoms are stationary; above this temperature the atoms vibrate back and forth

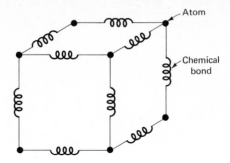

Figure 2-14. Atoms and the bonds between them.

Atom

Chemical bond

about the positions shown. If you visualize all the atoms of Fig. 2-14 springing back and forth randomly, you will have a rough idea of what goes on inside any material above absolute zero temperature.[1]

When free electrons flow through a load, they collide with atoms. The greater the number of collisions, the greater the resistance. As the surrounding temperature rises, the atoms vibrate faster and the number of collisions increases. (This idea is similar to an electric fan. You can easily pass an object through the vanes when the fan is off, but not when it is on.) Therefore, as a general rule, the resistance of most materials increases with temperature.

The *temperature coefficient* α of a material is the percent change in resistance with temperature. For instance, the temperature coefficient of copper is

$$\alpha = 0.4 \text{ percent per degree rise}$$

or equivalently,

$$\alpha = 4 \text{ percent for each } 10° \text{ rise}$$

Therefore, if a copper conductor has a resistance of 100 Ω at 20 °C, it has the following resistances:

100 Ω at 20°C
104 Ω at 30°C
108 Ω at 40°C
112 Ω at 50°C
etc.

Table 2-5 shows temperature coefficient of some common materials. All coefficients are positive except for carbon. With carbon resistors, the value of resistance *decreases*

[1] Heat energy is related to the random vibrations of these atoms. When you touch a hot body, the heat you feel is the vibration of these atoms.

TABLE 2-5. TEMPERATURE COEFFICIENTS

MATERIAL	α FOR A 10° RISE, %
Silver	4
Copper	4
Aluminum	4
Tungsten	5
Carbon	−0.5
Manganin	0
Constantan	0

approximately 0.5 percent for each 10° rise. A carbon resistor with 1000 Ω at 20°C becomes 995 Ω at 30°C, 990 Ω at 40°C, 985 Ω at 50°C, and so on.

The values given in Table 2-5 are rough approximations intended only for estimating resistance changes. Accurate values depend on many factors, such as the purity of the material, the insulating cover, etc. (Engineering handbooks and manufacturers' catalogs give precise temperature coefficients and conditions.)

Manganin and constantan are alloys of other metals; their main asset is a temperature coefficient of approximately zero. This is useful in wirewound resistors because it prevents the resistance from changing too much with temperature.

SUMMARY OF FORMULAS

DEFINING

$$I = \frac{Q}{t} \quad \text{(steady flow)} \tag{2-1}$$

$$W = fd \quad \text{(constant force)} \tag{2-2}$$

$$V = \frac{W}{Q} \tag{2-3}$$

$$R = \frac{V}{I} \tag{2-5}$$

EXPERIMENTAL

$$\frac{V}{I} = \text{constant} \quad \text{(Ohm's law)}$$

DERIVED

$$V = RI \qquad (2\text{-}6b)$$

$$I = \frac{V}{R} \qquad (2\text{-}6c)$$

$$R = \rho\, \frac{l}{A} \qquad (2\text{-}7)$$

$$R_1 = \frac{\rho_1}{\rho_2}\, R_2 \qquad (2\text{-}8)$$

Problems

2-1.   In half an hour, 18,000 C flow through an electric clothes dryer. What is the current?

2-2.   The battery of a small transistor radio supplies 5.4 C during each 10-min period of time. What does the current equal?

2-3.   The current through a soldering iron equals 0.5 A. How many coulombs does this represent during each second?

2-4.   Figure 2-15a shows the face of an ammeter. The reading is in milliamperes and represents the current through a transistor. How many coulombs pass through the transistor in 1 s? In 1 h?

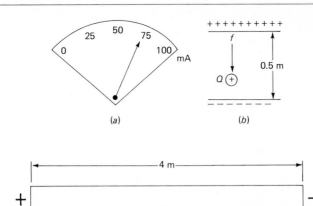

Figure 2-15.

2-5. The constant electric field of Fig. 2-15b exerts a downward force of 6 N on the positive charge shown. How much work has to be done to lift the charge from the lower plate to the upper plate?

2-6. If the positive charge of Fig. 2-15b is subject to a downward push of 10 N, how much work can this charge do if it falls from the upper to the lower plate?

2-7. The force on the positive charge of Fig. 2-15b is 12 N. What is the potential energy of this charge when it is located at the upper plate? What is the potential energy of this charge when it is halfway between the plates?

2-8. In Fig. 2-15b, the charge equals 8 C and the downward force is 48 N. What is the voltage between the two plates?

2-9. A negative charge of 5 C has 35 J more potential energy at the right end of Fig. 2-15c than at the left end. What is the voltage across the conductor?

2-10. Each 0.5 C passing through a flashlight bulb loses 1.5 J of potential energy. What is the voltage across the bulb?

2-11. Suppose you measure 15 V across a transistor. How much potential energy does each coulomb lose when it passes through the transistor? Each 20 C?

2-12. Each free electron in Fig. 2-15c is subject to a force of $8(10^{-20})$ N when the voltage across the conductor equals 2 V. What does the force equal when
  a. The voltage is doubled
  b. $V = 6$ V
  c. $V = 9$ V

2-13. The load of Fig. 2-16a has 5 V across it and 0.02 A through it. What is the resistance of the load?

2-14. In Fig. 2-16b the load has 12 V across it and a current of 3 μA (microamperes). How much resistance does the load have?

2-15. The voltage to the typical household is 115 V. If the current equals 50 A, what is the resistance?

2-16. An electric motor has 120 V across it and a current of 8 A through it. What is its resistance?

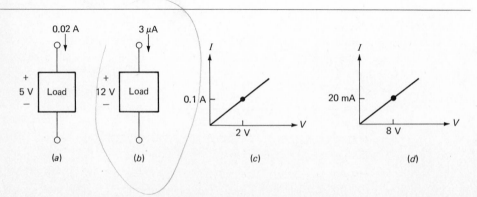

Figure 2-16.

8-17-16

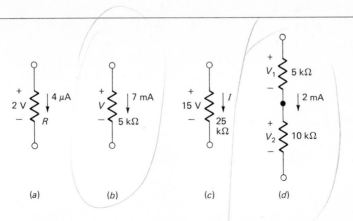

Figure 2-17.

(a)   (b)   (c)   (d)

2-17. Your skin has resistance. The value depends on its thickness and moisture. Suppose the voltage across a sample of skin is 1.5 V and a current of 5 μA results. What is the resistance?

2-18. In Fig. 2-16c, what is the current corresponding to 4 V? 8 V?

2-19. If we have 16 V across the load whose graph is shown in Fig. 2-16d, how much current is there through the load?

2-20. Figure 2-16c shows the voltage-current graph of a load. If we measure a current of 0.3 A through the load, what is the voltage across the load?

2-21. What is the resistance of a load whose voltage-current graph looks like Fig. 2-16c? Like Fig. 2-16d?

2-22. What resistance does the load of Fig. 2-17a have?

2-23. What is the voltage across the load shown in Fig. 2-17b?

2-24. How much current is there through the resistor of Fig. 2-17c?

2-25. The current through both resistors of Fig. 2-17d is 2 mA. Calculate the voltage across each resistor.

2-26. The resistance in Fig. 2-17a is a linear resistance. How much current is there if we change the voltage to 14 V?

2-27. If we triple the current in Fig. 2-17d, what are the new voltages?

2-28. The voltage across the resistor of Fig. 2-17c is changed from 15 to 45 V. What is the new value of the current?

2-29. The current in Fig. 2-17d is doubled. What are the new values of V_1 and V_2?

2-30. A piece of copper wire has a resistance of 0.05 Ω. If we double the length of this wire and keep the diameter the same, what is the new resistance?

2-31. A piece of copper wire 3 m in length has a resistance of 0.001 Ω. How much resistance does an equal length of copper wire have if the *diameter* is reduced by a factor of 2?

2-32. A conductor has a resistivity of 0.02, a length of 0.3 m, and a cross-sectional area of $5(10^{-7})$ m². Calculate the resistance.

Figure 2-18.

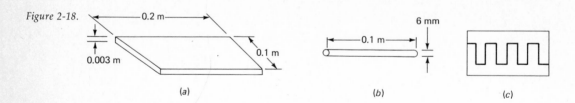

(a) (b) (c)

2-33. Figure 2-18a shows part of an aluminum chassis.
  a. Calculate the resistance from the left to the right end.
  b. Calculate the resistance from the front to the back end.

2-34. The cylindrical rod in Fig. 2-18b is made out of copper. The diameter is 6 mm (millimeters) and the length is 0.1 m. What is the resistance? If the rod were made out of glass, what would its resistance be?

2-35. If you break a light bulb and look at the filament, you will find a very fine piece of tungsten wire wound into a shape like a spring. Suppose the stretched-out length of the filament is 1.5 m and the cross-sectional area is 10^{-10} m². What is the resistance of the filament?

2-36. The length of the thick-film resistor in Fig. 2-18c is 0.15 m, and the cross-sectional area is $2(10^{-6})$ m². Calculate the resistance for a resistivity of 0.04.

2.37. How much resistance does 6 in of AWG-16 copper wire have? Of AWG-32 copper wire?

2-38. What resistance does 3 ft of AWG-24 silver wire have? Of aluminum wire? A mica rod of the same length and diameter?

2-39. A carbon resistor has a value of 5000 Ω at 20°C. What is its resistance at 50°C? At 0°C?

2-40. An aluminum conductor has a resistance of 10 Ω at 20°C. What resistance does it have at −30°C? At 100°C?

2-41. The temperature coefficients given in Table 2-5 are valid over several hundred degrees Celsius. For changes beyond this, the coefficients may change significantly. Nevertheless, in this problem we are going to use a temperature coefficient of 5 percent for tungsten no matter how large the temperature change. In other words, the resistivity of tungsten will increase 5 percent for a 10° rise, 50 percent for a 100° rise, 500 percent for a 1000° rise, and so forth.

When off, a light bulb has a resistance of 10 Ω. But when on, its resistance equals 100 Ω. The large increase in resistance takes place because of the tremendous increase in temperature necessary to make the filament incandescent. If the temperature of the tungsten filament is 20°C when off, what is it when the power is on?

2-42. The thick-film resistor of Fig. 2-18c has a temperature coefficient of 50 ppm (parts per million) per degree; this is equivalent to 0.005 percent per degree rise, or 0.05 percent per 10° rise. If the resistance is 1000 Ω at 20°C, what is it at 100°C?

ANSWERS TO TESTS

2-1. *d, a, b, c, a*
2-2. *c, a, b, d, a, c, d*
2-3. *a, d, c*
2-4. *c, b, b, d, b, d*
2-5. *c, d, a, d, d*
2-6. *d, a, b, d, a*

3. Series and Parallel Circuits

This chapter introduces two experimental formulas and uses them to analyze *series* and *parallel* circuits.

3-1. SERIES CIRCUIT

An electric circuit is a connection of sources and loads with one or more paths for charge flow. The first things to learn about circuits are

> *What is a series circuit?*
> *How does a switch work?*

Definition

A *series circuit* is one with only a single path for charges to flow through. Figure 3-1a is an example. Conventional flow is from the positive battery terminal through both resistors to the negative battery terminal. (Equivalently, electron flow is the opposite way.) The charges in Fig. 3-1a have only a single path they can follow in getting from one battery terminal to the other; this is why the circuit is classified as a series circuit.

Figure 3-1b is another example of a series circuit. Even though the circuit has two sources and four resistors, the charges still have only a single path they can follow when moving from one battery terminal to another. In other words, any electron that passes through R_4 must pass through R_3, R_2, and R_1.

Figure 3-1. Series circuits.

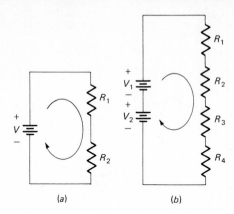

(a)   (b)

The switch

The principle behind a *switch* is simple enough; it is used to open or close the path for charge flow. Figure 3-2 shows an open switch; the path for charge flow is broken, and the current is zero. If we close the switch, however, the circuit becomes a series circuit, and charges can flow from one battery terminal to the other.

EXAMPLE 3-1.
Some Christmas tree lights are wired in series. Imagine 12 light bulbs in series across a source. What happens if one of the bulbs burns out?

SOLUTION.
If a bulb burns out, its filament opens. This interrupts the path for charge flow. Because all bulbs are in series, all the light bulbs go out.

Test 3-1

1. Series is the opposite of
    (*a*) circuit   (*b*) one path   (*c*) more than one path
    (*d*) several sources .... ( )

Figure 3-2. Circuit with switch.

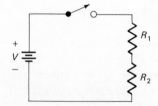

2. A series circuit always has one
   (a) voltage   (b) path   (c) source   (d) load .... ( )
3. Which of these belongs least?
   (a) one path   (b) one source   (c) series .... ( )

3-2. PARALLEL CIRCUIT

Figure 3-3a shows a *parallel circuit,* one whose loads are connected across the same voltage. Now there are two paths charges can flow through in getting from one battery terminal to the other. In the discussion to follow, find the answers to

> *What is an equipotential point?*
> *Why is the voltage the same for all loads?*

Parallel resistances

Figure 3-3b is another example of a parallel circuit; this time five resistors are connected across the source. Conventional flow into point A will split into two components; some of the charges flow down through R_1, and the rest flow to point B. At point B, the flow again divides. In this way, each of the five resistors gets a share of the total flow. After the charges have passed through the resistors, they combine and flow into the negative battery terminal as shown.

Connecting wires

The resistances of connecting wires are almost always small enough to neglect in electronic circuits. For instance, in Fig. 3-3b the wire between point A and point B is

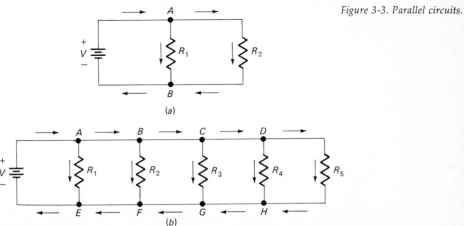

Figure 3-3. Parallel circuits.

short and has a resistance much less than 1 Ω. Typical loads in electronic circuits have resistances much greater than 1 Ω. For this reason, we can approximate the resistance of connecting wires as zero.

Equipotential points

Whenever the resistance between two points is zero, the voltage between the points is zero. This is true because

$$V = RI = 0 \times I = 0$$

No matter what the current, zero resistance results in zero voltage.

In Fig. 3-3b if we treat the resistance of connecting wires as zero, there is zero voltage between points A and B, points B and C, and so on. Because of this, charges have the same potential energy anywhere along the upper path. Whenever charges have the same potential energy along a path, the entire path is called an *equipotential point*. In Fig. 3-3b, this means the tops of all resistors are connected to an equipotential point.

Similarly, charges anywhere along the E-F-G-H path have the same potential energy. Therefore, the bottoms of all the resistors connect to another equipotential point. Because of this, Fig. 3-3b has two equipotential points. In fact, any parallel circuit has only *two equipotential points*.

One voltage

In a parallel circuit the same voltage appears across all loads, because these loads are connected between the same pair of equipotential points. If 5 V is across R_1 in Fig. 3-3b, 5 V must also be across R_2, R_3, and so on. The word "parallel" therefore is synonymous with "one voltage." A parallel circuit is a one-voltage circuit because the same voltage is across each load.

Since each load in a parallel circuit has the same voltage across it, you can remove one of the loads without disturbing the current in other loads. Most Christmas tree lights are wired in parallel, so that if one bulb burns out, the others stay on.

Test 3-2

1. A parallel circuit always has how many equipotential points?
   (*a*) one   (*b*) two   (*c*) three   (*d*) more than three .... ( )
2. The voltage is the same across all loads in a parallel circuit because the circuit has only two
   (*a*) paths   (*b*) equipotential points   (*c*) loads   (*d*) currents .... ( )
3. Series is to parallel as one path is to
   (*a*) one path   (*b*) two paths   (*c*) one voltage
   (*d*) none of these .... ( )

4. The following words can be rearranged to form a sentence: ONE IMPLIES PARAL-
   LEL VOLTAGE. Is the sentence
   (a) true   (b) false .... ( )
5. Which belongs least?
   (a) one voltage   (b) two equipotential points   (c) parallel
   (d) series .... ( )

3-3. KIRCHHOFF'S CURRENT LAW

Before the twentieth century, electric fluid was widely used because the electron had
not yet been discovered. In this preelectron era, Kirchhoff found two important experimental formulas; one for current, the other for voltage.

  This section is about one of the laws Kirchhoff discovered. Your goal is to learn the
answers to:

What is Kirchhoff's current law?
Why is the current law always true?

What it says

  Kirchhoff's *current law* says the sum of currents into a point equals the sum of currents out of the point. For instance, in Fig. 3-4a charges flow into and out of point A as
shown. The current into A equals I_1, and the current out of A is I_2. Kirchhoff's current
law says

$$I_1 = I_2$$

  Another example. In Fig. 3-4b charges flow into point A, where they split and
travel along two different paths. The current into A equals I_1, and the sum of currents
out of A is $I_2 + I_3$. In this case, Kirchhoff's current law says

$$I_1 = I_2 + I_3$$

  As still another example, in Fig. 3-4c the sum of currents into point A equals $I_1 + I_2$;
the sum of currents out of A is $I_3 + I_4$. Kirchhoff's current law tells us

$$I_1 + I_2 = I_3 + I_4$$

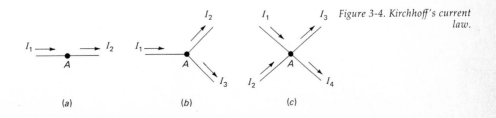

Figure 3-4. Kirchhoff's current law.

(a)   (b)   (c)

We can summarize Kirchhoff's current law by the following formula:

$$\Sigma \text{ currents in} = \Sigma \text{ currents out} \qquad (3\text{-}1)$$

where the Greek letter Σ (*sigma*) is shorthand for "the sum of." Equation (3-1) says the sum of currents in equals the sum of currents out.

Although simple, Kirchhoff's current law is crucial in circuit analysis. For one thing, it allows us to write an equation for each point in an electric circuit; using these equations, we can derive formulas relating the currents and voltages in the circuit.

Why it must be true

Kirchhoff's current law is an experimental formula; countless experiments have proved it, and this is the reason it is called a law. However, since the discovery of the electron, we can see why the current law must be true. If 1 million electrons flow into a point in 1 s, 1 million electrons must flow out of this point in 1 s; otherwise, electrons would be created or destroyed at the point. In an electric circuit, no electrons are created or destroyed; therefore, the rate of flow into each point must equal the rate of flow out of the point.

EXAMPLE 3-2.
What is the value of I in Fig. 3-5a?

SOLUTION.
Apply the current law to point A to get

$$7 = 3 + I$$

Now, solve for I to get

$$I = 4 \text{ A}$$

In Fig. 3-5a, when we visualize I equal to 4 A, we can see a total of 7 A going into point A and a total of 7 A coming out of point A.

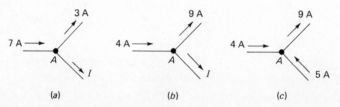

Figure 3-5. Examples of the current law.

EXAMPLE 3-3.
Solve for I in Fig. 3-5b.

SOLUTION.
With Kirchhoff's current law,

$$4 = 9 + I$$

Solving for I,

$$I = -5 \text{ A}$$

What does the negative sign mean? It means the true direction of conventional current is opposite that shown in the circuit. In other words, in Fig. 3-5b the true direction of conventional current I is opposite that shown. If we redraw the circuit showing true directions, we get Fig. 3-5c. Here we see a total of 9 A flowing into the point and 9 A flowing out of the point.

Whenever possible, we draw the true direction of current in circuit diagrams, using the rule that conventional current is from plus to minus in all loads. Occasionally at the beginning of a problem, it's impossible to tell which end of a load is positive. For this reason, in some problems we may guess the wrong direction of current. Kirchhoff's current law automatically catches errors like this because a negative sign turns up when solving for the unknown current.

Whenever you solve an equation and get a negative value of current, all you have to do is visualize or redraw the original circuit with this current in the opposite direction.

Test 3-3

1. Kirchhoff's current law is an example of which kind of formula?
   (*a*) defining   (*b*) experimental   (*c*) derived   .... ( )
2. The current law is true because electrons
   (*a*) can be created at a point   (*b*) can disappear at a point
   (*c*) can flow into a point
   (*d*) can neither be created nor destroyed at a point   .... ( )
3. The following words can be rearranged to form a sentence: COMES GOES IN OUT WHAT. With relation to the current law, the sentence is
   (*a*) true   (*b*) false   .... ( )

3-4. KIRCHHOFF'S VOLTAGE LAW

Kirchhoff's voltage law says the sum of voltages around a closed path is zero. In other words, start at any point in a circuit and go around a path that returns to the starting point; you will find all voltages add to zero.

This section answers the following questions:

What are the steps in applying the voltage law?
Why must the voltage law be true?

How to apply Kirchhoff's voltage law

Here are the key steps for applying Kirchhoff's voltage law to a circuit:

1. Select any point in the circuit as a starting point. Imagine you are going to walk clockwise around a path that returns to this starting point.
2. As you arrive at each source or load, write down the first sign you see (+ or −) and the magnitude of the voltage.
3. When you arrive back at the starting point, equate all voltages to zero.

To understand each of these steps, we will go through a few examples.

One source and one load

In Fig. 3-6a suppose we pick A as a starting point. Walking around the circuit in a clockwise direction, the first sign is plus and the magnitude of the voltage is V_2. So, we write

$$+V_2$$

Continuing in a clockwise direction, we next arrive at the source where the first sign is minus and the magnitude of voltage is V_1. Adding $-V_1$ to the other voltage gives

$$+V_2 - V_1$$

After walking through the source and arriving back at the starting point, we equate all voltages to zero to get

$$+V_2 - V_1 = 0$$

This is the Kirchhoff voltage equation for the closed path of Fig. 3-6a. It says the difference of V_2 and V_1 is zero.

If we transpose V_1 in the foregoing equation, we get this relation between V_1 and V_2:

$$V_1 = V_2$$

This tells us the source voltage equals the load voltage in Fig. 3-6a. The result makes sense, because the source and load are between the same pair of equipotential points.

Figure 3-6b shows a simple way to visualize the equality of V_1 and V_2. The elevator lifts balls from the ground level to the upper level; this increases their potential energy. When the balls fall back to the ground, they lose their potential energy. The potential

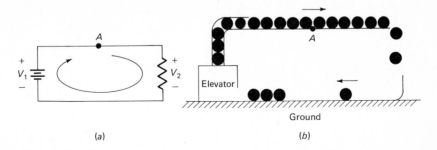

Figure 3-6. (a) *Kirchhoff's voltage law.* (b) *Mechanical analogy.*

(a)   *(b)*

energy lost equals the potential energy gained when the balls travel from A to the ground and back up to A.

One source and two loads

As another example of the use of Kirchhoff's voltage law, look at Fig. 3-7a. Starting at point A and going around the circuit in a clockwise direction, we first write

$$+V_2$$

then add $+V_3$ to get

$$+V_2 + V_3$$

then add $-V_1$ to get

$$+V_2 + V_3 - V_1$$

We then arrive back at the starting point and can equate all voltages to zero to get

$$+V_2 + V_3 - V_1 = 0$$

Transposing V_1 gives

$$V_1 = V_2 + V_3$$

This says source voltage equals the sum of load voltages. The result makes sense, because the energy gained by charges passing through the source equals the energy lost by these charges passing through the loads.

Two sources and two loads

In Fig. 3-7b, summing voltages around the circuit gives

$$V_3 + V_4 - V_2 - V_1 = 0$$

Figure 3-7. Examples of
Kirchhoff's voltage law.

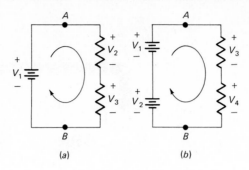

(a)   (b)

which rearranges into

$$V_1 + V_2 = V_3 + V_4$$

This says the sum of all source voltages equals the sum of all load voltages.

Again, the result makes sense. The total energy gained by charges passing through the sources should equal the total energy lost by these charges passing through the loads. (Charges passing through a source gain energy, because they move from a point of low potential energy to a point of high potential energy.)

Why the voltage law must be true

In general, Kirchhoff's voltage law says

$$\Sigma \text{ voltages around a closed path} = 0 \qquad (3\text{-}2)$$

Because voltage is the difference in potential energy per unit charge, it makes no difference how complicated the circuit is. As long as we go around a closed path, the voltages must add to zero because we are returning to the same potential-energy point.

Kirchhoff's voltage law combined with the current law allows us to write equations relating voltages and currents in electric circuits. Starting with these equations, we can derive all kinds of useful formulas.

EXAMPLE 3-4.
What is the value of V in Fig. 3-8a?

SOLUTION.
Summing voltages around the circuit gives

$$3 + 4 - V = 0$$

Solving for V,

$$V = 7 \text{ V}$$

In Fig. 3-8a, when we visualize V equal to 7 V, we see that the source voltage is the sum of the load voltages.

This is a simple example of using Kirchhoff's voltage law to get an equation which is then solved for the unknown voltage. In any circuit, each *loop* (closed path) gives one voltage equation. Therefore, whenever you know all the voltages in a loop except one, you can solve for the unknown voltage.

EXAMPLE 3-5.
Solve for V in Fig. 3-8b.

SOLUTION.
With Kirchhoff's voltage law,

$$3 + V - 12 = 0$$

which rearranges into

$$V = 9 \text{ V}$$

In Fig. 3-8b, V equal to 9 V means the source voltage equals the sum of the load voltages.

EXAMPLE 3-6.
Solve for V_1 and V_2 in Fig. 3-8c.

SOLUTION.
You can choose any closed path, because the voltage law holds true for each closed path. To minimize the work, always choose a path with the fewest unknowns. For instance, to get an equation with V_1, choose path A-B-C-A which gives

$$3 + V_1 - 12 = 0$$

Figure 3-8. Summing voltages around a loop.

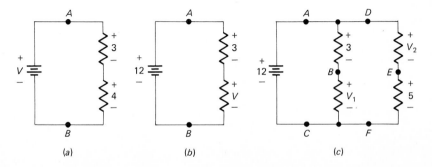

(a) (b) (c)

Solving for V_1 gives

$$V_1 = 9 \text{ V}$$

To get V_2, choose path D-E-F-C-A-D which gives

$$V_2 + 5 - 12 = 0$$

Solving for V_2 results in

$$V_2 = 7 \text{ V}$$

We will apply Kirchhoff's voltage law many times in this book. It is important to remember we can choose any closed path, usually the one with the fewest unknowns.

Test 3-4

1. Similar to the current law, the voltage law is
    (*a*) a defined formula   (*b*) an experimental formula
    (*c*) a derived formula   (*d*) none of these .... ( )
2. You are also allowed to sum voltages in a counterclockwise direction if you wish. This being the case, you must write down
    (*a*) the first sign of each voltage   (*b*) the second sign
    (*c*) either sign, provided you are consistent .... ( )
3. Kirchhoff's voltage law is true because
    (*a*) the starting and ending points correspond to the same potential energy
    (*b*) charges are neither created nor destroyed
    (*c*) voltages are neither created nor destroyed   (*d*) none of these .... ( )
4. The current law is to the voltage law as a point is to a
    (*a*) load   (*b*) circuit   (*c*) equipotential point   (*d*) loop .... ( )

3-5. SERIES-CIRCUIT ANALYSIS

With Kirchhoff's current and voltage laws you can learn more about series circuits. For series circuits the key questions are

Why is current the same at all points?
Why do voltage ratios equal resistance ratios?
What is equivalent resistance?

Same current at all points

Here is an important idea about a series circuit: current has the same value at any point in the circuit. This follows logically from Kirchhoff's current law. In Fig. 3-9*a*, the current law gives

$$I_1 = I_2$$

which says the current into A equals the current out of A. Furthermore, because the current out of one point is the current into the next point, we conclude current has the same value at all points. If the current is 5 A at point A, it must also be 5 A at B, C, and every other point in Fig. 3-9a.

Voltage ratios equal resistance ratios

Another important property of a series circuit is this: the ratio of the load voltages equals the ratio of the load resistances. If a resistance is twice as much as another resistance, it will have twice as much voltage across it. Or if a resistance is 10 times as large as another resistance, it will have 10 times as much voltage.

Here's the proof. In Fig. 3-9b, the current has the same value in R_1 and R_2; therefore, the voltages are

$$V_1 = R_1 I$$

$$V_2 = R_2 I$$

The ratio of these voltages is

$$\frac{V_1}{V_2} = \frac{R_1 I}{R_2 I}$$

or

$$\frac{V_1}{V_2} = \frac{R_1}{R_2} \tag{3-3}$$

Figure 3-9. Voltage ratios equal resistance ratios.

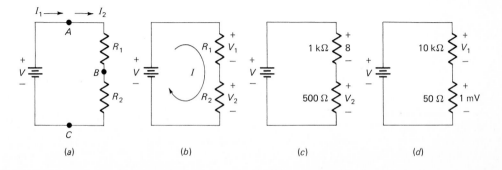

(a)   (b)   (c)   (d)

As an example, substituting the values of Fig. 3-9c into Eq. (3-3) gives

$$\frac{8}{V_2} = \frac{1000}{500}$$

or
$$V_2 = 4 \text{ V}$$

Visualizing V_2 as 4 V in Fig. 3-9c, we see the 1-kΩ resistance has 8 V and the 500-Ω resistance has 4 V, a 2-to-1 ratio for voltages and resistances.

As another example, the 50-Ω resistance of Fig. 3-9d has 1 mV across it. With Eq. (3-3),

$$\frac{V_1}{0.001} = \frac{10{,}000}{50} = 200$$

or
$$V_1 = 0.2 \text{ V} = 200 \text{ mV}$$

In Fig. 3-9d, when V_1 equals 200 mV, the voltage ratio is 200, the same as the resistance ratio.

Equivalent resistance

The *equivalent resistance* between two points in a series circuit equals the sum of the resistances between the points. In Fig. 3-10a the equivalent resistance between A and B is

$$R = 4000 + 2000 = 6000 \ \Omega$$

This equivalent resistance results in the same current as the two separate resistances. In other words, the current in Fig. 3-10b has the same value as the current in Fig. 3-10a.

Here is the proof. In Fig. 3-10c, we can sum voltages around the circuit to get

$$V_1 + V_2 - V = 0$$

or
$$V = V_1 + V_2 \tag{3-4a}$$

Furthermore, each load voltage in Fig. 3-10c is

$$V_1 = R_1 I$$

and
$$V_2 = R_2 I$$

Substituting these expressions into Eq. (3-4a) gives

$$V = R_1 I + R_2 I = (R_1 + R_2)I$$

or
$$I = \frac{V}{R_1 + R_2} \tag{3-4b}$$

This says the current in Fig. 3-10c equals the source voltage divided by the sum of the resistances.

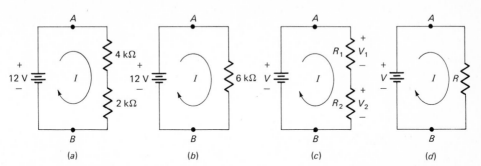

Figure 3-10. Example of equivalent resistance.

In Fig. 3-10d, the current is

$$I = \frac{V}{R}$$

When

$$R = R_1 + R_2$$

the current in Fig. 3-10d is the same as the current in Fig. 3-10c. In other words, as far as the current is concerned, the equivalent resistance of Fig. 3-10d acts the same as the two separate resistances of Fig. 3-10c.

By a proof similar to what we have just gone through, we can show the following. If a series circuit has n resistances, the equivalent resistance that results in the same current is

$$R = R_1 + R_2 + \cdots + R_n \qquad (3\text{-}5)$$

In words, Eq. (3-5) says the equivalent resistance between two points equals the sum of all the resistances between the points.

The symbol $\Rightarrow$ is shorthand for "is equivalent to." Figure 3-11a shows how to visualize Eq. (3-5) in circuit form. The circuit on the left is equivalent to the circuit on the right when R equals the sum of the individual resistances. (The dashes in Fig. 3-11a mean there are resistances between R_2 and R_n.)

If a series circuit has $R_1 = 20\ \Omega$, $R_2 = 50\ \Omega$, and $R_3 = 10\ \Omega$, the equivalent resistance is 80 Ω. If the source voltage in this circuit is 8 V, the current equals

$$I = \frac{V}{R} = \frac{8}{80} = 0.1 \text{ A} = 100 \text{ mA}$$

Generalization

Figure 3-11a uses a battery for its voltage source. Later we discuss other kinds of voltage sources. The idea of equivalent resistance does not depend on the kind of volt-

Figure 3-11. Equivalent resistance.

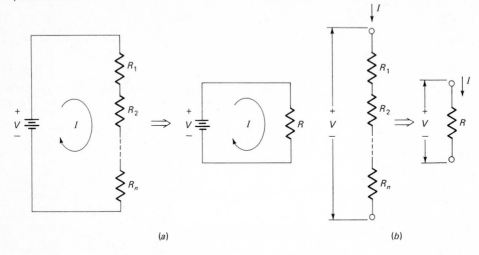

(a) (b)

age source used. For this reason, the way to visualize equivalent resistance is shown in Fig. 3-11*b*. Whenever we see a string of series resistances between two points, we can mentally lump all these into a single resistance equal to the sum of individual resistances. This equivalent resistance has the same *V/I* ratio as the string of resistances.

EXAMPLE 3-7.
Calculate the value of *I* in Fig. 3-12*a*.

SOLUTION.
The equivalent resistance is

$$R = R_1 + R_2 + R_3 = 2 + 3 + 1 = 6 \ \Omega$$

Figure 3-12*b* shows the equivalent circuit. With Ohm's law,

$$I = \frac{V}{R} = \frac{12}{6} = 2 \ \text{A}$$

Therefore, the current in the original circuit of Fig. 3-12*a* is 2 A.

EXAMPLE 3-8.
Prove that 2 A is the correct current in Fig. 3-12*a* by summing voltages around the circuit.

SOLUTION.
Whenever you have any doubt about the value of current in a series circuit, you can always check the value by working out the individual voltages and making sure they

add up to the source voltage. In Fig. 3-12a, when I equals 2 A,

$$V_1 = R_1 I = 2 \times 2 = 4 \text{ V}$$

$$V_2 = R_2 I = 3 \times 2 = 6 \text{ V}$$

$$V_3 = R_3 I = 1 \times 2 = 2 \text{ V}$$

The sum of these load voltages does equal the source voltage (12 V). Therefore, the value of I must be correct; any other value would violate Kirchhoff's voltage law.

EXAMPLE 3-9.
Figure 3-12c shows the series circuit analyzed in the two preceding examples. Prove the voltage ratios equal the resistance ratios.

SOLUTION.
In Fig. 3-12c, the voltage ratio of the upper resistances is

$$\frac{V_1}{V_2} = \frac{4}{6} = \frac{2}{3}$$

which does equal the resistance ratio

$$\frac{R_1}{R_2} = \frac{2}{3}$$

The two lower resistances have a voltage ratio of

$$\frac{V_2}{V_3} = \frac{6}{2} = 3$$

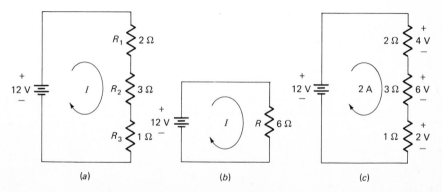

Figure 3-12. Calculating equivalent resistance.

(a)   (b)   (c)

and a resistance ratio of

$$\frac{R_2}{R_3} = \frac{3}{1} = 3$$

The top and bottom resistances have a voltage ratio of

$$\frac{V_1}{V_3} = \frac{4}{2} = 2$$

and a resistance ratio of

$$\frac{R_1}{R_3} = \frac{2}{1} = 2$$

Therefore, in all three possibilities, the voltage ratio equals the resistance ratio.

EXAMPLE 3-10.
What is the equivalent resistance between A and B in Fig. 3-13a? In Figure 3-13c?

SOLUTION.
Equivalent resistance between two points is the sum of the series resistances between the points. In Fig. 3-13a, the sum of the resistances is 8 kΩ. Figure 3-13b shows this equivalent resistance.

The equivalent resistance between the AB terminals of Fig. 3-13c equals 6 kΩ. Figure 3-13d shows this equivalent resistance.

EXAMPLE 3-11.
Replace each branch of Fig. 3-13e by its equivalent resistance.

Figure 3-13. Simplifying circuits with equivalent resistance.

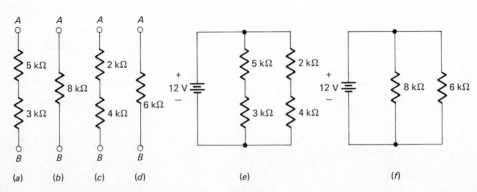

SOLUTION.

The branch with 5 kΩ and 3 kΩ has an equivalent resistance of 8 kΩ; the branch with 2 kΩ and 4 kΩ has an equivalent resistance of 6 kΩ.

Figure 3-13f shows the circuit with each branch replaced by its equivalent resistance. Note the new circuit is a parallel circuit with two resistances.

EXAMPLE 3-12.

A string of Christmas tree lights is wired in series. If there are 12 bulbs with a resistance of 100 Ω each, what is the equivalent resistance? If 120 V is applied to the string, what is the current?

SOLUTION.

The equivalent resistance equals

$$R = 12 \times 100 = 1.2 \text{ k}\Omega$$

The current is

$$I = \frac{V}{R} = \frac{120}{1200} = 0.1 \text{ A} = 100 \text{ mA}$$

EXAMPLE 3-13.

Two thick-film resistors are in series and have values of 500 Ω and 1 kΩ. What is the equivalent resistance? If 30 mV is applied to these resistances, what is the resulting current?

SOLUTION.

The two resistances act like a single resistance of

$$R = 500 + 1000 = 1.5 \text{ k}\Omega$$

The current equals

$$I = \frac{V}{R} = \frac{0.03}{1500} = 20 \text{ }\mu\text{A}$$

Test 3-5

1. The main reason current is the same at all points in a series circuit is
    (a) Kirchhoff's voltage law
    (b) current out of any point equals current into the next point
    (c) voltages change around the loop
    (d) voltage ratios equal resistance ratios .... ( )
2. Equation (3-3) says the ratio of two voltages equals the ratio of the corresponding resistances. We know this is true because the formula is
    (a) defined   (b) experimental   (c) derived .... ( )

3. Equivalent resistance in a series circuit has no relation to
   (a) voltage source used   (b) sum of resistances   (c) V/I ratio
   (d) individual resistances .... ( )
4. Which of the following does not belong?
   (a) equivalent resistance   (b) total resistance
   (c) same voltage-to-current ratio   (d) same voltage .... ( )
5. Ten volts are across a string of three resistances with values of 2 Ω, 6 Ω, and 10 Ω.
   Which of the following is closest to the current?
   (a) 5 A   (b) 0.5 A   (c) 0.6 A   (d) 0.4 A .... ( )

3-6. PARALLEL-CIRCUIT ANALYSIS

Parallel circuits also have special properties you need to know about. Again, Kirchhoff's current and voltage laws give the starting equations for deriving these properties. Try to learn all you can about parallel circuits, especially

> *How are the currents related?*
> *What is the product-over-sum rule?*
> *How do you combine more than two resistances?*

Same voltage across all loads

The voltage across all loads in a parallel circuit has the same value. For instance, in Fig. 3-14a

$$V = V_1 = V_2$$

We proved this earlier by the following argument: the voltages have to be the same because each load is between the same pair of equipotential points as the source.

Another way to prove the voltage is the same in a parallel circuit is by applying Kirchhoff's voltage law. In Fig. 3-14a, summing voltages around the left loop gives

$$V_1 - V = 0$$

or
$$V = V_1$$

Similarly, summing voltages around the right loop gives

$$V_2 - V_1 = 0$$

or
$$V_1 = V_2$$

It follows that

$$V = V_1 = V_2$$

By a similar proof, we can show

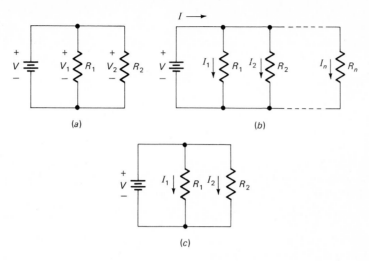

Figure 3-14. Analyzing parallel circuits.

$$V = V_1 = V_2 = \cdots = V_n$$

for a parallel circuit with n loads.

Currents

Since voltage has the same value across all loads, the current in each load of Fig. 3-14b is

$$I_1 = \frac{V}{R_1}$$

$$I_2 = \frac{V}{R_2}$$

$$\cdot$$
$$\cdot$$
$$\cdot$$

$$I_n = \frac{V}{R_n}$$

Furthermore, by applying Kirchhoff's current law, the total current from the source equals the sum of the load currents. In symbols,

$$I = I_1 + I_2 + \cdots + I_n$$

Current ratios equal inverse resistance ratios

Another important property of a parallel circuit is this: the ratio of load currents equals the inverse ratio of the load resistances. If a resistance is twice as much as another resistance, it has only half the current. If a resistance is 10 times as large as another resistance, it has only one-tenth the current.

Here is the proof. In Fig. 3-14c, the voltage is the same across R_1 and R_2; therefore, the load currents are

$$I_1 = \frac{V}{R_1}$$

and

$$I_2 = \frac{V}{R_2}$$

The ratio of these two equations is

$$\frac{I_1}{I_2} = \frac{V/R_1}{V/R_2}$$

or

$$\frac{I_1}{I_2} = \frac{R_2}{R_1} \tag{3-6}$$

This derived formula says the ratio of currents equals the inverse ratio of resistances.

As an example, in Fig. 3-15a the current through R_1 is 1 mA; therefore, with Eq. (3-6),

$$\frac{I_1}{I_2} = \frac{R_2}{R_1}$$

$$\frac{0.001}{I_2} = \frac{500}{1000}$$

or

$$I_2 = 0.002 \text{ A} = 2 \text{ mA}$$

Visualizing I_2 equal to 2 mA in Fig. 3-15a, we see the 500-Ω resistance has twice as much current as the 1000-Ω resistance.

Figure 3-15. Current ratio equals the reciprocal of resistance ratio.

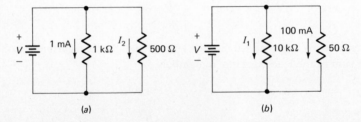

(a)   (b)

As another example, the 50-Ω resistance of Fig. 3-15b has 100 mA through it. With Eq. (3-6),

$$\frac{I_1}{0.1} = \frac{50}{10,000} = \frac{1}{200}$$

or
$$I_1 = 0.5 \text{ mA}$$

In Fig. 3-15b, when I_1 equals 0.5 mA, the current through the 50-Ω resistance is 200 times larger than the current through the 10-kΩ resistance.

In general, no matter how many loads are in parallel, the ratio of any two load currents equals the inverse ratio of the load resistances.

Equivalent resistance

In parallel-circuit analysis you often use the *equivalent resistance* between two points. Unlike a series circuit, do not add parallel resistances. Instead, you must use a more complicated rule for combining parallel resistances. What follows is the derivation of this rule.

In Fig. 3-16a, the current law applied to point A gives

$$I = I_1 + I_2 \tag{3-7a}$$

Next, note

$$I_1 = \frac{V}{R_1}$$

and
$$I_2 = \frac{V}{R_2}$$

Substituting these expressions into Eq. (3-7a) gives

$$I = \frac{V}{R_1} + \frac{V}{R_2}$$

which factors into

$$I = \left(\frac{1}{R_1} + \frac{1}{R_2}\right) V \tag{3-7b}$$

In Fig. 3-16b, the current equals

$$I = \frac{V}{R} \tag{3-7c}$$

where R is the equivalent resistance, the single resistance that produces the same value of I as exists in the original circuit of Fig. 3-16a. For the two I's to be equal, the right-hand members of Eqs. (3-7b) and (3-7c) must be equal. Therefore,

$$\frac{1}{R} = \frac{1}{R_1} + \frac{1}{R_2} \tag{3-8a}$$

Figure 3-16. Equivalent resistance.

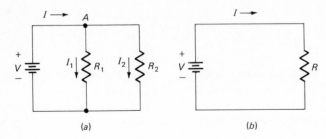

(a)   (b)

This equation says the reciprocal of the equivalent resistance equals the sum of reciprocals of the parallel resistances.

Using a common denominator, we can rewrite Eq. (3-8a) as

$$\frac{1}{R} = \frac{R_1 + R_2}{R_1 R_2}$$

And by inverting both sides,

$$R = \frac{R_1 R_2}{R_1 + R_2} \tag{3-8b}$$

This result is a good example of a derived formula, first discussed in Chap. 1. We began by applying Kirchhoff's current law to get Eq. (3-7a). Rearranging this gave Eq. (3-7b). Since the load current produced by an equivalent resistance equals the sum of the individual load currents, we were able to arrive at Eq. (3-8a). Finally, we algebraically manipulated Eq. (3-8a) to get Eq. (3-8b). There are many steps between the starting equation and the final equation; however, as long as each step is valid, the final result is valid.

And what does the final result say? It says the equivalent resistance equals the *product* of parallel resistances divided by the *sum* of these resistances. Because Eq. (3-8b) is used so often, we call it the *product-over-sum rule*.

Examples

Suppose two 5-kΩ resistances are in parallel as shown in Fig. 3-17a. The equivalent resistance is

$$R = \frac{5000 \times 5000}{5000 + 5000} = \frac{25,000,000}{10,000} = 2500 \; \Omega$$

If a 3-kΩ resistance is in parallel with a 6-kΩ resistance, the equivalent resistance equals

$$R = \frac{3000 \times 6000}{3000 + 6000} = \frac{18,000,000}{9000} = 2000 \ \Omega$$

as shown in Fig. 3-17b.

When 100 Ω is in parallel with 300 Ω, the equivalent resistance is

$$R = \frac{100 \times 300}{100 + 300} = \frac{30,000}{400} = 75 \ \Omega$$

as illustrated by Fig. 3-17c.

In general, therefore, whenever two resistances are in parallel, we can calculate the equivalent resistance by the product-over-sum rule.

Several parallel resistances

How do you calculate the equivalent resistance of more than two parallel resistances? There are two ways. First, combine two resistances at a time until you have reduced the original circuit to a single resistance. For instance, given the circuit of Fig. 3-18a, combine the 3-kΩ and 6-kΩ resistances into an equivalent 2-kΩ resistance as shown in Fig. 3-18b. In turn, these 2-kΩ resistances have an equivalent resistance of 1-kΩ (Fig. 3-18c).

The second method for combining more than two parallel resistances is this. By a proof similar to one given earlier,

$$\frac{1}{R} = \frac{1}{R_1} + \frac{1}{R_2} + \cdots + \frac{1}{R_n}$$

Figure 3-17. Examples of equivalent resistance.

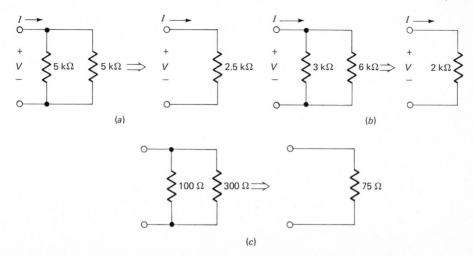

(a)   (b)

(c)

Figure 3-18. Three parallel
resistances combined to two, and
then one.

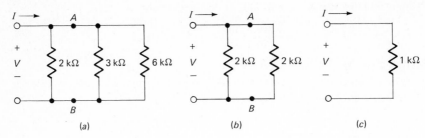

(a) (b) (c)

which rearranges into

$$R = \frac{1}{1/R_1 + 1/R_2 + \cdots + 1/R_n} \tag{3-9}$$

This derived formula says you add the reciprocals of the parallel resistances; then the reciprocal of the sum equals the equivalent resistance.

Equation (3-9) is useful when you have an electronic slide rule or calculator to work with. You can quickly add the reciprocal of each resistance to get the denominator. The equivalent resistance equals the reciprocal of this sum. For instance, suppose a circuit has 2 Ω, 3 Ω, and 6 Ω in parallel (Fig. 3-19a). With Eq. (3-9),

$$R = \frac{1}{1/2 + 1/3 + 1/6} = \frac{1}{0.5 + 0.333 + 0.167}$$

$$= 1 \ \Omega$$

Therefore, the V/I ratio of Fig. 3-19b is equal to the V/I ratio of Fig. 3-19a.

Figure 3-19. Combining three
parallel resistances by reciprocal
rule.

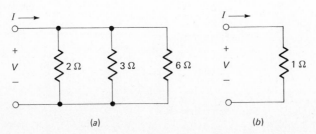

(a) (b)

EXAMPLE 3-14.
Calculate the value of I in Fig. 3-20a.

SOLUTION.
One way of finding I is to replace 3 kΩ and 6 kΩ by an equivalent resistance of 2 kΩ (calculated earlier). The current in the equivalent circuit of Fig. 3-20b is

$$I = \frac{V}{R} = \frac{12}{2000} = 6 \text{ mA}$$

This is the same value of current as in the original circuit (Fig. 3-20a).

EXAMPLE 3-15.
Work out the value of I in Fig. 3-20a without using equivalent resistance.

SOLUTION.
A direct attack on Fig. 3-20a is to calculate I_1 and I_2. Then, using Kirchhoff's current law, you get the value of I. In Fig. 3-20a,

$$I_1 = \frac{12}{3000} = 4 \text{ mA}$$

and

$$I_2 = \frac{12}{6000} = 2 \text{ mA}$$

Figure 3-20c shows these current values. Applying Kirchhoff's current law to point A gives

$$I = 4 \text{ mA} + 2 \text{ mA}$$
$$= 6 \text{ mA}$$

EXAMPLE 3-16.
Verify that the ratio of the load currents in Fig. 3-20c equals the inverse ratio of the load resistances.

Figure 3-20. Calculating currents in parallel circuits.

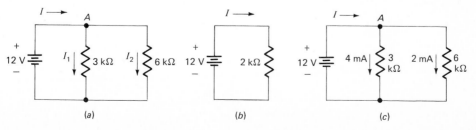

| (a) | (b) | (c) |

SOLUTION.

All we have to do is show that

$$\frac{I_1}{I_2} = \frac{R_2}{R_1}$$

in Fig. 3-20c. Using the given values,

$$\frac{I_1}{I_2} = \frac{4 \text{ mA}}{2 \text{ mA}} = 2$$

and

$$\frac{R_2}{R_1} = \frac{6 \text{ k}\Omega}{3 \text{ k}\Omega} = 2$$

Therefore, I_1/I_2 does equal R_2/R_1.

EXAMPLE 3-17.

A set of Christmas tree lights is wired in parallel. If there are 20 bulbs with a resistance of 10 kΩ each, what is the equivalent resistance? If 100 V is across the bulbs, what is the total current? What is the current in each bulb?

SOLUTION.

With Eq. (3-9),

$$R = \frac{1}{1/10{,}000 + 1/10{,}000 + \cdots + 1/10{,}000}$$
$$= \frac{1}{0.0001 + 0.0001 + \cdots + 0.0001}$$

There are 20 terms in the denominator, each equal to 0.0001. Therefore,

$$R = \frac{1}{20 \times 0.0001} = \frac{1}{0.002}$$
$$= 500 \ \Omega$$

The total current equals

$$I = \frac{100}{500} = 0.2 \text{ A} = 200 \text{ mA}$$

The current in each bulb is

$$I_n = \frac{100}{10{,}000} = 0.01 \text{ A} = 10 \text{ mA}$$

Test 3-6

1. In a parallel circuit, the total current
   (a) has the same value at all points
   (b) is the difference in load currents
   (c) is smaller than any load current
   (d) is greater than any load current .... ( )
2. Two loads are in parallel with 3 A in each. The total current equals
   (a) 0   (b) 3 A   (c) 6 A   (d) 9 A .... ( )
3. The equivalent resistance of two parallel resistances is always
   (a) greater than the larger   (b) greater than either
   (c) less than either   (d) more than the smaller .... ( )
4. Equation (3-9) is
   (a) a defined formula   (b) an experimental formula
   (c) a derived formula .... ( )
5. The current through the equivalent resistance of a parallel circuit is always
   (a) less than any load current   (b) greater than any load current
   (c) smaller than the largest load current   (d) none of these .... ( )

3-7. SHORTCUTS FOR PARALLEL RESISTANCES

Getting the equivalent resistance in a series circuit is easy; merely add the resistances. Calculating the equivalent resistance in a parallel circuit is harder, because the formulas are more complicated. In electronics you have to analyze many series and parallel circuits. For this reason, the shortcuts of this section can save time and effort. In particular, you will learn

> *The fast way to combine equal parallel resistances*
> *The lower and upper limits for unequal parallel resistances*
> *The cancel-and-restore rule*

Equal resistances

If two parallel resistances are equal, their equivalent resistance is half the value of either. The proof is next to obvious. With the product-over-sum rule,

$$R = \frac{R_1 R_1}{R_1 + R_1} = \frac{R_1{}^2}{2R_1}$$

or

$$R = \frac{R_1}{2} \qquad\qquad\qquad (3\text{-}10a)$$

Figure 3-21. Rules for parallel resistances.

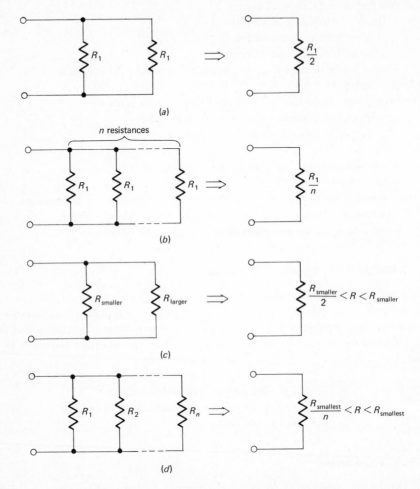

Figure 3-21*a* summarizes this rule. When two 10-kΩ resistors are in parallel, the equivalent resistance equals 5 kΩ. With 300 Ω in parallel with 300 Ω, the equivalent resistance is 150 Ω.

With Eq. (3-9) we can derive this formula for *n* equal resistances in parallel:

$$R = \frac{R_1}{n} \tag{3-10b}$$

Figure 3-21*b* summarizes this rule in circuit form. If ten 8-kΩ resistances are in parallel, the equivalent resistance is 800 Ω. Or when twenty 10-kΩ resistances are in parallel, the equivalent resistance equals 500 Ω.

Unequal resistances

The total current in a parallel circuit equals the sum of the individual currents; therefore, the total current is always greater than any individual current. Because of this, the equivalent resistance is always smaller than the smallest parallel resistance. For instance, given 3 kΩ, 5 kΩ, 2 kΩ, and 7 kΩ, we automatically know the equivalent resistance is less than 2 kΩ (the smallest).

When two unequal resistances are in parallel, the equivalent resistance is less than the smaller. In symbols,

$$R < R_{\text{smaller}} \qquad\qquad (3\text{-}11a)$$

Furthermore, when the larger resistance is only slightly more than the smaller, the equivalent resistance is only slightly more than $R_{\text{smaller}}/2$. In symbols,

$$\frac{R_{\text{smaller}}}{2} < R \qquad\qquad (3\text{-}11b)$$

Combining the two inequalities just given,

$$\frac{R_{\text{smaller}}}{2} < R < R_{\text{smaller}} \qquad\qquad (3\text{-}11c)$$

Figure 3-21c illustrates this useful rule. Given two unequal resistances in parallel, the equivalent resistance must be less than the smaller resistance, but more than half this value. As an example, given 3 kΩ in parallel with 7 kΩ, the smaller resistance is 3 kΩ; half of this is 1.5 kΩ; therefore, the equivalent resistance is between 1.5 Ω and 3 kΩ. Similarly, if 5 kΩ are in parallel with 12 kΩ, the equivalent resistance is between 2.5 Ω and 5 kΩ.

When n unequal resistances are in parallel, the checking rule is

$$\frac{R_{\text{smallest}}}{n} < R < R_{\text{smallest}} \qquad\qquad (3\text{-}12)$$

Figure 3-21d illustrates this rule. If 3 kΩ, 5 kΩ, 2 kΩ, and 7 kΩ are in parallel, the smallest resistance is 2 kΩ; with n equal to 4, $R_{\text{smallest}}/4$ is 500 Ω. Therefore, the equivalent resistance is between 500 Ω and 2 kΩ.

Cancel-and-restore rule

Parallel-resistance calculations can be simplified as follows. Cancel zeros from each resistance before you set up the product-over-sum calculation; then, restore the same number of zeros to the answer.

If 3000 Ω are in parallel with 6000 Ω, cancel three zeros from each resistance to get 3 and 6; canceling three zeros is equivalent to moving the decimal point three places to the left. Set up the product-over-sum calculation as

$$R = \frac{3 \times 6}{3 + 6} = \frac{18}{9} = 2$$

Now restore three zeros to this answer to get

$$R = 2000 \ \Omega$$

Restoring three zeros means moving the decimal point three places to the right. The final answer of 2000 Ω is the equivalent resistance of 3000 Ω in parallel with 6000 Ω.

As another example, suppose 400 Ω and 600 Ω are in parallel. Cancel two zeros (move the decimal point two places to the left) to get 4 and 6. Then,

$$R = \frac{4 \times 6}{4 + 6} = \frac{24}{10} = 2.4$$

Restore two zeros (move the decimal point two places to the right) to get

$$R = 240 \ \Omega$$

When several resistances are in parallel, cancel zeros from each resistance before using Eq. (3-9); then restore the same number of zeros to the answer. If 400 Ω, 600 Ω, and 1200 Ω are in parallel, cancel two zeros and set up Eq. (3-9) as follows:

$$R = \frac{1}{1/4 + 1/6 + 1/12} = \frac{1}{0.25 + 0.167 + 0.083}$$

$$= 2$$

Restore two zeros to get the final answer

$$R = 200 \ \Omega$$

EXAMPLE 3-18.
Two light bulbs with a resistance of 100 Ω each are in parallel. What is the equivalent resistance? If 40 of these light bulbs are in parallel, what is the equivalent resistance?

SOLUTION.
With two equal resistances,

$$R = \frac{100}{2} = 50 \ \Omega$$

When 40 of these bulbs are in parallel,

$$R = \frac{100}{40} = 2.5 \ \Omega$$

EXAMPLE 3-19.
The equivalent resistance of Fig. 3-22a has to lie between what two values? What is the equivalent resistance?

SOLUTION.
The smaller resistance is 10 kΩ; half of this is 5 kΩ; therefore, the equivalent resistance is between 5 kΩ and 10 kΩ.

With the product-over-sum rule and the cancel-and-restore rule,

$$R = \frac{1 \times 3}{1 + 3} = \frac{3}{4} = 0.75$$

Restoring zeros gives

$$R = 7500 \ \Omega = 7.5 \ k\Omega$$

EXAMPLE 3-20.

What range must the equivalent resistance of Fig. 3-22b be in? What is the equivalent resistance?

SOLUTION.

The smallest resistance is 10 kΩ; one-third of this is 3.33 kΩ; therefore, the equivalent resistance is between 3.33 kΩ and 10 kΩ.

With Eq. (3-9) and the cancel-and-restore rule,

$$R = \frac{1}{1/1 + 1/3 + 1/5} = \frac{1}{1 + 0.333 + 0.2}$$

$$= 0.652$$

Restoring zeros gives

$$R = 6520 \ \Omega = 6.52 \ k\Omega$$

The answer is between 3.33 kΩ and 10 kΩ, as it should be.

3-8.   COMBINATION CIRCUITS

Often, circuits combine series and parallel connections; we call these *combination circuits.* By applying rules for series and parallel circuits, we can reduce combination circuits to an equivalent resistance.

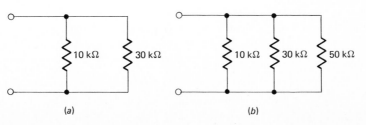

Figure 3-22. Examples of calculating equivalent resistance.

(a)   (b)

*Figure 3-23. Reducing a
combination circuit to a single
resistance.*

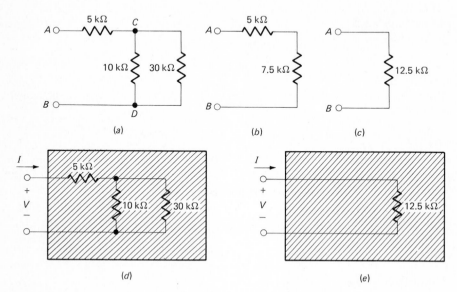

(a)   (b)   (c)

(d)   (e)

Examples

Figure 3-23a shows a circuit that is neither series nor parallel; yet it combines both ideas. The 10 kΩ and 30 kΩ are in parallel because they are between the same pair of equipotential points (C and D). Combining these two resistances, we get 7.5 kΩ, as shown in Fig. 3-23b.

The 5-kΩ resistance of Fig. 3-23b is in series with the 7.5-kΩ resistance; we can combine these into a 12.5-kΩ resistance (Fig. 3-23c). This 12.5-kΩ resistance is the equivalent resistance of the original circuit (Fig. 3-23a). Stated another way, the V/I ratio for the AB terminals has the same value. In terms of measurements, an ohmmeter would read 12.5 kΩ when connected across the AB terminals of Fig. 3-23a, b, or c.

In fact, if the original circuit and the equivalent resistance are inside two identical sealed boxes (Figs. 3-23d and e), there is no way to know which is which because both have exactly the same V/I ratio. If we apply 12.5 V to each sealed box, 1 mA is the value of current to each box. Sealed boxes like this are often called *black boxes*.

Here is another example of reducing a combination circuit to an equivalent resistance. In Fig. 3-24a, the left branch has 6 kΩ in series with 14 kΩ. We can combine these two resistances to get Fig. 3-24b. Now 20 kΩ are in parallel with 60 kΩ. Canceling four zeros gives

$$R = \frac{2 \times 6}{2 + 6} = 1.5$$

Restoring four zeros gives

$$R = 15{,}000 \ \Omega = 15 \ k\Omega$$

as shown in Fig. 3-24c.

Again, realize Figs. 3-24a and c have ohmmeter and black-box equivalence; an ohmmeter reads 15 kΩ for either circuit; put both circuits inside black boxes, and you can't tell one from the other, at least not by voltage and current measurements.

General idea

Here are the steps for reducing any combination circuit to an equivalent resistance:

1. Examine the circuit, looking for series and parallel connections.
2. Replace each series section by its equivalent resistance, and each parallel section by its equivalent resistance.
3. If the resulting circuit has more than one resistance, repeat steps 1 and 2 until you arrive at a single resistance.

This three-step procedure works with almost all the combination circuits you are likely to run into.

EXAMPLE 3-21.
Work out the equivalent resistance between the *AB* terminals of Fig. 3-25a.

SOLUTION.
The 10-kΩ and 30-kΩ resistances are in parallel; the 6-kΩ and 14-kΩ resistances are in series. Combining these parallel and series sections gives Fig. 3-25b.

In Fig. 3-25b, 5 kΩ and 7.5 kΩ are in series; 20 kΩ and 60 kΩ are in parallel. Combining resistances gives Fig. 3-25c.

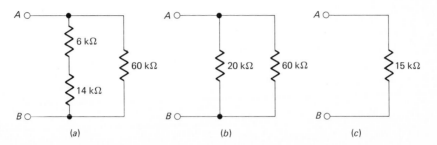

Figure 3-24. Equivalent resistance of a combination circuit.

Figure 3-25.

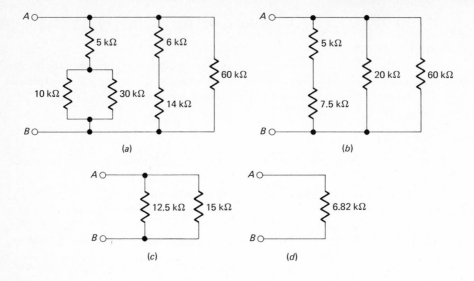

(a)

(b)

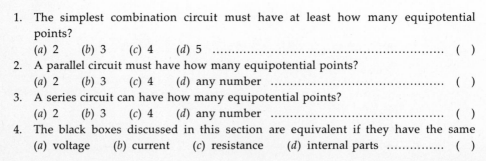

(c)

(d)

Finally, in Fig. 3-25c, 12.5 kΩ and 15 kΩ are in parallel. So,

$$R = \frac{12.5 \times 15}{12.5 + 15} = 6.82$$

After restoring zeros,

$$R = 6.82 \text{ k}\Omega$$

as shown in Fig. 3-25d.

Therefore, an ohmmeter across the *AB* terminals of Fig. 3-25a reads 6.82 kΩ.

Test 3-7

1. The simplest combination circuit must have at least how many equipotential points?
   (*a*) 2 (*b*) 3 (*c*) 4 (*d*) 5 .... ( )
2. A parallel circuit must have how many equipotential points?
   (*a*) 2 (*b*) 3 (*c*) 4 (*d*) any number .... ( )
3. A series circuit can have how many equipotential points?
   (*a*) 2 (*b*) 3 (*c*) 4 (*d*) any number .... ( )
4. The black boxes discussed in this section are equivalent if they have the same
   (*a*) voltage (*b*) current (*c*) resistance (*d*) internal parts .... ( )

SUMMARY OF FORMULAS

EXPERIMENTAL

$$\Sigma \text{ currents in} = \Sigma \text{ currents out} \tag{3-1}$$

$$\Sigma \text{ voltages around a closed path} = 0 \tag{3-2}$$

DERIVED

$$\frac{V_1}{V_2} = \frac{R_1}{R_2} \quad \text{(series)} \tag{3-3}$$

$$R = R_1 + R_2 + \cdots + R_n \quad \text{(series)} \tag{3-5}$$

$$\frac{I_1}{I_2} = \frac{R_2}{R_1} \quad \text{(parallel)} \tag{3-6}$$

$$R = \frac{R_1 R_2}{R_1 + R_2} \quad \text{(parallel)} \tag{3-8b}$$

$$R = \frac{1}{1/R_1 + 1/R_2 + \cdots + 1/R_n} \quad \text{(parallel)} \tag{3-9}$$

$$\frac{R_{\text{smaller}}}{2} < R < R_{\text{smaller}} \quad \text{(parallel)} \tag{3-12}$$

Problems

3-1. As mentioned in Chap. 2, AWG-22 copper wire is used a great deal in transistor radios, TV receivers, and other electronic equipment. Normally, the current in this connecting wire is kept less than 1 A to avoid excessive self-heating.

What is the resistance of a 3-in piece of this wire? If the current is pushed to the normal limit, what is the maximum voltage drop across this piece of connecting wire?

3-2. How many equipotential points are there in Fig. 3-26a? In Fig. 3-26b?

3-3. Calculate the value of I in Fig. 3-26c.

3-4. Work out the value of I in Fig. 3-26d. Is the true direction of I away or toward point A?

3-5. In Fig. 3-26a, the current from the positive source terminal equals 8 mA. A downward current of 3 mA is in R_1. What is the magnitude and direction of the current in R_2?

3-6. Figure 3-27a shows a *Wheatstone bridge*, a circuit analyzed in Chap. 4. With Kirchhoff's current law, you can answer these questions:

Figure 3-26.

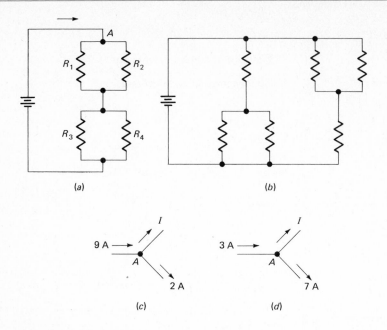

(a)   (b)

(c)   (d)

a. What is the value of I_1?
b. What is the value of I_2?
c. The value of I_3?
d. I_4?

3-7.   Apply Kirchhoff's current law to Fig. 3-27b to find the value of I_1, I_2, I_3, I_4, I_5, and I_6.

3-8.   Calculate the value of V in the series circuit of Fig. 3-28a.

3-9.   In Fig. 3-28b, work out the value of V_1. Next, work out the value of V_2.

3-10. What is the value of V in Fig. 3-28c?

Figure 3-27.

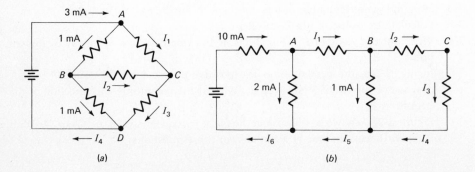

(a)   (b)

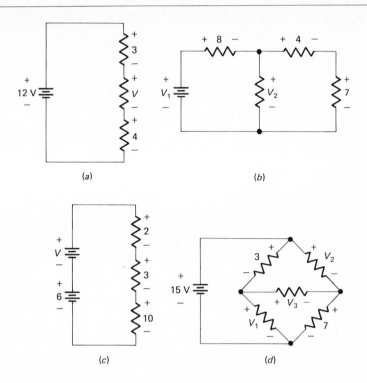

Figure 3-28.

(a)

(b)

(c)

(d)

3-11. In Fig. 3-28d, calculate each of the following:
   a. The value of V_1
   b. V_2
   c. V_3
3-12. Use Eq. (3-3) to work out the value of V_1 in Fig. 3-29a. Next, what is the value of V? The value of I?
3-13. Calculate the values of V_1, V_2, and V in Fig. 3-29b.
3-14. What is the value of I in Fig. 3-29c? Calculate the voltage across each resistance.
3-15. Work out the value of I for the circuit of Fig. 3-29d. How much voltage is there across each resistance? What is the total source voltage?
3-16. What is the equivalent resistance in each of the following:
   a. Fig. 3-29a
   b. Fig. 3-29b
   c. Fig. 3-29c
   d. Fig. 3-29d
3-17. What is the equivalent resistance between the AB terminals of Fig. 3-30a?
3-18. In Fig. 3-30b, what is the equivalent resistance between the AB terminals? Between the CD terminals? The EF terminals?

Figure 3-29.

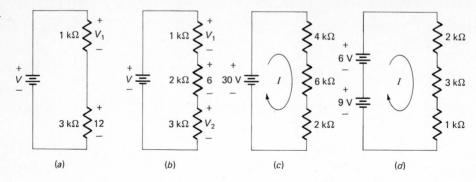

(a)   (b)   (c)   (d)

3-19.   Calculate the equivalent resistance between the AB terminals in Fig. 3-30c.
3-20.   A hundred light bulbs each with a resistance of 1 Ω are in series. What is the
        equivalent resistance?   100Ω
3-21.   A wirewound resistor has 1000 turns of AWG-22 copper wire. Each turn has a
        circumference of 1 in. Visualize each turn as a series resistance; then, you can
        see there are 1000 resistances in series. What is the equivalent resistance?
3-22.   Use Eq. (3-6) to work out the value of I_1 in Fig. 3-31a. What is the value of I?
3-23.   With Eq. (3-6), calculate the value of I_1 in Fig. 3-31b. What is the value of I_3?
        Of I?

Figure 3-30.

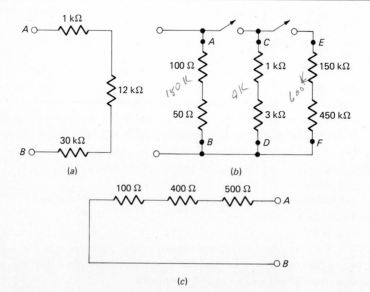

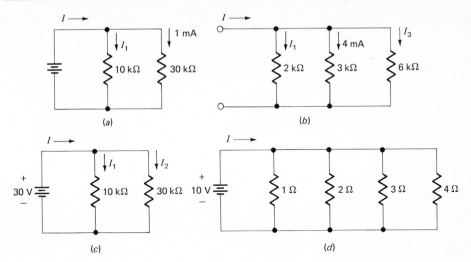

Figure 3-31.

3-24. Calculate the value of I in Fig. 3-31c using the equivalent resistance. Also, work out the values of I_1 and I_2.

3-25. Using equivalent resistance, calculate the value of I in Fig. 3-31d. Next, find the values of currents in the parallel resistances.

3-26. What is the equivalent resistance in each of the following:
   a. Fig. 3-31a
   b. Fig. 3-31b
   c. Fig. 3-31c
   d. Fig. 3-31d

3-27. Do not work out the equivalent resistance in this problem. Instead, give the minimum and maximum possible values for the equivalent resistance for
   a. Fig. 3-32a
   b. Fig. 3-32b
   c. Fig. 3-32c

3-28. Repeat the previous problem, but use
   a. Fig. 3-32d
   b. Fig. 3-32e

3-29. Two 10-MΩ resistances are in parallel. What is the equivalent resistance?

3-30. A pair of 50-kΩ resistances are connected in parallel. What is the equivalent resistance?

3-31. Work out the equivalent resistance between the AB terminals of Fig. 3-32a.

3-32. What is the equivalent resistance from A to B in Fig. 3-32b?

3-33. Calculate the equivalent resistance of Fig. 3-32c.

3-34. What resistance would an ohmmeter read from A to B in Fig. 3-32d?

Figure 3-32.

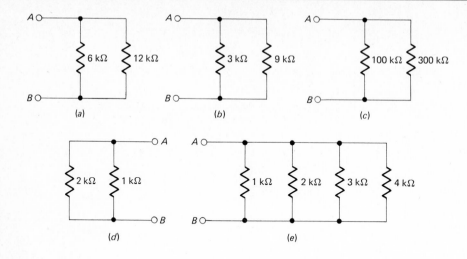

(a)   (b)   (c)

(d)   (e)

3-35. Work out the equivalent resistance between A and B in Fig. 3-32e.

3-36. Fifty light bulbs, each with a resistance of 200 Ω, are in parallel. What is the equivalent resistance?

3-37. Each load resistance in an ordinary house is in parallel across the power line. A seven-room house has the following resistances for its rooms: 5 Ω, 10 Ω, 20 Ω,

Figure 3-33.

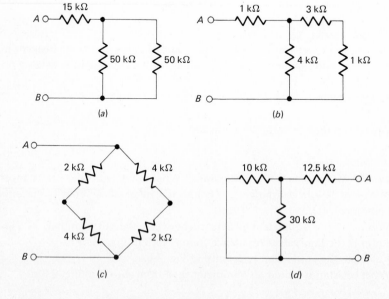

(a)   (b)

(c)   (d)

20 Ω, 30 Ω, 30 Ω, and 30 Ω. What is the equivalent resistance for the entire household?

3-38. A 1-kΩ resistance is in parallel with a 1-MΩ resistance. What is the equivalent resistance? Does the answer approach the value of the smaller or larger resistance?

3-39. What is the equivalent resistance between A and B in Fig. 3-33a?

3-40. Calculate the equivalent resistance from A to B in Fig. 3-33b.

3-41. Work out the equivalent resistance between the AB terminals of Fig. 3-33c.

3-42. What is the equivalent resistance from A to B in Fig. 3-33d?

ANSWERS TO TESTS

3-1. c, b, b

3-2. b, b, c, a, d

3-3. b, d, a

3-4 b, c, a, d

3-5. b, c, a, d, c

3-6. d, c, c, c, b

3-7. b, a, d, c

4. Three Theorems

A *theorem* is a mathematically provable statement. In this book, the word "theorem" is used only when something important is said about electrical quantities. This chapter discusses three theorems. The first is a shortcut, not really essential, but useful in many circuits. The next two theorems help us rearrange and simplify circuits.

In fact, the second and third theorems of this chapter are so crucial that without them it is impossible to have a deep understanding of circuits. These theorems are so vital we discuss them intensively in this chapter and apply them constantly in later chapters. Furthermore, when you get to electronics, you will find the last two theorems of this chapter are the keys to transistor-circuit analysis.

4-1. VOLTAGE-DIVIDER THEOREM

The voltage from a source may be more than you need. To reduce voltage, you can use a *voltage divider* (see Fig. 4-1a). V is the *input* voltage and V_2 is the *output* voltage. Since charges lose potential energy passing through R_1, the output voltage is less than the input.

The two main questions this section answers are

> *What is the voltage-divider theorem?*
> *How does a potentiometer work?*

Formula for output voltage

Here is how to derive a formula for V_2. In Fig. 4-1a,

$$V_2 = R_2 I$$

and
$$V = (R_1 + R_2)I$$

The ratio of these two equations is

$$\frac{V_2}{V} = \frac{R_2}{R_1 + R_2}$$

or
$$\frac{V_2}{V} = \frac{R_2}{R} \qquad\qquad (4\text{-}1a)$$

where $R = R_1 + R_2$, the equivalent resistance between the input terminals. Multiplying both sides of Eq. (4-1a) by V gives

$$V_2 = \frac{R_2}{R} V \qquad\qquad (4\text{-}1b)$$

Equation (4-1a) echos an idea learned earlier about series circuits: the ratio of the voltages equals the ratio of the resistances. In the series connection of Fig. 4-1a, V_2 is the voltage across R_2, and V is the voltage across R; the voltage ratio V_2/V equals the resistance ratio R_2/R.

Equation (4-1b) is an alternative formula that gives output voltage in terms of resistance ratio and input voltage. In fact, this equation summarizes the *voltage-divider theorem*: the output voltage of a voltage divider equals the resistance ratio R_2/R times the input voltage.

As an example, 10 V drives the voltage divider of Fig. 4-1b. R_2 equals 7 kΩ and R equals 10 kΩ; therefore,

$$V_2 = \frac{R_2}{R} V = \frac{7000}{10,000} 10 = 7 \text{ V}$$

As another example, Fig. 4-1c shows an input of 15 V. R_2 is 2 kΩ and R is 10 kΩ.

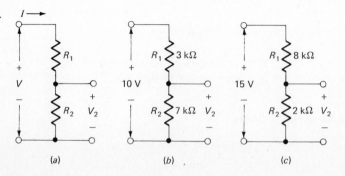

Figure 4-1. Voltage divider.

(a)   (b)   (c)

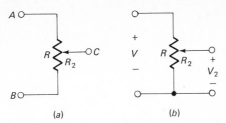

Figure 4-2. Potentiometer.

So,

$$V_2 = \frac{R_2}{R} V = \frac{2000}{10,000} \, 15 = 3 \text{ V}$$

Potentiometer

A *potentiometer* is a resistor with three terminals instead of two (Fig. 4-2a). The resistance between A and B equals R; the resistance between C and B equals R_2. Terminal C is called a *wiper* because it can slide up or down, tapping off different values of resistance. When the wiper is moved all the way up, R_2 equals R. If the wiper is moved all the way down, R_2 is zero.

The volume control for a transistor radio is a potentiometer used as a voltage divider (Fig. 4-2b). Input voltage V is an electrical signal; output voltage V_2 is eventually converted into the sound we hear. Since

$$V_2 = \frac{R_2}{R} V$$

we can control the volume or strength of the sound by moving the wiper.

Adjustable voltage dividers like Fig. 4-2b are widely used in radio, TV, measuring instruments, etc.

EXAMPLE 4-1.
In Fig. 4-3a, what are the minimum and maximum output voltages? The output voltage when the wiper is in the middle?

SOLUTION.
With the wiper all the way down, we tap off zero resistance and the output voltage equals

$$V_2 = \frac{R_2}{R} V = \frac{0}{R} V = 0$$

When the wiper is at the top, R_2 equals R and the output voltage is

$$V_2 = \frac{R_2}{R} V = \frac{R}{R} V = V = 2 \text{ V}$$

When the wiper is in the middle, R_2 equals half of R; the output voltage therefore equals

$$V_2 = \frac{R_2}{R} V = \frac{5000}{10,000} 2 = 1 \text{ V}$$

EXAMPLE 4-2.

What is the equivalent resistance between the input terminals of Fig. 4-3b?

SOLUTION.

The resistances are in series. So,

$$R = 900,000 + 90,000 + 9000 + 1000$$
$$= 1,000,000 \ \Omega = 1 \text{ M}\Omega$$

EXAMPLE 4-3.

In Fig. 4-3b, when the switch is moved to different positions, different output voltages result. Calculate the output voltage for each position A through E.

SOLUTION.

Similar to Example 4-1, R_2 equals zero means V_2 is zero; R_2 equals R means V_2 is equal to V. Therefore, position A gives a 10-V output and position E gives a zero output.

Figure 4-3. (a) Continuously variable divider. (b) Step divider.

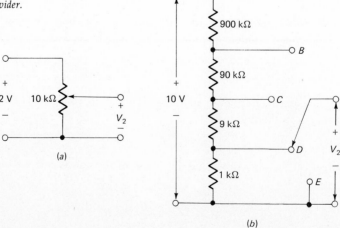

In position D, R_2 equals 1 kΩ and

$$V_2 = \frac{R_2}{R} V = \frac{1000}{1{,}000{,}000} 10 = 0.01 \text{ V} = 10 \text{ mV}$$

For position C, R_2 is 10 kΩ and

$$V_2 = \frac{10{,}000}{1{,}000{,}000} 10 = 0.1 \text{ V} = 100 \text{ mV}$$

Position B has an R_2 of 100 kΩ and

$$V_2 = \frac{100{,}000}{1{,}000{,}000} 10 = 1 \text{ V}$$

The output voltage changes in *decade* steps, by a factor of 10: 10 mV, 100 mV, 1 V, and 10 V. Many electronic instruments use decade-step voltage dividers like Fig. 4-3b.

Test 4-1

1. A voltage divider like Fig. 4-1a always has an output
   (*a*) less than the input   (*b*) greater than the input
   (*c*) equal to the input   (*d*) none of these .... ( )
2. The voltage-divider theorem is true because R_1 and R_2 have the same
   (*a*) current   (*b*) voltage   (*c*) resistance .... ( )
3. The output of a potentiometer is always
   (*a*) less than the input   (*b*) greater than the input
   (*c*) equal to or less than the input   (*d*) none of these .... ( )

4-2. THEVENIN QUANTITIES

When analyzing an electronic circuit, you seldom have to find all voltages and currents in the circuit; most of the time you will be after the voltage or current for a single resistance. When this is the case, *Thevenin's theorem* is often the easiest way to a solution. Without doubt, this theorem tops the list for importance; experienced technicians and engineers have used it thousands of times.

Section 4-3 states the Thevenin theorem and the conditions that must be satisfied when you apply it. Before you can understand the theorem, you first need to learn the answers to

> *What is Thevenin voltage?*
> *What is Thevenin resistance?*
> *What is a Thevenin circuit?*

Figure 4-4. Thevenin quantities.

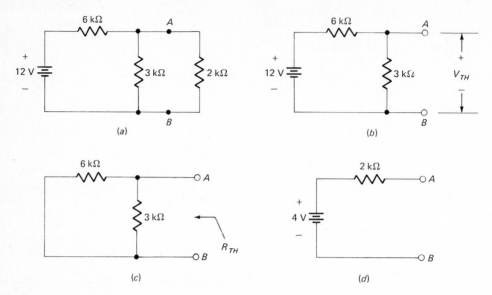

(a) (b) (c) (d)

Thevenin voltage

The *Thevenin voltage* between a pair of terminals is defined as the voltage that results when the load between these terminals is opened. For instance, Fig. 4-4*a* shows a 2-kΩ load between the *AB* terminals. If the load is opened (removed), the circuit reduces to Fig. 4-4*b*. By definition, the voltage appearing between the *AB* terminals of this *open-load circuit* is called the Thevenin voltage V_{TH}.

Figure 4-4*b* is a voltage divider. So, in this particular case, the Thevenin voltage equals

$$V_{TH} = \frac{R_2}{R} V = \frac{3000}{9000} 12 = 4 \text{ V}$$

Thevenin resistance

The *Thevenin resistance* between a pair of terminals is defined as the resistance between these terminals when the load is open and the source is reduced to zero. In Fig. 4-4*b*, the load is open; if we now visualize the source reduced to zero, the circuit simplifies to Fig. 4-4*c*. By definition, the resistance between the *AB* terminals of this *zero-source circuit* is called the Thevenin resistance. (Note: reducing a voltage source to zero is the same as replacing it by zero resistance, because $R = 0$ means $V = RI = 0$.)

The 6-kΩ resistor of Fig. 4-4c is in parallel with the 3-kΩ resistor, because both resistors are between the same pair of equipotential points. Therefore, in this particular case, the Thevenin resistance equals

$$R_{TH} = \frac{6000 \times 3000}{6000 + 3000} = 2 \text{ k}\Omega$$

Thevenin circuit

A *Thevenin circuit* is defined as a circuit that has a voltage source of V_{TH} in series with a resistance of R_{TH}. For instance, we have found the Thevenin voltage and Thevenin resistance between the AB terminals of Fig. 4-4a; the values are

$$V_{TH} = 4 \text{ V}$$

$$R_{TH} = 2 \text{ k}\Omega$$

Figure 4-4d is the Thevenin circuit with these values.

EXAMPLE 4-4.
Figure 4-5a shows a circuit with the load already removed. Work out the values of V_{TH} and R_{TH}. Show the Thevenin circuit.

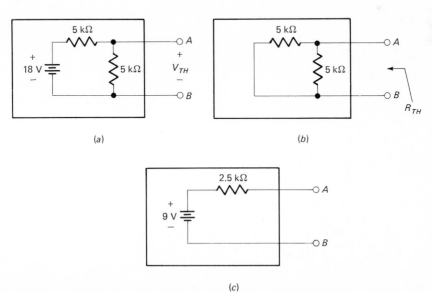

Figure 4-5. Applying Thevenin's theorem.

(a)

(b)

(c)

SOLUTION.
Since Fig. 4-5*a* is a voltage divider with equal resistors, the Thevenin voltage equals half the source voltage. So,

$$V_{TH} = 9 \text{ V}$$

With the source reduced to zero, the circuit looks like Fig. 4-5*b*. Two 5-kΩ resistors in parallel produce a Thevenin resistance of

$$R_{TH} = 2.5 \text{ k}\Omega$$

The Thevenin circuit is an equivalent circuit with a 9-V source and a 2.5-kΩ series resistor as shown in Fig. 4-5*c*.

EXAMPLE 4-5.
The load between the *AB* terminals of Fig. 4-6*a* has already been removed. What is the Thevenin voltage? The Thevenin resistance? The Thevenin circuit?

SOLUTION.
There is no voltage across the 2-kΩ resistor because there is no current in it. As a result, the voltage between the *AB* terminals equals the voltage across the 3-kΩ resistor (Kirchhoff's voltage law). Since the 6-kΩ resistor and the 3-kΩ resistor form a voltage divider,

Figure 4-6. Another example of Thevenin's theorem.

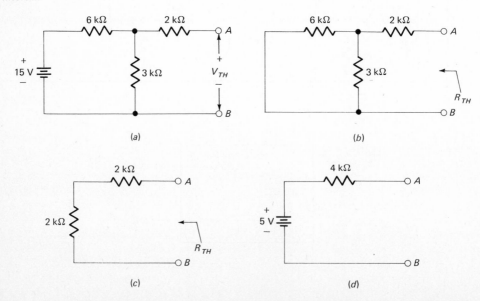

$$V_{TH} = \frac{R_2}{R} V = \frac{3000}{9000} 15 = 5 \text{ V}$$

Next, get R_{TH}. Figure 4-6b shows the circuit with the source reduced to zero. The 6-kΩ resistor and the 3-kΩ resistor are in parallel, because they are between the same pair of equipotential points. When these two resistors are combined, the circuit simplifies to Fig. 4-6c. The Thevenin resistance between the AB terminals is

$$R_{TH} = 2 \text{ kΩ} + 2 \text{ kΩ} = 4 \text{ kΩ}$$

Figure 4-6d shows the Thevenin circuit; it's an equivalent circuit with a 5-V source in series with a 4-kΩ resistor.

EXAMPLE 4-6.
Figure 4-7a shows a 10-kΩ load between the AB terminals. Work out the Thevenin voltage, the Thevenin resistance, and show the Thevenin circuit.

SOLUTION.
First, open the 10-kΩ resistor to get Fig. 4-7b. The parallel connection of 6 kΩ and 3 kΩ is equivalent to 2 kΩ. Therefore, the Thevenin voltage is

$$V_{TH} = \frac{R_2}{R} V = \frac{1000}{3000} 24 = 8 \text{ V}$$

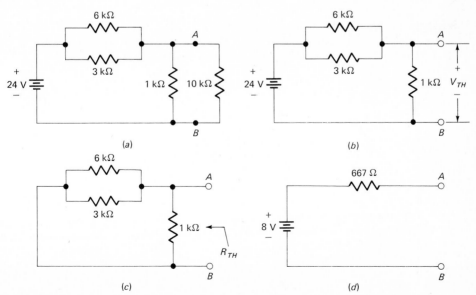

Figure 4-7. Applying Thevenin's theorem.

Second, reduce the source to zero to get Fig. 4-7c. The 6 kΩ and 3 kΩ have an equivalent resistance of 2 kΩ; therefore, the Thevenin resistance is

$$R_{TH} = \frac{2000 \times 1000}{2000 + 1000} = 667 \ \Omega$$

Figure 4-7d shows the Thevenin circuit. In this equivalent circuit the 8-V source is in series with the 667-Ω resistor.

Test 4-2

1. To get the Thevenin voltage, you
   (a) short the load   (b) reduce the source to zero   (c) open the load
   (d) none of these .... ( )
2. In calculating the Thevenin voltage, you often use
   (a) the zero-source circuit   (b) an ohmmeter   (c) an ammeter
   (d) the voltage-divider theorem .... ( )
3. A Thevenin circuit with a load connected between its AB terminals has how many equipotential points?
   (a) 1   (b) 2   (c) 3   (d) 4 .... ( )
4. Which of these does not belong?
   (a) Thevenin voltage   (b) zero source   (c) Thevenin resistance .... ( )

4-3. THEVENIN'S THEOREM

During your reading of this section, learn the answers to

What is Thevenin's theorem?
What two conditions must be satisfied?

The theorem

Suppose you have a complicated circuit and want the current or voltage for only one resistor in this circuit. Nothing prevents you from drawing or visualizing this resistor as a load resistor R_L connected between a pair of AB terminals (Fig. 4-8a); the rest of the complicated circuit is left of the AB terminals.

Thevenin (1883) was the first to prove the following: the circuit left of the AB terminals (Fig. 4-8a) acts like a voltage source V_{TH} in series with a resistance R_{TH} (Fig. 4-8b). This important statement is called *Thevenin's theorem*. To put it another way, a Thevenin circuit like Fig. 4-8b produces the same load current and load voltage as exists in the original circuit of Fig. 4-8a.

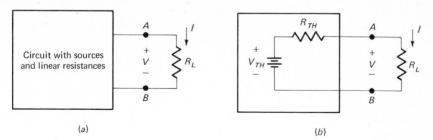

Figure 4-8. Thevenin's theorem.

(a)   (b)

Linearity and independent sources

The proof of Thevenin's theorem is too advanced to reproduce here. In the proof, however, are two crucial conditions that must be satisfied when you apply the theorem.

First, all resistances left of the *AB* terminals in the original circuit (Fig. 4-8a) must be *linear*. (A 5-kΩ resistor has to have a constant value of 5 kΩ even though its voltage and current vary.) Unless this condition is satisfied, you cannot use Thevenin's theorem. The linearity condition does not apply to R_L; it can be linear or nonlinear.

Second, only *independent sources* are reduced to zero when finding R_{TH}; an independent source is one whose value is independent of any load voltage or load current in the original circuit. This book uses only independent sources; so when you find R_{TH}, reduce all sources to zero. (Example 4-7 discusses a dependent source.)

Analogy to logarithms

If you are not familiar with logarithms yet, skip to Example 4-7. Given two numbers M and N, you can find the product by adding the logarithms of the numbers and taking the antilogarithm of the sum. Sometimes, this indirect method is easier than direct multiplication. Figure 4-9a illustrates the idea.

There's no mathematical relation between logarithms and Thevenin's theorem. But there is this similarity: if you are after the current or voltage of only one resistor in a complicated circuit, it may be easier to get the Thevenin circuit and then calculate the desired current or voltage. Figure 4-9b shows this indirect approach: first you get the Thevenin circuit; second you connect the load resistor and the calculate the current or voltage.

In many cases, the indirect method of using a Thevenin circuit turns out to be much easier and faster than direct methods based on Kirchhoff's laws.

EXAMPLE 4-7.
Are the sources of Fig. 4-10a and b dependent or independent sources?

Figure 4-9. Analogy of Thevenin's theorem to logarithms.

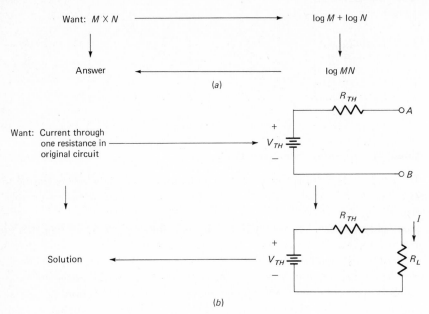

(a)

(b)

Figure 4-10. Dependent and independent sources.

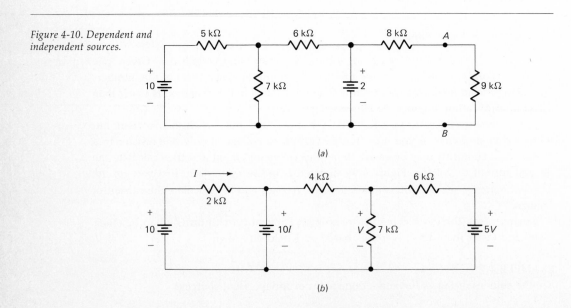

(a)

(b)

SOLUTION.

The value of an independent source does not depend on any load current or load voltage in the circuit. The sources in Fig. 4-10*a* have values of 10 V and 2 V; these values are constants and do not depend on voltage or current elsewhere in the circuit. Therefore, the sources are independent sources.

In Fig. 4-10*b*, the source on the left has a value of 10 V, which in no way depends on other voltages or currents; therefore, the left source is an independent source.

But the middle and right sources of Fig. 4-10*b* are different. *I* is the current through the 2-kΩ resistor; therefore, the 10*I* source has a value that does depend on the current in another part of the circuit. Also, *V* is the voltage across the 7-kΩ resistor; therefore, the 5*V* source has a value that depends on another voltage. The 10*I* and 5*V* sources are examples of dependent sources.

This is the last time you will encounter dependent voltage sources in this book; the rest of the book uses only independent sources, so no problems arise when reducing sources to zero. Later when you get into electronics, you will again see dependent sources; when you apply Thevenin's theorem to these circuits, remember to reduce only the independent sources to zero; leave the dependent sources alone.

EXAMPLE 4-8.

Find the values of *V* and *I* in Fig. 4-11*a* using Thevenin's theorem.

SOLUTION.

First, remove the load resistance to get the open-load circuit of Fig. 4-11*b*. In this open-load circuit, no charges flow through the 1-kΩ resistance; therefore, no voltage can be across the 1-kΩ resistance. Kirchhoff's voltage law implies V_{TH} equals the voltage across the 3-kΩ resistance. With the voltage-divider theorem,

$$V_{TH} = \frac{R_2}{R} V = \frac{3000}{9000} 12 = 4 \text{ V}$$

Second, reduce the 12-V source to zero to get the zero-source circuit of Fig. 4-11*c*. The 6-kΩ resistance is in parallel with the 3-kΩ resistance; these two combine into the equivalent 2-kΩ resistance of Fig. 4-11*d*. The Thevenin resistance between the terminals is

$$R_{TH} = 3 \text{ k}\Omega$$

Third, draw the Thevenin circuit with the load connected (Fig. 4-11*e*). In this circuit

$$I = \frac{4}{8000} = 0.5 \text{ mA}$$

The load voltage equals

$$V = R_L I = 5000 \times 0.0005 = 2.5 \text{ V}$$

Figure 4-11. Example of Thevenin's theorem.

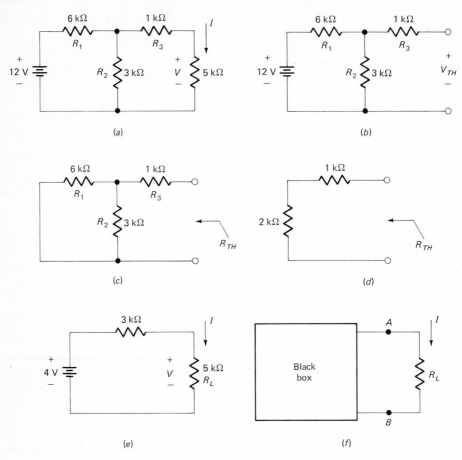

(a)

(b)

(c)

(d)

(e)

(f)

If you were to analyze the original circuit (Fig. 4-11a), you would get exactly the same load current and load voltage for the 5-kΩ resistor.

EXAMPLE 4-9.

A black box has a load resistance of R_L connected to its output terminals as shown in Fig. 4-11f. When the load is opened, a voltmeter reads 15 V between the *AB* terminals. When all sources inside the black box are reduced to zero, an ohmmeter measures 10 kΩ between the *AB* terminals. What is the Thevenin circuit for the black box?

SOLUTION.

No matter what's inside the black box, it acts the same as a 15-V source in series with a 10-kΩ resistance. In other words, no matter what you connect between the *AB* terminals, you get exactly the same value of load current with the black box or the Thevenin circuit.

The point is you can measure V_{TH} and R_{TH} in a built-up circuit; some circuits are extremely complicated, and it may be easier to measure the Thevenin quantities than to calculate them.

EXAMPLE 4-10.

Solve for the value of I in Fig. 4-12a.

Figure 4-12. Applying Thevenin's theorem more than once.

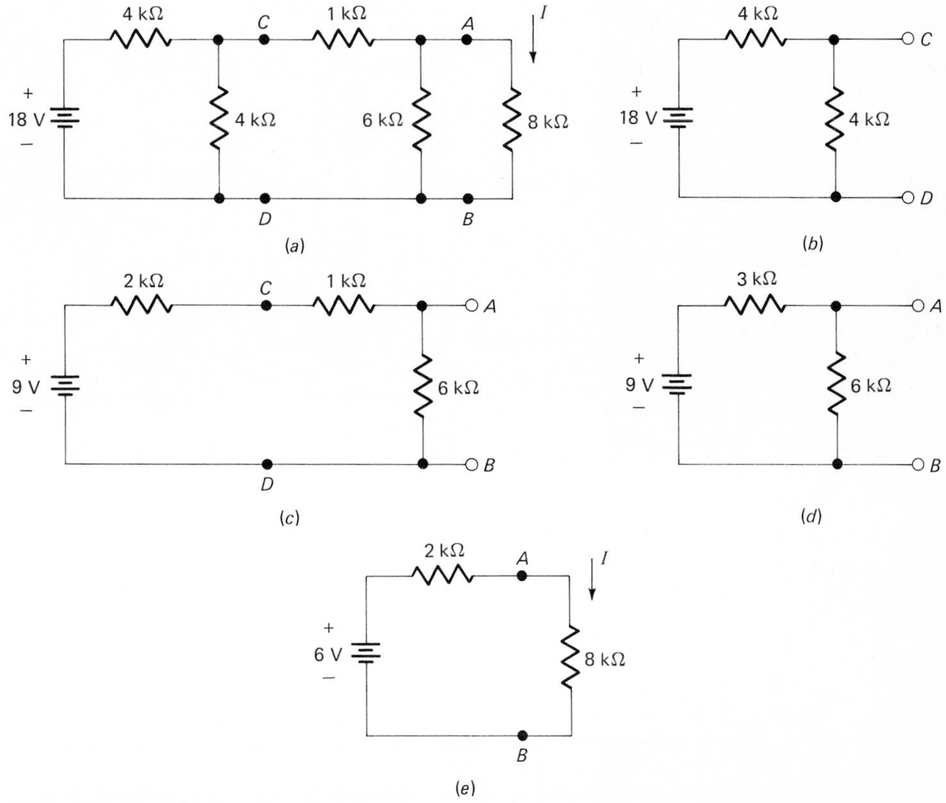

SOLUTION.

The key idea in this example is to apply Thevenin's theorem more than once. The first application of Thevenin's theorem is at the *CD* terminals. When we open the *CD* terminals, we get the open-load circuit of Fig. 4-12*b*. The equal resistances give a V_{TH} of 9 V. With a zero source, 4 kΩ in parallel with 4 kΩ gives an R_{TH} of 2 kΩ. Therefore, we can replace the original circuit left of the *CD* terminals by the Thevenin circuit shown in Fig. 4-12*c*.

The resulting circuit is still not as simple as it can be. Combining the 2-kΩ and 1-kΩ resistances gives the 3-kΩ resistance of Fig. 4-12*d*. Applying the voltage-divider theorem to Fig. 4-12*d* gives

$$V_{TH} = \frac{6000}{9000} \, 9 = 6 \text{ V}$$

and the Thevenin resistance is

$$R_{TH} = \frac{3000 \times 6000}{3000 + 6000} = 2 \text{ kΩ}$$

Figure 4-12*e* shows the final circuit with the load connected. In this series circuit,

$$I = \frac{6}{10,000} = 0.6 \text{ mA}$$

Remember the point of this problem: you can apply Thevenin's theorem more than once in the same problem.

Test 4-3

1. You cannot use Thevenin's theorem if the
   (*a*) load is nonlinear   (*b*) load is linear   (*c*) load is dependent
   (*d*) internal resistances are nonlinear .... ( )
2. A source is dependent if it
   (*a*) has a numerical value   (*b*) has a fixed value
   (*c*) has a variable value
   (*d*) is controlled by a current or voltage elsewhere in the circuit .... ( )
3. Which of the following belongs least?
   (*a*) controlled   (*b*) fixed   (*c*) numerical   (*d*) independent .... ( )
4. In working out a problem, you are allowed to use Thevenin's theorem how many times?
   (*a*) 1   (*b*) 2   (*c*) 3   (*d*) any number .... ( )

4-4. OTHER WAYS TO GET THEVENIN RESISTANCE

Opening the load terminals is usually practical. In other words, we can look at a schematic diagram and mentally disconnect the load resistance during the calculation of

V_{TH}. Likewise, given a built-up circuit, we usually can disconnect the load and measure V_{TH} with a voltmeter.

Measuring R_{TH} is not always straightforward. With some built-up circuits, you cannot reduce the sources to zero. The sources may be inside a sealed box with no way of adjusting them to zero. Therefore, you need alternative ways of finding R_{TH}. This section is about two other widely used methods for calculating or measuring Thevenin resistance.

In what follows, your job is to learn

What is the shorted-load method?
What is the matched-load method?

Shorted-load method

Given any two-terminal black box with sources and linear resistances, we know it acts the same as a voltage source in series with an equivalent resistance (see Fig. 4-13a). The Thevenin circuit behind the *AB* terminals produces exactly the same load current as the original circuit, regardless of what is connected to the *AB* terminals. This is the whole point of the Thevenin theorem.

This being the case, we can even connect a *short* (a load with zero resistance) between the load terminals as shown in Fig. 4-13b. From now on, this special case is called the *shorted-load condition*. The resulting load current is designated I_{SL}, where the subscripts stand for shorted load.

In Fig. 4-13b,

$$I_{SL} = \frac{V_{TH}}{R_{TH}} \qquad\qquad (4\text{-}2)$$

If $V_{TH} = 10$ V and $R_{TH} = 2$ kΩ, the shorted-load current is

$$I_{SL} = \frac{10}{2000} = 5 \text{ mA}$$

Figure 4-13. Shorted-load method.

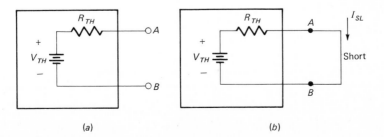

(a)   (b)

Or given a V_{TH} of 6 V and an R_{TH} of 20 Ω, the shorted-load current is

$$I_{SL} = \frac{6}{20} = 0.3 \text{ A}$$

If you have the values of V_{TH} and I_{SL}, you can get R_{TH} by using

$$R_{TH} = \frac{V_{TH}}{I_{SL}} \tag{4-3}$$

This formula is derived from Eq. (4-2). It says the Thevenin resistance equals the ratio of the Thevenin voltage to the shorted-load current. So, if the values of V_{TH} and I_{SL} are measured for a built-up circuit, the value of R_{TH} can be calculated.

Here's an example of the shorted-load method. If you build the circuit of Fig. 4-14a, you can measure the voltage between the AB terminals; it will equal

$$V_{TH} = 4 \text{ V}$$

Next, you can short the AB terminals as shown in Fig. 4-14b. If you measure the current through this short, you will find

$$I_{SL} = 2 \text{ mA}$$

Figure 4-14. Example of shorted-load method.

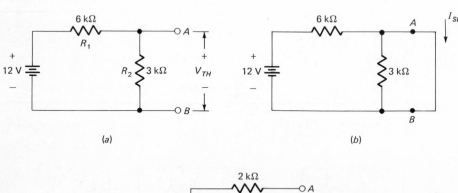

(a)   (b)

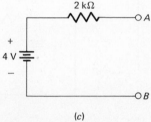

(c)

Once you have V_{TH} and I_{SL}, you can calculate R_{TH} with Eq. (4-3):

$$R_{TH} = \frac{V_{TH}}{I_{SL}} = \frac{4}{0.002} = 2 \text{ k}\Omega$$

Therefore, Fig. 4-14a has a Thevenin circuit like Fig. 4-14c.

A word of caution: the shorted-load method is usually safe with electronic circuits. As a guide, an I_{SL} less than 1 A is almost always safe. If you have no idea how large I_{SL} may be, don't put a short across the load terminals of a built-up circuit.

Matched-load method

A *rheostat* is a variable resistor. One way to make a rheostat is to connect the wiper to one end of a potentiometer (see Fig. 4-15a). Moving the wiper up or down changes the load resistance; in turn, this changes the load voltage.

A *matched load* is one whose resistance equals the Thevenin resistance:

$$R_{\text{match}} = R_{TH} \qquad (4-4)$$

If the Thevenin resistance of a circuit is 5 kΩ, a matched load has a resistance of 5 kΩ. When the rheostat of Fig. 4-15b is adjusted to R_{match}, the load voltage equals half of the Thevenin voltage:

$$V_L = 0.5V_{TH} \qquad (4-5)$$

You can measure the Thevenin resistance as follows. Given a black box, connect a rheostat across its load terminals and adjust the resistance to get a load voltage of half the Thevenin voltage. Then disconnect the rheostat and measure its resistance with an ohmmeter. The reading on the ohmmeter equals R_{TH}.

As an example, the circuit of Fig. 4-16a has these Thevenin values:

$$V_{TH} = 4 \text{ V}$$

$$R_{TH} = 2 \text{ k}\Omega$$

Figure 4-15. Matched-load method.

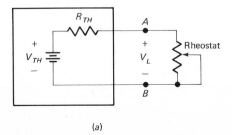

(a)

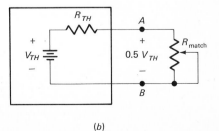

(b)

Figure 4-16. Example of matched-load method.

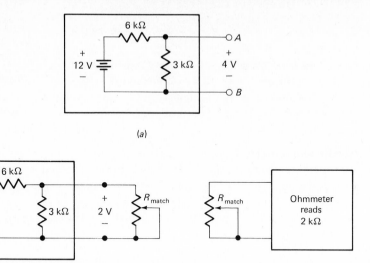

(a)

(b)   (c)

Suppose the circuit is inside a black box and you have no idea what the circuit is. With a voltmeter between the load terminals of Fig. 4-16a, you would read 4 V; this is the Thevenin voltage. Next, you can connect a rheostat between the load terminals and adjust resistance to get a load voltage of 2 V (Fig. 4-16b). When you disconnect the rheostat and measure its resistance (Fig. 4-16c), the ohmmeter will read 2 kΩ. In this way, you are finding the Thevenin resistance by measurement rather than calculation.

Summary

Now we have three methods for finding R_{TH}: the zero-source method, the shorted-load method, and the matched-load method. All are important. When analyzing schematic diagrams, choose whichever method involves the least amount of work (usually the zero-source method).

When you have builtup circuits, you can measure the Thevenin quantities. Sometimes, measurement of R_{TH} is impossible with the zero-source method, because you cannot physically reduce the sources to zero. In this case, use the shorted-load or matched-load method.

EXAMPLE 4-11.

A flashlight battery is a good example of a sealed box (Fig. 4-17a). When a voltmeter is connected between the terminals, it reads 1.5 V (Fig. 4-17b). When an ammeter is connected between the terminals, it reads 1.5 A (Fig. 4-17c). What is the Thevenin circuit for the flashlight battery?

SOLUTION.

Voltmeters have very high resistances. Because of this, the voltmeter of Fig. 4-17b reads the Thevenin voltage of the battery to a close approximation. As shown in Fig. 4-17b, $V_{TH} = 1.5$ V.

Ammeters have very low resistances. Because of this, the ammeter of Fig. 4-17c reads the shorted-load current to a close approximation.

With Eq. (4-3),

$$R_{TH} = \frac{V_{TH}}{I_{SL}} = \frac{1.5 \text{ V}}{1.5 \text{ A}} = 1 \text{ } \Omega$$

Figure 4-17d shows the Thevenin circuit for a flashlight battery (these are typical values).

In earlier discussions, we treated the Thevenin resistance of batteries as zero. This is usually a good approximation because the R_{TH} of the battery is normally much

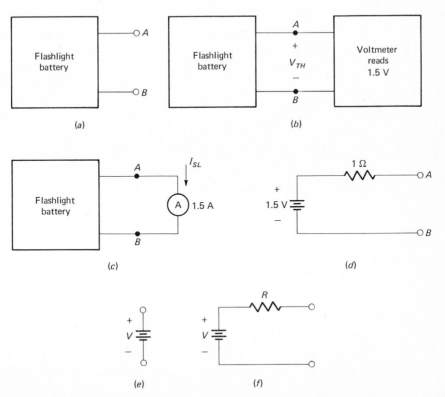

Figure 4-17. Thevenin circuit of flashlight battery.

smaller than the load resistances connected to the battery. From now on, whenever we use the symbol of Fig. 4-17e, we mean the source has zero Thevenin resistance. If the Thevenin resistance is important in a discussion or problem, we will draw it separately as shown in Fig. 4-17f.

This example treated voltmeter resistance as infinite and ammeter resistance as zero. This is how a good voltmeter and ammeter should act, that is, the voltmeter should not disturb the voltage across the open terminals, and the ammeter should appear like a short. When you need exact answers, however, you have to include the effects of voltmeter and ammeter resistances (see Chap. 7).

EXAMPLE 4-12.
Figure 4-18a shows the schematic symbol of a *solar cell,* a device that produces an output voltage when sunlight strikes it. (Solar cells are often used on satellites.) The voltmeter of Fig. 4-18b reads 2.5 V, and the ammeter of Fig. 4-18c reads 100 mA. What is the Thevenin circuit for the solar cell?

SOLUTION.
Again, approximate the voltmeter resistance as infinite and the ammeter resistance as zero. In Fig. 4-18b,

$$V_{TH} = 2.5 \text{ V}$$

In Fig. 4-18c,

$$I_{SL} = 100 \text{ mA}$$

Figure 4-18. Thevenin circuit of solar cell.

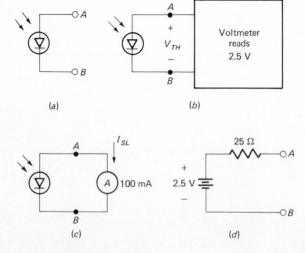

(a)

(b)

(c)

(d)

The ratio of these quantities equals the Thevenin resistance:

$$R_{TH} = \frac{2.5}{0.1} = 25 \ \Omega$$

Figure 4-18d is the Thevenin circuit for the solar cell.

EXAMPLE 4-13.
Suppose a black box contains the circuit of Fig. 4-19a. If you put a short between the AB terminals, what does I_{SL} equal? If you adjust a rheostat to get half the Thevenin voltage, what resistance does the rheostat have?

SOLUTION.
Figure 4-19b shows the Thevenin circuit left of the AB terminals. Thevenin's theorem tells us Fig. 4-19b produces the same load effects as the original circuit of Fig. 4-19a.
    When a short is connected between the load terminals (Fig. 4-19c), the resulting current is

$$I_{SL} = \frac{V_{TH}}{R_{TH}} = \frac{5}{1000} = 5 \ \text{mA}$$

When a matched load is connected between the AB terminals (Fig. 4-19d), it has the same value as R_{TH}; therefore,

$$R_{\text{match}} = R_{TH} = 1 \ \text{k}\Omega$$

Figure 4-19. Shorted-load current and matched load.

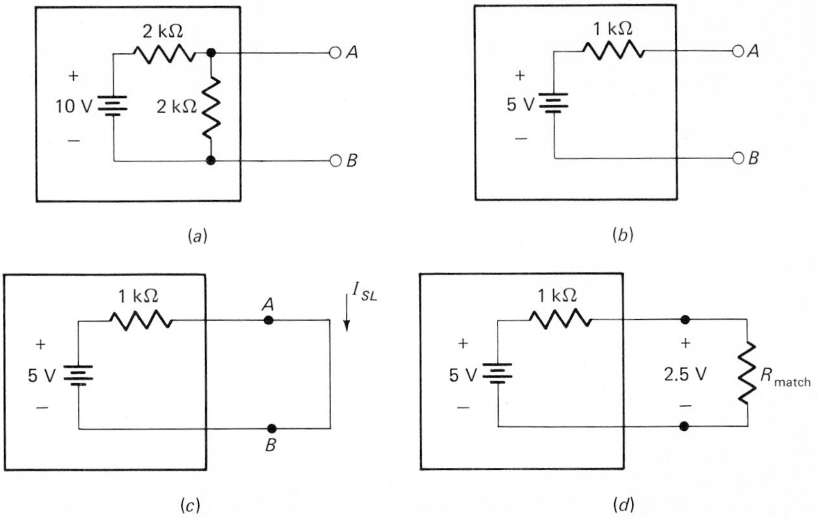

(a)

(b)

(c)

(d)

Figure 4-20. Applying Thevenin's
theorem.

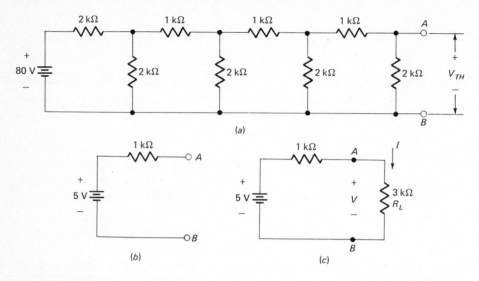

(a)

(b)   (c)

EXAMPLE 4-14.
If you build the circuit of Fig. 4-20a, you will measure a V_{TH} of 5 V. If you also connect a
rheostat between the AB terminals, you will find that a 1-kΩ load resistance drops the
voltage to 2.5 V. What is the Thevenin circuit to the left of the AB terminals?

SOLUTION.
This is easy. We are given V_{TH} equal to 5 V. We are also given R_{match} equals 1 kΩ.
Therefore, R_{TH} is 1 kΩ.
    Figure 4-20b summarizes the solution; this is the Thevenin circuit for the original
circuit.

EXAMPLE 4-15.
If you connect a 3-kΩ load resistance between the AB terminals of Fig. 4-20a, what is
the load current? The load voltage?

SOLUTION.
Figure 4-20c shows the 3-kΩ resistance connected to the Thevenin circuit. The current is

$$I = \frac{5}{1000 + 3000} = 1.25 \text{ mA}$$

and the voltage is

$$V = R_L I = 3000 \times 0.00125 = 3.75 \text{ V}$$

The values of I and V in Fig. 4-20c are the same as when a 3-kΩ load resistance is connected between the AB terminals of the original circuit (Fig. 4-20a). If the circuit is built up, you can verify this by measurements.

Test 4-4

1. I_{SL} always equals
   (a) a nonzero value   (b) the load current with a shorted source
   (c) the ratio of the Thevenin voltage to the Thevenin resistance   .... ( )
2. In the shorted-load method for getting R_{TH}, you
   (a) measure R_{TH} with an ohmmeter and a shorted source
   (b) short the load and measure the load voltage   (c) calculate R_{TH} .... ( )
3. With a shorted load, a Thevenin circuit has how many equipotential points?
   (a) 1   (b) 2   (c) 3   (d) any number .... ( )
4. The shorted-load method works
   (a) experimentally   (b) on paper   (c) either   (d) neither   .... ( )
5. A shorted load is to a matched load as zero voltage is to
   (a) Thevenin voltage   (b) Thevenin resistance   (c) source voltage
   (d) half Thevenin voltage .... ( )
6. A matched load always
   (a) equals Thevenin resistance   (b) has half the V_{TH} across it
   (c) drops the load voltage to half the open-load voltage
   (d) all of the foregoing .... ( )
7. The current through a matched load equals
   (a) twice the shorted-load current   (b) half the shorted-load current
   (c) V_{TH}/R_{TH} (d) none of these .... ( )
8. A short is defined as
   (a) zero resistance   (b) infinite resistance   (c) zero current
   (d) nonzero voltage .... ( )

4-5. THE WHEATSTONE BRIDGE

Figure 4-21a shows a *Wheatstone bridge,* a circuit used a great deal in instruments and other applications. R_1 and R_2 form a voltage divider on the left side of the bridge; R_3 and R_4 are a voltage divider on the right side of the bridge. Usually, a load resistance bridges the path between the AB terminals (Fig. 4-21b). As shown, current I is in the load resistance.

The big questions are

When is a Wheatstone bridge balanced?
How do you find load current in a bridge?

Open-load condition

When there is no load resistance between the AB terminals, we have the open-load circuit of Fig. 4-21a. The voltage divider on the left produces a voltage across R_2 of

$$V_2 = \frac{R_2}{R_1 + R_2} V \qquad (4\text{-}6a)$$

The voltage divider on the right sets up a voltage across R_4 of

$$V_4 = \frac{R_4}{R_3 + R_4} V \qquad (4\text{-}6b)$$

Figure 4-21c shows these voltages.

To get an important formula, sum the voltages around the A-B-C-A loop of Fig. 4-21c:

$$V_{AB} + V_4 - V_2 = 0$$

or $\qquad\qquad\qquad V_{AB} = V_2 - V_4 \qquad\qquad\qquad (4\text{-}7)$

This says the voltage between the AB terminals for the open-load condition equals the difference of V_2 and V_4, where V_2 and V_4 are given by Eqs. (4-6a) and (4-6b).

For instance, if V_2 is 5 V and V_4 is 2 V, then

$$V_{AB} = 5 - 2 = 3 \text{ V}$$

Figure 4-21. Wheatstone bridge.

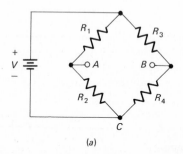

(a)

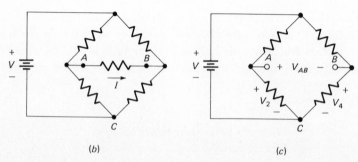

(b)   (c)

On the other hand, if V_2 equals 4 V and V_4 is 6 V,

$$V_{AB} = 4 - 6 = -2 \text{ V}$$

As mentioned earlier, when a negative sign turns up in an answer for current or voltage, it means the current or voltage is opposite the direction shown on the original diagram. Therefore, in Fig. 4-21c a value of -2 V for V_{AB} means the true polarity of voltage is opposite that shown.

Balanced bridge

There is an important special case. When the voltage on the left side equals the voltage on the right side of Fig. 4-21c, $V_2 = V_4$ and

$$V_{AB} = V_2 - V_4 = 0$$

This special case is called *balance,* and we refer to the circuit as a *balanced bridge.*
When the bridge is balanced,

$$V_2 = V_4$$

or with Eqs. (4-6a) and (4-6b),

$$\frac{R_2}{R_1 + R_2} V = \frac{R_4}{R_3 + R_4} V$$

Dividing both sides by V leaves

$$\frac{R_2}{R_1 + R_2} = \frac{R_4}{R_3 + R_4}$$

By algebra, this rearranges into

$$\frac{R_1}{R_2} = \frac{R_3}{R_4} \qquad \text{(balance)} \tag{4-8}$$

Equation (4-8) is called the *balanced-bridge formula.* It tells us a bridge is balanced when the ratio of the resistances on the left side equals the ratio of the resistances on the right side.

Many instruments use the balanced-bridge condition. Among other things, we can use a balanced bridge to measure unknown resistances to a high degree of accuracy. Example 4-18 discusses this further.

Unbalanced and loaded bridge

Figure 4-22a shows a Wheatstone bridge drawn with vertical lines instead of diagonal ones (for convenience). A load resistance R_L is between the AB terminals. Unless

Figure 4-22. Applying Thevenin's theorem to Wheatstone bridge.

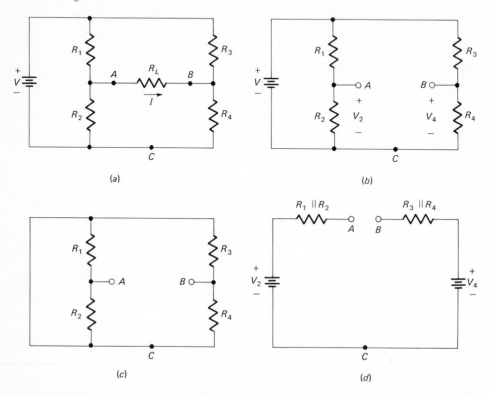

(a)

(b)

(c)

(d)

the bridge is balanced, there is a voltage across the load and a current through it. We are going to derive a formula for I by applying Thevenin's theorem.

We begin by removing the load as shown in Fig. 4-22b. In this open-load circuit, there are two voltage dividers. The voltage divider on the left has a Thevenin voltage of

$$V_2 = \frac{R_2}{R_1 + R_2} V$$

When the source is reduced to zero, we get the zero-source circuit shown in Fig. 4-22c. The Thevenin resistance between the AC terminals is the parallel of R_1 and R_2:

$$R_{TH} = R_1 \parallel R_2$$

where the vertical lines $\parallel$ are shorthand for "in parallel with." We calculate the value of $R_1 \parallel R_2$ by the product-over-sum rule. Figure 4-22d shows the Thevenin circuit for the left voltage divider.

Similarly, the right voltage divider has a Thevenin voltage of

$$V_4 = \frac{R_4}{R_3 + R_4} V$$

and a Thevenin resistance of $R_3 \parallel R_4$. Figure 4-22d shows the Thevenin circuit for the right voltage divider.

The final step in deriving a formula for I is to reconnect the load resistance between the AB terminals as shown in Fig. 4-23. This series circuit produces exactly the same value of I as the original Wheatstone bridge (Fig. 4-22a). Summing voltages around the series circuit of Fig. 4-23 gives

$$R_L I + (R_3 \parallel R_4)I + V_4 - V_2 + (R_1 \parallel R_2)I = 0$$

Rearranging and solving for I,

$$I = \frac{V_2 - V_4}{R_1 \parallel R_2 + R_L + R_3 \parallel R_4} \tag{4-9}$$

Equation (4-9) is not really complicated. It says the current equals the difference of the open-load voltages on each side of the bridge divided by the sum of $R_1 \parallel R_2$, R_L, and $R_3 \parallel R_4$. After you go through a few examples, finding the current in an unbalanced bridge becomes easy.

The unbalanced Wheatstone-bridge problem shows the power of the Thevenin theorem. Try finding a formula for current in a different way, and you will have a problem on your hands. There are other ways, of course. But they are incredibly difficult compared to the simplicity of applying Thevenin's theorem to each side of the bridge.

EXAMPLE 4-16.
What is the value of I in Fig. 4-24a?

SOLUTION.
Remove the 4-kΩ load resistance to get the open-load circuit of Fig. 4-24b. The left voltage divider produces a Thevenin voltage of

$$V_2 = \frac{6000}{3000 + 6000} \, 12 = 8 \text{ V}$$

Figure 4-23. Thevenin circuit for Wheatstone bridge.

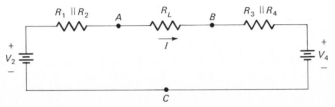

Figure 4-24. Solving an unbalanced-bridge problem.

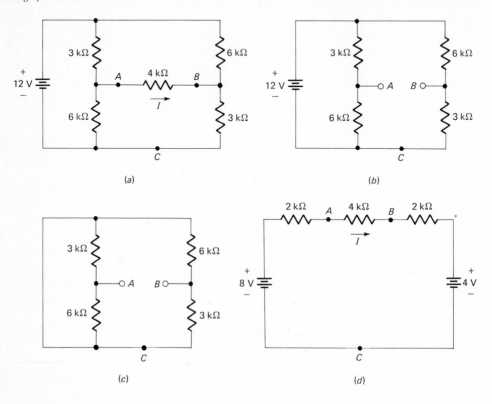

(a)

(b)

(c)

(d)

The right voltage divider produces a Thevenin voltage of

$$V_4 = \frac{3000}{6000 + 3000}\, 12 = 4 \text{ V}$$

When the source is reduced to zero, we get the zero-source circuit of Fig. 4-24c. The equivalent resistance between the *AC* terminals is 3 kΩ in parallel with 6 kΩ; therefore, the left voltage divider has a Thevenin resistance of

$$R_1 \parallel R_2 = 3000 \parallel 6000 = 2 \text{ k}\Omega$$

Similarly, the equivalent resistance between the *BC* terminals is

$$R_3 \parallel R_4 = 6000 \parallel 3000 = 2 \text{ k}\Omega$$

Figure 4-24d shows both Thevenin circuits with the load reconnected between the *AB* terminals. The difference between V_2 and V_4 is 4 V. The equivalent resistance of the series circuit is 8 kΩ. Therefore,

$$I = \frac{4}{8000} = 0.5 \text{ mA}$$

In the future, you can solve for the load current in an unbalanced bridge by going through all the steps as we did in the example, or you can substitute directly in Eq. (4-9).

EXAMPLE 4-17.
Repeat the preceding example, but use an ammeter of zero resistance in place of R_L.

SOLUTION.
Figure 4-25a shows the Wheatstone bridge with an ammeter between the AB terminals, and Fig. 4-25b is the equivalent circuit. Because the ammeter has zero resistance, we can redraw the circuit as shown in Fig. 4-25c. The current is

$$I = \frac{8 - 4}{2000 + 2000} = \frac{4}{4000} = 1 \text{ mA}$$

Therefore, 1 mA is in the ammeter of Fig. 4-25a.

EXAMPLE 4-18.
Figure 4-26 shows a *Wheatstone-bridge ohmmeter*, an instrument used to measure resistance accurately. When measuring an unknown resistance, you adjust the rheostat until the bridge balances.

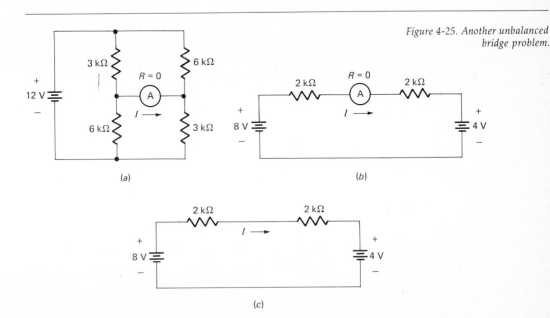

Figure 4-25. Another unbalanced bridge problem.

(a)

(b)

(c)

Figure 4-26. Measuring resistance with a Wheatstone bridge.

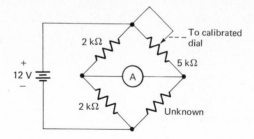

If the bridge balances when the rheostat equals 5 kΩ, what value does the unknown resistance have?

SOLUTION.

The key equation is the balanced formula:

$$\frac{R_1}{R_2} = \frac{R_3}{R_4}$$

Substituting the values given in Fig. 4-26,

$$\frac{2000}{2000} = \frac{5000}{R_4}$$

Solving for R_4 gives

$$R_4 = 5 \text{ k}\Omega$$

In a commercial Wheatstone-bridge ohmmeter, when you adjust R_3 to get bridge balance (zero ammeter current), you automatically are turning a calibrated dial that points to the value of the unknown resistance. In other words, you don't have to calculate the value of R_4 as we did in this example; it's indicated automatically on a commercial Wheatstone-bridge ohmmeter.

Test 4-5

1. An unloaded bridge must be balanced if
    (*a*) load voltage is zero   (*b*) load current is zero
    (*c*) load resistance is infinite   (*d*) none of these .... ( )
2. A loaded bridge must be balanced if
    (*a*) load current is zero   (*b*) load voltage is zero
    (*c*) load resistance is zero   (*d*) none of these .... ( )
3. A bridge with a shorted load must be balanced if
    (*a*) load voltage is zero   (*b*) load resistance is zero
    (*c*) load current is zero   (*d*) none of these .... ( )

4. An unloaded bridge must be unbalanced if
    (*a*) load current is not zero   (*b*) load voltage is not zero
    (*c*) load current is zero   (*d*) none of these .... ( )
5. A loaded bridge must be unbalanced if
    (*a*) load current is not zero   (*b*) load voltage is zero
    (*c*) load resistance does not match Thevenin resistance .... ( )
6. We found load current in an unbalanced bridge by applying Thevenin's theorem
    how many times?
    (*a*) 1   (*b*) 2   (*c*) 3   (*d*) any number .... ( )

4-6. BASIC IDEA OF SUPERPOSITION THEOREM

Whenever a circuit has more than one source, the *superposition theorem* may reduce the
work of finding currents and voltages. When you study electronics, you will find the
superposition theorem very helpful in understanding how transistor circuits work.

This section begins the study of the superposition theorem for a circuit with two
sources. In the discussion that follows, dig out the answers to

> *What is a first-source circuit?*
> *What is a second-source circuit?*
> *How do you apply the superposition theorem?*

First-source and second-source circuits

Figure 4-27*a* shows a circuit with two voltage sources. To keep the sources distinct,
we call the one on the left the *first source*, and the one on the right the *second source*.
Suppose we are after the current I through the lower 3-kΩ resistor. There are several
ways to find the value of I. This section describes the superposition method.

The *first-source circuit* is the new circuit you get when you reduce all sources except
the first to zero. Figure 4-27*b* shows the first-source circuit. I_1 is the current through the
3-kΩ resistor in this new circuit.

Similarly, if you reduce all sources to zero except the second, the resulting circuit
will be known as the *second-source circuit*. Figure 4-27*c* shows the second-source circuit. Notice that I_2 is the current through the 3-kΩ resistor in this new circuit.

Relation to original current

We are after the current I in the original circuit of Fig. 4-27*a*. The superposition
theorem says I equals the algebriac sum of I_1 and I_2. So, instead of calculating the value
of I directly from the original circuit, we can calculate the values of I_1 and I_2 in the first-
source and second-source circuits; the sum of these individual currents is the original
current I.

Figure 4-27. Superposition theorem. (a) Original circuit. (b) First-source circuit. (c) Second-source circuit.

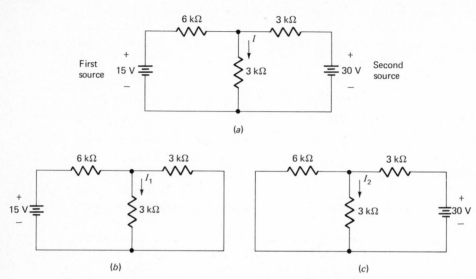

(a)

(b)

(c)

The superposition theorem splits the original two-source problem into two one-source problems. Instead of having to solve a difficult two-source problem, we can solve two simple one-source problems.

If we carry out the work, we will find I_1 equals 1 mA in Fig. 4-27b and $I_2 = 4$ mA in Fig. 4-27c. With the superposition theorem,

$$I = I_1 + I_2 = 1 \text{ mA} + 4 \text{ mA}$$
$$= 5 \text{ mA}$$

Therefore, 5 mA is the current through the 3-kΩ resistance of the original circuit.

The next section states the superposition theorem in formal terms as it applies to any circuit. At that moment, all you need to remember about a two-source circuit is this:

1. The first-source circuit is the original circuit with all sources reduced to zero except the first.
2. The second-source circuit is the original circuit with all sources reduced to zero except the second.

EXAMPLE 4-19.

Figure 4-28a shows the equivalent circuit derived in Example 4-16. Draw the first-source and second-source circuits.

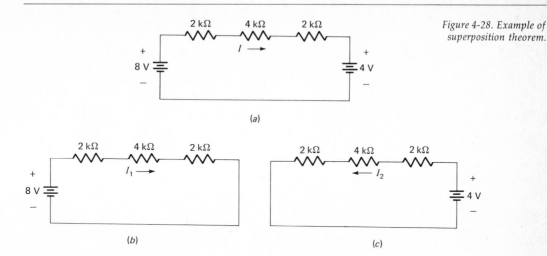

Figure 4-28. Example of superposition theorem.

SOLUTION.

When we reduce all sources except the first to zero, we get Fig. 4-28b, the first-source circuit. Note the true direction of I_1 must be to the right, because conventional flow is from the positive battery terminal to the negative.

To get the second-source circuit, reduce all sources except the second to zero as shown in Fig. 4-28c. Here, the true direction of I_2 is to the left because conventional current is from the positive to the negative battery terminal.

EXAMPLE 4-20.

Given the circuit of Fig. 4-29a, show the first-source and second-source circuits.

SOLUTION.

Call the upper source the *first source,* and the lower one the *second source.* Figure 4-29b is the first-source circuit; Fig. 4-29c is the second-source circuit.

Figure 4-29. Another example of superposition.

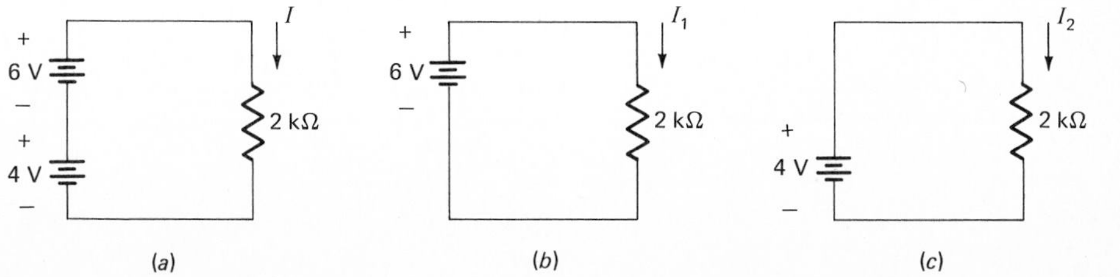

Test 4-6

1. Reducing a voltage source to zero is the same as
   (*a*) opening it   (*b*) shorting it   (*c*) eliminating it
   (*d*) none of these .... ( )
2. How many sources are there in the first-source circuit?
   (*a*) 1   (*b*) 2   (*c*) 3   (*d*) any number .... ( )
3. How many resistances are there in the first-source circuit?
   (*a*) 1   (*b*) 2   (*c*) 3   (*d*) any number .... ( )
4. Which belongs least in the following?
   (*a*) original current   (*b*) Thevenin voltage
   (*c*) first-source and second-source circuits   (*d*) algebraic sum .... ( )
5. The following words can be rearranged to form a sentence: ORIGINAL CURRENT
   SUM IS ALGEBRAIC NOT. With respect to the superposition theorem, this is
   (*a*) true   (*b*) false .... ( )

4-7. SUPERPOSITION THEOREM

When the original circuit has many sources, and we are after the current through a load
resistance, we can visualize or draw the circuit as shown in Fig. 4-30*a*. To the right of
the *AB* terminals is the load resistance whose current we are after. To the left of the *AB*
terminals is the rest of the circuit with *n* sources and any connection of linear resistances.

*Figure 4-30. Superposition
theorem.*

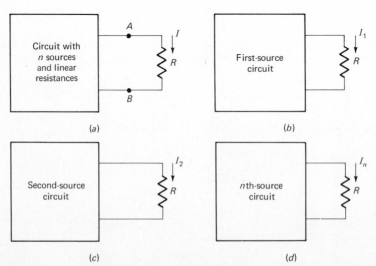

(*a*)   (*b*)

(*c*)   (*d*)

If we reduce all sources to zero except the first, we get the first-source circuit of Fig. 4-30b. The load current in this first-source circuit is I_1.

Reducing all sources to zero except the second gives the second-source circuit (Fig. 4-30c). The load current in this new circuit equals I_2.

Continuing like this for each source, we finally arrive at the last or nth source. In the nth-source circuit of Fig. 4-30d, the load current equals I_n. Therefore, given an original circuit with n sources, we get n new circuits, each with one source.

The remainder of this section answers these questions:

> *How do you apply the superposition theorem to a multisource circuit?*
> *What two conditions must be satisfied?*
> *What does algebraic sum mean?*

The theorem

The superposition theorem says current I in the original circuit equals the algebraic sum of currents $I_1, I_2, \ldots, I_n$ produced by the sources taken one at a time. In symbols,

$$I = I_1 + I_2 + \cdots + I_n \qquad (4\text{-}10)$$

The superposition theorem reminds us of a divide-and-conquer strategy. We start with an original circuit having n sources. A direct attack means solving for I with all sources in the circuit *at the same time.* But with the superposition theorem, we divide the original problem into n simpler problems. By solving each simpler problem and summing the individual currents, we get current I in the original circuit. Figure 4-31 summarizes this divide-and-conquer stategy.

Two conditions

The proof of the superposition theorem is too advanced to give in this book. In the proof, however, are two important conditions that must be satisfied when you apply the theorem.

First, *all* resistances in the circuit including the load must be *linear;* otherwise, you cannot use the superposition theorem.

Second, in getting the new circuits, you reduce only *independent* sources to zero. If

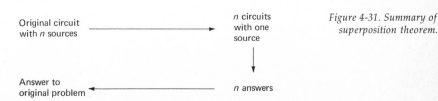

Figure 4-31. Summary of superposition theorem.

there are any dependent sources (Example 4-7), you leave these alone. As mentioned earlier, this book uses only independent sources, so no difficulty arises. When you get to electronics, however, you will come up against dependent sources; to get each one-source circuit, you reduce all independent sources to zero except the one of interest; leave all dependent sources alone.

Algebraic sum

In Eq. (4-10), use the algebraic sum. This means taking the direction of individual currents into account. When an individual current is in the same direction shown for the original current, add the magnitude of the individual current. But when the individual current is opposite the direction shown for the original current, subtract its magnitude.

Figure 4-32a shows an original circuit with current I down. Suppose this circuit contains two sources. If the first-source circuit (Fig. 4-32b) has a downward current of 5

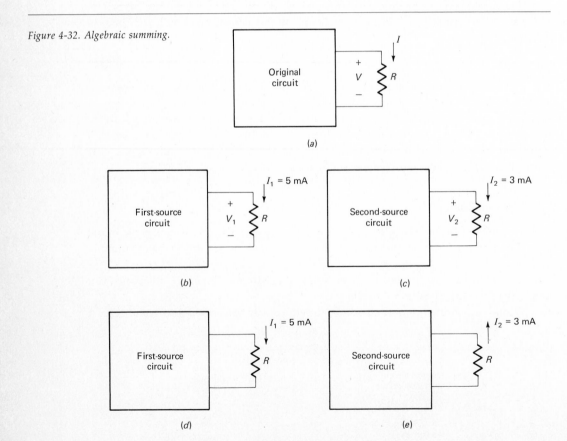

Figure 4-32. Algebraic summing.

(a)

(b)

(c)

(d)

(e)

mA and the second-source circuit (Fig. 4-32c) has a downward current of 3 mA, the original current is

$$I = 5 \text{ mA} + 3 \text{ mA} = 8 \text{ mA}$$

We have added individual currents, because they are in the same direction as the original current; therefore, each individual current increases the total rate of flow.

On the other hand, if it turns out that I_1 is 5 mA downward (Fig. 4-32d) and I_2 is 3 mA upward (Fig. 4-32e), then

$$I = 5 \text{ mA} - 3 \text{ mA} = 2 \text{ mA}$$

I_2 opposes or reduces the effect of I_1; for this reason, we subtract its magnitude.

Voltages

We have given the superposition theorem for currents. The same idea applies to voltages. In other words, the voltage across the load resistance in the original circuit equals the algebraic sum of the individual load voltages produced by the sources taken one at a time. In symbols,

$$V = V_1 + V_2 + \cdots + V_n \qquad (4\text{-}11)$$

For instance, in Fig. 4-32b and c, if $V_1 = 5$ V and $V_2 = 3$ V, then $V = 8$ V.

EXAMPLE 4-21.
Calculate the value of I in Fig. 4-33a using the superposition theorem. Check the answer by another method.

SOLUTION.
Figure 4-33b shows the first-source circuit; the current is

$$I_1 = \frac{6}{2000} = 3 \text{ mA}$$

Figure 4-33. Example 4-21.

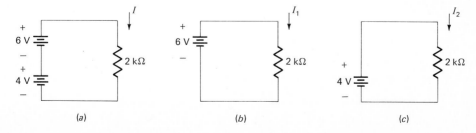

(a)   (b)   (c)

In the second-source circuit of Fig. 4-33c, the current is

$$I_2 = \frac{4}{2000} = 2 \text{ mA}$$

Since the individual currents are in the same direction as shown for the original current, add magnitudes to get

$$I = 3 \text{ mA} + 2 \text{ mA}$$
$$= 5 \text{ mA}$$

We can check this answer directly. In Fig. 4-33a, the voltage across the 2-kΩ resistance is the sum of source voltages (Kirchhoff's voltage law). Therefore, the current equals

$$I = \frac{10}{2000} = 5 \text{ mA}$$

This proves the superposition theorem for the simple circuit of Fig. 4-33a.

EXAMPLE 4-22.
Example 4-16 converted an unbalanced Wheatstone bridge into the series circuit of Fig. 4-34a. Find the value of I by applying the superposition theorem.

SOLUTION.
Figure 4-34b shows the first-source circuit; the current equals

$$I_1 = \frac{8}{8000} = 1 \text{ mA}$$

Figure 4-34. Example 4-22.

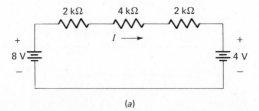

(a)

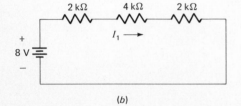

(b)

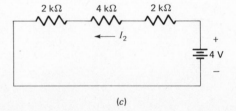

(c)

In the second-source circuit of Fig. 4-34c,

$$I_2 = \frac{4}{8000} = 0.5 \text{ mA}$$

Since I_1 is in the same direction as I, and I_2 is in the opposite direction,

$$I = 1 \text{ mA} - 0.5 \text{ mA}$$
$$= 0.5 \text{ mA}$$

This answer found by superposition is the same as the answer found earlier by a different method. Therefore, this proves the superposition theorem for the original circuit of Fig. 4-34a.

EXAMPLE 4-23.
Calculate the value of I in Fig. 4-35a.

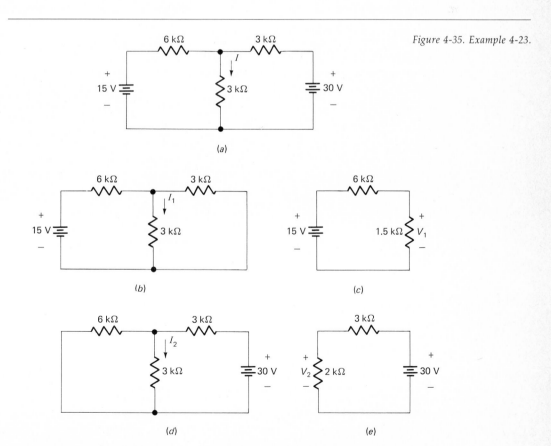

Figure 4-35. Example 4-23.

(a)

(b)

(c)

(d)

(e)

SOLUTION.
Figure 4-35*b* shows the first-source circuit. Since 3 kΩ are in parallel with 3 kΩ, we get the 1.5-kΩ resistance shown in Fig. 4-35*c*. With the voltage-divider theorem,

$$V_1 = \frac{1500}{6000 + 1500}\ 15 = 3 \text{ V}$$

In Fig. 4-35*b*, 3 V is across the middle 3-kΩ resistance; the current through this resistance is

$$I_1 = \frac{3}{3000} = 1 \text{ mA}$$

Figure 4-35*d* shows the second-source circuit; the parallel combination of 6 kΩ and 3 kΩ is the 2-kΩ resistance of Fig. 4-35*e*. Using the voltage-divider theorem,

$$V_2 = \frac{2000}{3000 + 2000}\ 30 = 12 \text{ V}$$

These 12 V are across the middle 3-kΩ resistance of Fig. 4-35*d*; therefore,

$$I_2 = \frac{12}{3000} = 4 \text{ mA}$$

Currents I_1 and I_2 are in the same direction as the original current I of Fig. 4-35*a*. So,

$$I = 1 \text{ mA} + 4 \text{ mA}$$
$$= 5 \text{ mA}$$

There is an alternative way to solve the problem. After finding V_1 and V_2, add these individual voltages to get the original voltage:

$$V = V_1 + V_2 = 3 + 12$$
$$= 15 \text{ V}$$

This is the voltage across the middle 3-kΩ resistance of the original circuit (Fig. 4-35*a*). Since this is the total voltage,

$$I = \frac{15}{3000} = 5 \text{ mA}$$

which agrees with the answer found by adding individual currents.

EXAMPLE 4-24.
Use the superposition theorem to find the Thevenin circuit left of the *AB* terminals of Fig. 4-36*a*.

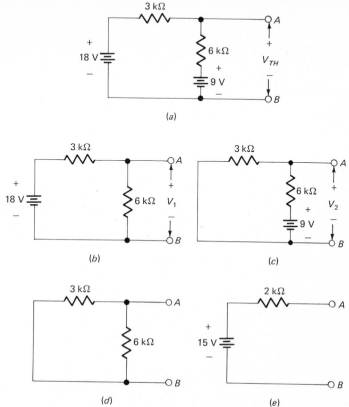

Figure 4-36. Example 4-24.

(a)

(b)   (c)

(d)   (e)

SOLUTION.

This chapter has discussed three theorems: the voltage-divider theorem, Thevenin's theorem, and the superposition theorem. In this example, we use all three theorems.

To get the Thevenin circuit for Fig. 4-36a, we have to work out V_{TH} and R_{TH}. To find V_{TH}, add the voltages produced across the AB terminals when the sources are taken one at a time. In the first-source circuit of Fig. 4-36b,

$$V_1 = \frac{6000}{3000 + 6000} \, 18 = 12 \text{ V}$$

In the second-source circuit of Fig. 4-36c,

$$V_2 = \frac{3000}{6000 + 3000} \, 9 = 3 \text{ V}$$

V_1 and V_2 have the same polarity as V_{TH}; therefore,

$$V_{TH} = V_1 + V_2 = 12 + 3$$
$$= 15 \text{ V}$$

This is the Thevenin voltage for the original circuit.

To get R_{TH}, reduce all sources to zero as shown in Fig. 4-36d. The equivalent resistance between the AB terminals is

$$R_{TH} = 3000 \parallel 6000$$
$$= 2 \text{ k}\Omega$$

Figure 4-36e shows the Thevenin circuit for Fig. 4-36a.

Test 4-7

1. Which of these least belongs?
   (a) all sources acting at the same time   (b) superposition theorem
   (c) divide-and-conquer strategy   (d) n simpler circuits .... ( )
2. Superposition is closest in meaning to
   (a) adding magnitudes   (b) adding or subtracting   (c) algebraic sum
   (d) summing .... ( )
3. You cannot use the superposition theorem if
   (a) the load is linear   (b) the load is independent
   (c) any resistance is linear   (d) any resistance is nonlinear .... ( )
4. You can use the superposition theorem provided
   (a) the load is nonlinear   (b) all sources are independent
   (c) all resistances are linear and you reduce only independent sources
   (d) all sources are independent and resistances are linear .... ( )
5. The algebraic sum of two currents is greater than either current only when the currents are in the
   (a) same direction   (b) opposite direction   (c) none of these .... ( )

4-8. REVIEW OF THE THREE THEOREMS

The three theorems of this chapter are among the most practical circuit theorems because they will help you solve an enormous range of problems.

The voltage-divider theorem is easy to understand and remember. It applies to any one-source series circuit, and says the output voltage across resistance R_2 equals R_2/R times the input voltage.

Thevenin's theorem is outstanding. It reduces a complicated circuit to a series circuit. Because of this, it eliminates all unnecessary information and lets you concentrate on the essential part of the problem.

The superposition theorem is useful when the original circuit has more than one source. This theorem divides a many-source circuit into simpler one-source circuits.

SUMMARY OF FORMULAS

DERIVED

$$V_2 = \frac{R_2}{R} V \tag{4-1b}$$

$$I_{SL} = \frac{V_{TH}}{R_{TH}} \tag{4-2}$$

$$\frac{R_1}{R_2} = \frac{R_3}{R_4} \quad \text{(balance)} \tag{4-8}$$

$$I = I_1 + I_2 + \cdot \cdot \cdot + I_n \quad \text{(superposition)} \tag{4-10}$$

Problems

4-1. In Fig. 4-37a, what is the output voltage when the wiper is all the way up? All the way down? At the middle position?

4-2. There are four switch positions in Fig. 4-37b. Calculate the output voltage for each.

4-3. Figure 4-37c shows a step voltage divider used in some voltmeters. What is the value of V_{out} in each switch position A through F?

4-4. In the open-load circuit of Fig. 4-38a, what is the value of V_{TH}? R_{TH}?

4-5. Figure 4-38b shows an open-load circuit. What does V_{TH} equal? R_{TH}?

4-6. What is the Thevenin circuit left of the AB terminals in Fig. 4-38c?

4-7. The wiper of Fig. 4-38d is in the middle of the 20-kΩ potentiometer. What does V_{TH} equal? R_{TH}?

4-8. A 75-kΩ resistance is connected between the AB terminals of Fig. 4-38a. What is the current through this resistance? The voltage across it?

4-9. When we connect a 60-kΩ resistance between the AB terminals of Fig. 4-38b, the voltage across these terminals decreases. What is the new value of the voltage?

4-10. A 12.5-kΩ resistance is attached to the AB terminals in Fig. 4-38c. What is the value of the load current? The load voltage?

Figure 4-37.

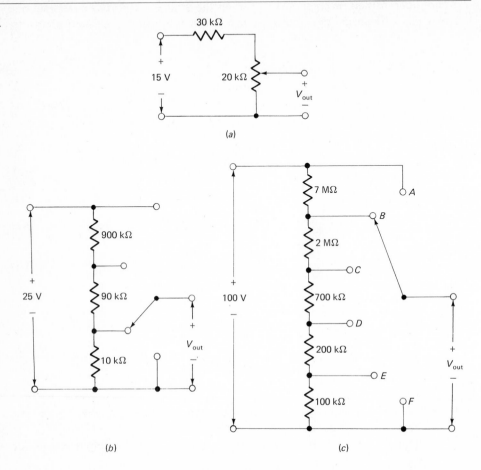

(a)

(b)   (c)

4-11.   The *AB* terminals of Fig. 4-38*d* are connected to an electronic circuit whose resistance is 10 kΩ. What is the voltage between the *AB* terminals when the wiper is at the middle? At the top?

4-12.   In the open-load circuit of Fig. 4-38*e*, what does V_{TH} equal? R_{TH}? If a 1-kΩ resistance is connected between the *AB* terminals, what is the current through it? The voltage across it?

4-13.   A penlight battery has a Thevenin voltage of 1.5 V. When the battery terminals are shorted, the resulting current is 2 A. What is the Thevenin resistance of this battery?

4-14.   The battery used in a transistor radio has an open-load voltage of 9 V. When a short is placed between the battery terminals, 1.25 A results. What is the Thevenin resistance of this battery?

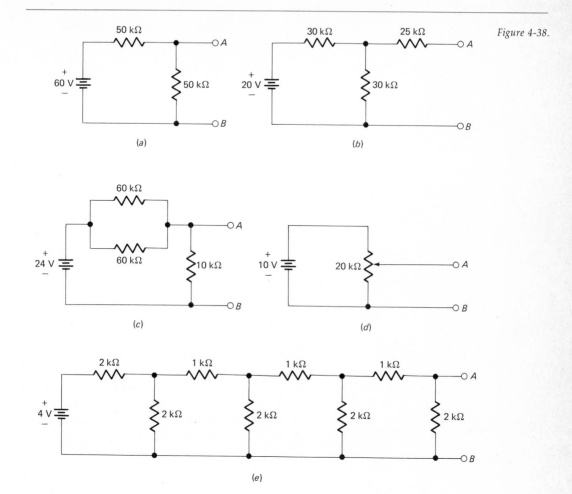

Figure 4-38.

4-15. The voltmeter of Fig. 4-39a has an infinite resistance, and the ammeter of Fig. 4-39b has a zero resistance. The voltmeter reads 100 mV; the ammeter reads 0.1 mA. What is the Thevenin resistance of the black box?

4-16. In Fig. 4-39a and b, the voltmeter looks like an open, and the ammeter like a short. The voltmeter reads 2 V, and the ammeter reads 1 mA. If a 3-kΩ resistance is connected between the AB terminals as shown in Fig. 4-39c, what is the current through it? The voltage across it?

4-17. A black box has an open-load voltage of 10 V and a shorted-load current of 2 mA. If a rheostat is connected between the AB terminals as shown in Fig. 4-39d, what is the resistance of the rheostat when the load voltage drops to 5 V? What is the resistance if the load voltage equals 7.5 V?

Figure 4-39.

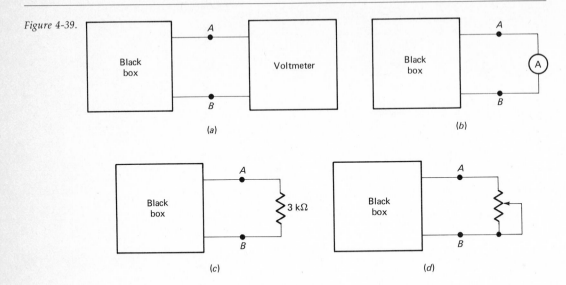

(a)   (b)

(c)   (d)

4-18. A solar cell has a Thevenin voltage of 6 V. When a rheostat is connected across the solar cell as shown in Fig. 4-40*a*, the load voltage drops to 3 V for a resistance of 50 Ω. What is the Thevenin resistance of the solar cell?

4-19. Figure 4-40*b* shows the schematic symbol for a *thermocouple*. This device is made by joining two dissimilar metals. A voltage appears between the *AB* terminals, and the value of this voltage depends on the temperature at the junction of the two metals. The higher the temperature, the greater the voltage.

Suppose we measure an open-load voltage across the thermocouple of 100 mV, and a shorted-load current of 0.2 mA. What is the Thevenin resistance of the thermocouple?

4-20. The open-load voltage of a black box equals 4 V. A load resistance of 8 kΩ results in a load voltage of 2 V. What is the value of shorted-load current? The current through a 32-kΩ load resistance?

4-21. What is the voltage across the 2-kΩ resistance of Fig. 4-41*a*? Across the 6-kΩ resistance? Between the *AB* terminals?

Figure 4-40.

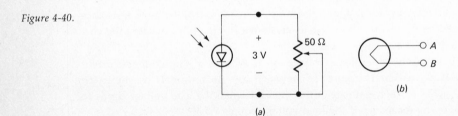

(a)

(b)

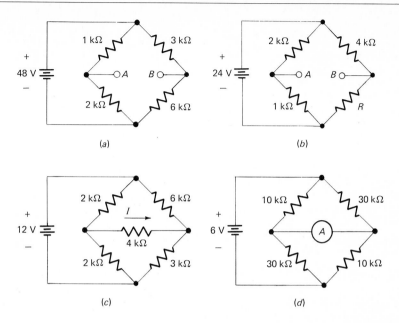

Figure 4-41.

(a)

(b)

(c)

(d)

4-22. In Fig. 4-41*b*, what is the voltage across the 1-kΩ resistance? If *R* equals 8 kΩ what is the voltage across *R*? What value of *R* balances the bridge?

4-23. Calculate the value of *I* in Fig. 4-41*c*.

4-24. The ammeter of Fig. 4-41*d* has such a low resistance that we will approximate it by a zero resistance. What does the ammeter read?

4-25. The *tolerance* of a resistor is the maximum percent error it can have from its specified value. For instance, a 100-Ω resistor with a tolerance of ±5 percent can have a value between 95 Ω and 105 Ω.

In Fig. 4-41*a*, the 1-kΩ resistance has a tolerance of ±1 percent. All other resistances have the values shown. What is the maximum voltage that may appear between the *AB* terminals?

4-26. Although an ammeter is usually approximated as a zero resistance, it actually has some resistance. Suppose the ammeter of Fig. 4-41*d* has a resistance of 50 Ω. What is the current through the ammeter? Does the ammeter resistance in this circuit have a small or a large effect on the load current?

4-27. In Fig. 4-41*b*, a short is placed between the *AB* terminals. Calculate the current for each of these:
  a. *R* = 1 kΩ.
  b. *R* = 2 kΩ.
  c. *R* = 3 kΩ.

4-28. Draw the first-source and second-source circuits for Fig. 4-42*c*.

4-29. Draw the first-source, second-source, and third-source circuits of Fig. 4-42*d*.

Figure 4-42.

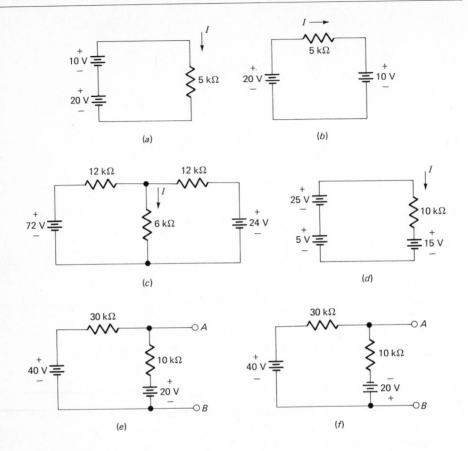

(a)

(b)

(c)

(d)

(e)

(f)

4-30. Draw the first-source and second-source circuits of Fig. 4-42e. Do the same for Fig. 4-42f.

4-31. Calculate the value of I in Fig. 4-42a using the superposition theorem. Check the answer by another method.

4-32. What is the value of I in Fig. 4-42b? The value of I_1 and I_2 in the first-source and second-source circuits?

4-33. Use the superposition theorem to find the value of I in Fig. 4-42c.

4-34. In Fig. 4-42d, work out the value of I with the superposition theorem. What is the voltage across the 10-kΩ resistance?

4-35. Work out the Thevenin voltage across the AB terminals of Fig. 4-42e by using the superposition theorem.

4-36. The 20-V source of Fig. 4-42f has the opposite polarity from the 20-V source of Fig. 4-42e. What is the Thevenin voltage between the AB terminals of Fig. 4-42e? Of Fig. 4-42f?

4-37. A 12.5-kΩ resistance is connected between the *AB* terminals of Fig. 4-42*e*. What is the current through this resistance? The voltage across it?

ANSWERS TO TESTS

4-1. *a, a, c*

4-2. *c, d, c, a*

4-3. *d, d, a, d*

4-4. *c, c, b, c, d, d, b, a*

4-5. *a, a, c, b, a, b*

4-6. *b, a, d, b, b*

4-7. *a, c, d, c, a*

5. More Theorems

We have defined three basic quantities: current, voltage, and resistance; discussed three crucial laws: Ohm's law, Kirchhoff's current law, and Kirchhoff's voltage law; examined three key theorems: the voltage-divider theorem, Thevenin's theorem, and the superposition theorem.

This chapter covers a few more theorems. These new theorems coupled with the material of earlier chapters will enable you to solve a wide variety of practical and important problems.

5-1. CONDUCTANCE THEOREM

In this section, find the answers to

What is conductance?
What does the conductance theorem say?
How do you combine parallel conductances?

Conductance

The *conductance* of a load equals the current through the load divided by the voltage across the load. In symbols, the defining formula is

$$G = \frac{I}{V} \qquad\qquad (5\text{-}1a)$$

where G = conductance
I = current
V = voltage

As an example, if the load current is 20 A and the load voltage is 4 V,

$$G = \frac{20 \text{ A}}{4 \text{ V}} = 5 \text{ A/V}$$

Or if I equals 12 A and V is 6 V,

$$G = \frac{12 \text{ A}}{6 \text{ V}} = 2 \text{ A/V}$$

The metric unit of measure for conductance is the *mho* defined as

$$1 \text{ mho} = 1 \text{ ampere per volt}$$

or
$$1 \text{ mho} = 1 \text{ A/V}$$

Therefore, in the preceding examples the answers are 5 mhos and 2 mhos.

Relation to resistance

Surely, you have noticed the relation between conductance and resistance. Conductance is the reciprocal of resistance. That is,

$$G = \frac{I}{V} = \frac{1}{V/I}$$

or
$$G = \frac{1}{R} \qquad\qquad (5\text{-}1b)$$

Therefore, a load with a resistance of 5 Ω has a conductance of

$$G = \frac{1}{5 \text{ Ω}} = 0.2 \text{ mhos}$$

Or a resistance of 8 kΩ is equivalent to a conductance of

$$G = \frac{1}{8000 \text{ Ω}} = 0.000125 \text{ mhos} = 125 \ \mu\text{mhos}$$

(Notice *mho* is *ohm* spelled backward; this is a reminder of the reciprocal relation between conductance and resistance.)

What the conductance theorem says

The conductance theorem says R can be replaced by $1/G$ in all formulas. For instance, the basic formula

$$I = \frac{V}{R}$$

can be written as

$$I = \frac{V}{1/G} = VG \qquad\qquad (5\text{-}1c)$$

So, given the conductance of a load, we can calculate the load current by the product of the load voltage and the conductance. If a load has a conductance of 5 mhos, a voltage of 10 V results in a current of

$$I = 10 \times 5 = 50 \text{ A}$$

Conductance is so simple it sometimes is confusing. Why bother with it? To begin with, some electronic instruments measure conductance rather than resistance. Because of this, occasionally the formula

$$I = VG$$

is more convenient to use than

$$I = \frac{V}{R}$$

Parallel conductances are additive

Perhaps the strongest argument for using conductance is this: the equivalent conductance of a parallel circuit is the *sum* of the parallel conductances. If 0.5 mho is in parallel with 0.25 mho (Fig. 5-1a), the equivalent conductance equals 0.75 mho (Fig. 5-1b).

The proof of this additive property of parallel conductances is simple. Earlier, we

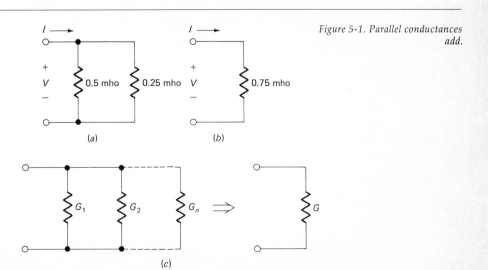

Figure 5-1. Parallel conductances add.

(a) (b) (c)

Figure 5-2. Example of conductances.

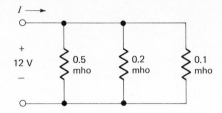

showed parallel resistances were related by

$$\frac{1}{R} = \frac{1}{R_1} + \frac{1}{R_2} + \cdot\,\cdot\,\cdot + \frac{1}{R_n} \tag{5-2}$$

With the conductance theorem, this becomes

$$G = G_1 + G_2 + \cdot\,\cdot\,\cdot + G_n \tag{5-3}$$

This says the equivalent conductance is the sum of the parallel conductances (Fig. 5-1c).

If you work a great deal with parallel circuits and if you measure conductance rather than resistance, you will find it easier to work with conductance. But in this book we prefer resistance, even in parallel circuits, because it minimizes confusion and is more common in industry. Only rarely do we have to use conductance.

EXAMPLE 5-1.
A *mhometer* reads 0.5 mho, 0.2 mho, and 0.1 mho. If these three conductances are connected across 12 V (see Fig. 5-2), what is the value of I?

SOLUTION.
The equivalent conductance is the sum of the individual conductances. In this case,

$$G = G_1 + G_2 + G_3 = 0.5 + 0.2 + 0.1 = 0.8 \text{ mhos}$$

Current equals voltage times conductance; therefore,

$$I = VG = 12 \times 0.8 = 9.6 \text{ A}$$

Test 5-1

1. What comes next in this series:
       RESISTANCE, CONDUCTANCE, V/I, _____
   (a) mhos   (b) ohms   (c) volts   (d) I/V .... ( )
2. Conductance is to resistance as 3 is to
   (a) 2   (b) three   (c) eerht   (d) 1/3 .... ( )
3. If we apply the conductance theorem to

$$K = \frac{R_1}{R_2}$$

we get $K =$

(a) $G_1 G_2$ (b) G_1/G_2 (c) G_2/G_1 (d) $1/G_1 G_2$ .... ( )

4. Conductances are added to get the equivalent parallel conductance when the loads are in

(a) series   (b) parallel   (c) either   (d) neither .... ( )

5. A 1-Ω resistance in series with a 2-Ω resistance results in an equivalent conductance of

(a) 3 mhos   (b) 1.5 mhos   (c) 0.333 mhos   (d) none of these .... ( )

5-2. IDEAL SOURCES

As you read about *ideal sources,* look for the answers to

> *What is an ideal voltage source?*
> *What is an ideal current source?*

Ideal voltage source

An *ideal voltage source* is one whose *Thevenin resistance is zero.* Because of this, the load resistance across the ideal voltage source has no effect on the voltage produced by the source. For instance, the ideal battery of Fig. 5-3a has an R_{TH} of zero; therefore, it produces a constant voltage V no matter what the setting of the wiper. From now on, the battery symbol of Fig. 5-3a always represents an ideal battery unless otherwise indicated.

Chapter 8 discusses other kinds of ideal voltage sources; these sources produce voltages that *vary with time.* Like constant sources, time-varying sources are still defined as ideal if R_{TH} is zero.

Figure 5-3. Types of sources.

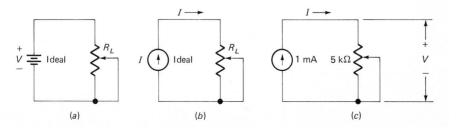

(a)   (b)   (c)

Ideal current source

An *ideal current source* is a source that produces a current *unaffected by load resistance*. Figure 5-3*b* shows the schematic symbol for an ideal current source. A current *I* comes out of this source, no matter what the value of load resistance. Even though the wiper is moved up and down, *I* does not change. Any source that produces a current whose value is unaffected by load resistance is classified as an ideal current source.

Ideal current sources may be *constant* or *time-varying*. When constant, the current source produces a current that does not change with time. Figure 5-3*c* is an example of an ideal and constant-current source. 1 mA comes out of the source no matter what the setting of the wiper. If the wiper is moved all the way up, the load voltage equals

$$V = R_L I = 0 \times 0.001 = 0$$

Move the wiper to the middle and the load voltage becomes

$$V = 2500 \times 0.001 = 2.5 \text{ V}$$

Or if it is moved all the way down,

$$V = 5000 \times 0.001 = 5 \text{ V}$$

Chapter 8 will discuss time-varying current sources. Similar to constant-current sources, time-varying current sources are ideal when they produce a current unaffected by load resistance.

The best example of a current source is the transistor. Electronics courses explain the physical action inside a transistor, and at that time you will understand how a current source works internally.

Besides being ideal or nonideal, constant or time-varying, current sources are also *dependent* or *independent*. A current source is dependent if its value depends on a current or voltage in another part of the circuit. If its value doesn't depend on other currents or voltages, it's independent. The current source of Fig. 5-3*c* is classified as ideal, constant, and independent. It's ideal because its value is unaffected by load resistance; it's constant because its value doesn't change with time; and it's independent because its value doesn't depend on other currents or voltages. (Example 5-3 discusses dependent and independent sources further.)

EXAMPLE 5-2.

Suppose the Thevenin circuit of a complicated circuit is a 1000-V source in series with a 10-MΩ resistor as shown in Fig. 5-4. With a 100-kΩ rheostat between the *AB* terminals, what is the minimum load voltage? The maximum?

SOLUTION.

With the wiper all the way up, the load resistance equals zero, and the load current is

$$I = \frac{V_{TH}}{R_{TH}} = \frac{1000}{10(10^6)} = 100 \ \mu\text{A}$$

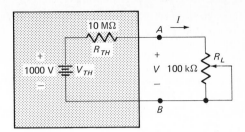

Figure 5-4. Approximate current source.

The load voltage equals

$$V = R_L I = 0 \times 100 \ \mu A = 0$$

Move the wiper all the way down, and the load resistance equals 100 kΩ. The load current then equals

$$I = \frac{V_{TH}}{R_{TH} + R_L} = \frac{1000}{10.1(10^6)} = 99 \ \mu A$$

and the load voltage is

$$V = R_L I = 100(10^3) \times 99(10^{-6}) = 9.9 \ V$$

So, the load voltage varies from 0 to 9.9 V, and the load current from 100 μA to 99 μA (essentially unchanged).

Here's the point. Hand the black box and rheostat (Fig. 5-4) over to someone without telling him what's inside the box. When he moves the wiper up and down, he notices the load current is essentially unaffected by the load resistance; only the load voltage changes. Because of this, he will approximate the black box as an ideal current source, something that puts out a current whose value is unaffected by the load resistance.

The key idea in Fig. 5-4 is to have a load resistance much smaller than the Thevenin resistance of the black box. When this is the case, changes in the load resistance have almost no effect on the equivalent series resistance. In symbols, the equivalent series resistance in Fig. 5-4 is

$$R = R_{TH} + R_L$$

or $$R \cong R_{TH} \quad \text{when} \quad R_L << R_{TH}$$

The symbol $\cong$ means "approximately equals," and the symbol $<<$ means "is much smaller than." We read the equation as R approximately equals R_{TH} when R_L is much smaller than R_{TH}.

Using transistors and other devices, we can build electronic circuits that act like the black box of Fig. 5-4, circuits whose Thevenin resistance is much larger than any practical load resistance.

EXAMPLE 5-3.

Current sources may be *dependent* or *independent* of a current or voltage in another part of the circuit. Classify the current sources of Figs. 5-5*a* and *b*.

SOLUTION.

In Fig. 5-5*a*, the current source has a value of 2 mA. First, notice that a change in the 1-kΩ load resistor has no effect on the 2-mA current source; the source will produce 2 mA no matter what the value of the load resistance. So the source is ideal. Second, notice the value of the source is a constant; it doesn't change with time. It's 2 mA now, and it will be 2 mA an hour from now. So the source is a constant source. Third, notice that current I in the left loop has no effect on the 2-mA source; therefore, the source is independent of I and all voltages in the circuit. The current source therefore is classified as ideal, constant, and independent.

The current source in Fig. 5-5*b* is different. To begin with, it has a value of 100I; this value is unaffected by the load resistance. Even if the 1-kΩ resistor is changed, the source still has a value of 100I; therefore, it's ideal. Next, realize that the 100I source has nothing to do with time; it equals 100I now, it will equal 100I an hour from now, and so on. So the source is constant with respect to time. Finally, notice that the source depends on a current in another part of the circuit; it depends on the current in the left loop. If anything causes the left-loop current to change, such as a change in the voltage source or in the left-loop resistors, I will change and this in turn will change 100I. This means the current source is dependent. So the current source of Fig. 5-5*b* is ideal, constant, and dependent.

For the remainder of this book, we concentrate on sources that are ideal and independent; they may be constant or time-varying. Remember: *ideal* means unaffected by load resistance, *constant* means unchanging in time, and *independent* means unaffected by currents or voltages in other parts of the circuit.

EXAMPLE 5-4.

Work out the voltage across the 1-kΩ load resistor of Figs. 5-5*a* and *b*.

Figure 5-5. Dependent and independent current sources.

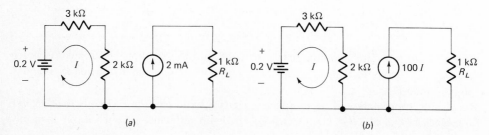

(a)   (b)

SOLUTION.
In Fig. 5-5*a*, the current source produces 2 mA. The resulting load voltage equals

$$V_L = R_L I = 1000 \times 0.002 = 2 \text{ V}$$

Figure 5-5*b* is a three-step problem. First, get the current in the left loop:

$$I = \frac{V}{R} = \frac{0.2}{5000} = 0.04 \text{ mA} = 40 \ \mu\text{A}$$

Second, work out the current in the right loop:

$$100I = 100 \times 0.04 \text{ mA} = 4 \text{ mA}$$

Third, calculate the load voltage:

$$V_L = R_L \times 100I = 1000 \times 0.004 = 4 \text{ V}$$

(This three-step calculation is common in transistor-circuit analysis.)

Test 5-2

1. Used with sources, the word "ideal" means
   (*a*) constant in time   (*b*) independent   (*c*) time-varying
   (*d*) unaffected by load resistance .... ( )
2. An ideal current source always puts out a current that is
   (*a*) constant in time   (*b*) unaffected by load resistance
   (*c*) independent of all other voltages and currents
   (*d*) uncontrolled by any other voltage or current   .... ( )
3. A constant current source produces a current that is
   (*a*) controlled   (*b*) constant in time   (*c*) dependent
   (*d*) dependent only on load resistance .... ( )
4. An ideal 4-mA current source drives a 5-kΩ resistor. The voltage across the current
   source is
   (*a*) time-varying   (*b*) zero   (*c*) 20 V   (*d*) dependent .... ( )

5-3. INCLUDING CURRENT SOURCES IN EARLIER THEOREMS

How does a current source change Thevenin's theorem? The superposition theorem? When you apply these theorems, reduce all independent sources to zero as before. With a voltage source this means $V = 0$. As discussed earlier, zero resistance means zero voltage. This is why reducing an ideal voltage source to zero is equivalent to replacing it by a *short*.

Reducing a current source to zero means $I = 0$; that is, zero current passes through the source. The only way this is possible is for the source to be open. In other words, reducing an ideal current source to zero is equivalent to replacing it by an *open* (infi-

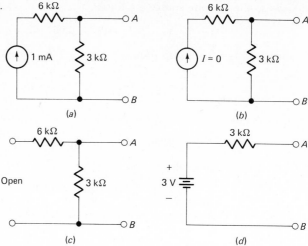

Figure 5-6. Thevenin's theorem.

nite resistance). For example, suppose we want the Thevenin circuit for Fig. 5-6a. The load resistance has already been removed, so all of the 1 mA must flow through 6 kΩ in series with 3 kΩ. The Thevenin voltage across the open-load terminals is

$$V_{TH} = 3000 \times 0.001 = 3 \text{ V}$$

To get R_{TH}, reduce the current source to zero as shown in Fig. 5-6b. The way to visualize $I = 0$ is by replacing the current source with an open (Fig. 5-6c). The equivalent resistance from A to B in this case equals

$$R_{TH} = 3 \text{ k}\Omega$$

Figure 5-6d shows the Thevenin circuit for Fig. 5-6a.

In summary, when applying Thevenin's theorem or the superposition theorem, reducing sources to zero is the same as

1. Replacing each voltage source by a short.
2. Replacing each current source by an open.

EXAMPLE 5-5.
From now on, *Thevenize* will mean "get the Thevenin circuit for." In Fig. 5-7a, Thevenize the circuit left of the *AB* terminals.

SOLUTION.
Figure 5-7b shows the original circuit after the load resistance is removed. The current source forces 5 mA through the 1-kΩ resistance because charges have nowhere else to

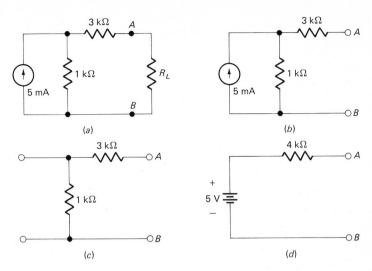

Figure 5-7. Applying Thevenin's theorem.

go in this open-load circuit. As a result, the Thevenin voltage is

$$V_{TH} = 1000 \times 0.005 = 5 \text{ V}$$

Next, reduce the current source to zero as shown in Fig. 5-7c. The equivalent resistance between the AB terminals is

$$R_{TH} = 3000 + 1000 = 4 \text{ k}\Omega$$

Figure 5-7d is the Thevenin circuit for Fig. 5-7b. Put either of these inside a black box and you can't tell one from the other. In other words, either circuit produces the same load current and voltage when R_L is connected to the load terminals.

EXAMPLE 5-6.
Thevenize Fig. 5-8a left of the AB terminals.

SOLUTION.
There are two sources; so we can use the superposition theorem to work out the Thevenin voltage between the AB terminals. Figure 5-8b shows the first-source circuit (the current source has been reduced to zero). With the voltage-divider theorem,

$$V_1 = \frac{4000}{8000} 10 = 5 \text{ V}$$

Figure 5-8c is the second-source circuit (the voltage source has been reduced to

Figure 5-8. Another example of Thevenin's theorem.

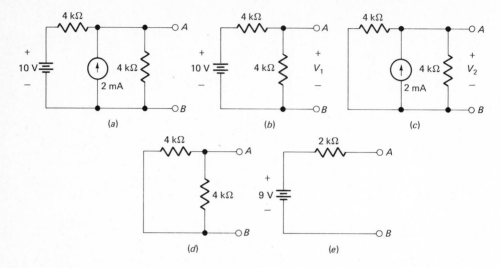

(a)   (b)   (c)

(d)   (e)

zero). Because the two resistances are equal, the current splits equally, resulting in 1 mA through each. It follows that

$$V_2 = 4000 \times 0.001 = 4 \text{ V}$$

The Thevenin voltage for the original circuit is the sum of the individual voltages:

$$V_{TH} = V_1 + V_2 = 5 + 4$$
$$= 9 \text{ V}$$

To get the Thevenin resistance of Fig. 5-8a, reduce both sources to zero: this means replacing the voltage source by a short and the current source by an open, as shown in Fig. 5-8d. The equivalent resistance between AB is

$$R_{TH} = 4000 \parallel 4000 = 2 \text{ k}\Omega$$

Figure 5-8e is the Thevenin circuit for Fig. 5-8a. Connect any load resistance across either circuit, and you get the same load current. For instance, with a short across the load terminals of Fig. 5-8e, the current is

$$I_{SL} = \frac{9}{2000} = 4.5 \text{ mA}$$

This is the same value of load current that would occur with a short between the AB terminals of the original circuit (Fig. 5-8a).

As you see, it's easier to work with the Thevenin circuit than the original circuit. The reason is simple: a loaded Thevenin circuit has only one loop, whereas the original circuit can have any number of loops; each extra loop makes it that much harder to calculate the load current.

5-4. NORTON'S THEOREM

What is Norton's theorem?

Norton's theorem is a useful alternative to Thevenin's theorem. For some circuits, we prefer Norton's theorem because it leads to easier calculations.

What did Norton prove? In Fig. 5-9a, the box contains sources and linear resistances. Norton proved this black box acts the same as a current source in parallel with a resistance (see Fig. 5-9b). He showed the current source has a value of I_{SL}, the shorted-load current of the original circuit, and that the resistance equals the Thevenin resistance. The equivalent circuit inside the box of Fig. 5-9b is called a *Norton circuit*.

To summarize, Norton's theorem says you can replace the contents of a black box (Fig. 5-9a) by a Norton circuit (Fig. 5-9b). To get the value of the current source, replace R_L by a short as shown in Fig. 5-9c; then, calculate or measure I_{SL}. To get the value of the parallel resistance, use any of the methods for finding R_{TH}: the zero-source method, the shorted-load method, or the matched-load method.

EXAMPLE 5-7.
What is the Norton circuit left of the *AB* terminals in Fig. 5-10a?

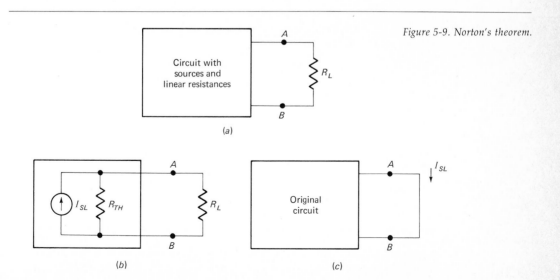

Figure 5-9. Norton's theorem.

(a)

(b)

(c)

Figure 5-10. Applying Norton's theorem.

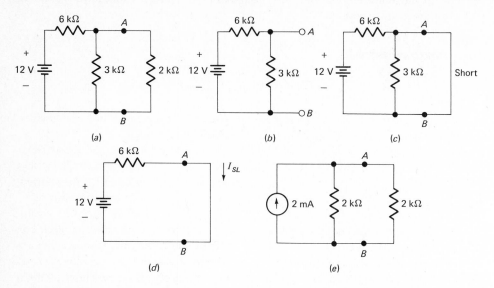

(a)   (b)   (c)

(d)   (e)

SOLUTION.

First, remove the 2-kΩ load resistance to get Fig. 5-10b. The Thevenin resistance of this circuit is 2 kΩ, found by the usual methods.

Second, replace R_L by a short as shown in Fig. 5-10c. With a short between the *AB* terminals, the load voltage drops to zero. Because of this, no current is in the 3-kΩ resistance. All the shorted-load current passes through the short, as shown in Fig. 5-10d. So,

$$I_{SL} = \frac{V}{R} = \frac{12}{6000} = 2 \text{ mA}$$

Figure 5-10e is the Norton circuit with the 2-kΩ load reconnected to the *AB* terminals. This circuit produces the same load current and voltage as the original circuit of Fig. 5-10a.

EXAMPLE 5-8.

Repeat the preceding example, but use this approach: first get the Thevenin circuit and then the Norton circuit.

SOLUTION.

This example shows an alternative way to find the Norton circuit. Look at Fig. 5-10b. The Thevenin voltage is

Figure 5-11. Converting a Thevenin circuit to a Norton circuit.

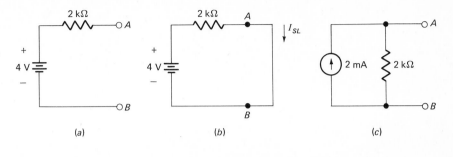

(a)   (b)   (c)

$$V_{TH} = \frac{R_2}{R}\, V = \frac{3000}{9000}\, 12 = 4 \text{ V}$$

and the Thevenin resistance is

$$R_{TH} = 6000 \parallel 3000 = 2 \text{ k}\Omega$$

Figure 5-11*a* shows the Thevenin circuit.

Now apply Norton's theorem to Fig. 5-11*a*. With a short across the load terminals, Fig. 5-11*b* gives

$$I_{SL} = \frac{4}{2000} = 2 \text{ mA}$$

The parallel resistance in a Norton circuit has the same value as the Thevenin resistance; therefore, Fig. 5-11*c* is the Norton circuit for Fig. 5-11*a* and for Fig. 5-10*b*.

The whole point is this. If you already have the Thevenin circuit worked out for some original circuit, you don't have to go back to the original circuit to work out the Norton circuit. You can get the Norton circuit directly from the Thevenin circuit.

EXAMPLE 5-9.
Convert the Norton circuit of Fig. 5-12*a* to a Thevenin circuit.

SOLUTION.
The load terminals of Fig. 5-12*a* are open; therefore, the Thevenin voltage is

$$V_{TH} = 5000 \times 0.003 = 15 \text{ V}$$

Next, open the current source; the equivalent resistance between the *AB* terminals is

$$R_{TH} = 5 \text{ k}\Omega$$

Therefore, Fig. 5-12*b* is the Thevenin circuit for Fig. 5-12*a*.

Figure 5-12. Applying Thevenin's theorem.

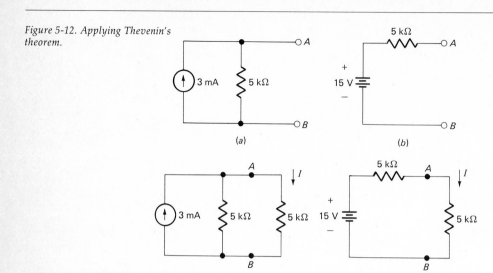

(a)   (b)

(c)   (d)

EXAMPLE 5-10.

A load resistance of 5 kΩ is connected between the *AB* terminals of Figs. 5-12*a* and *b*. Calculate the load current and voltage in each circuit.

SOLUTION.

Figure 5-12*c* shows the Norton circuit with a 5-kΩ load resistance. The current out of the source must split equally between the two resistances. So, the load current is

$$I = 1.5 \text{ mA}$$

and the load voltage is

$$V = 5000 \times 0.0015 = 7.5 \text{ V}$$

Figure 5-12*d* shows the Thevenin circuit with a 5-kΩ load resistance. In this case, the load current equals

$$I = \frac{15}{5000 + 5000} = 1.5 \text{ mA}$$

and the load voltage is

$$V = 5000 \times 0.0015 = 7.5 \text{ V}$$

The point is this. Given an original circuit, we can replace it by a Thevenin circuit or a Norton circuit to find the load current; either gives the same answer. Because of this, we have the option of using whichever is easier in a particular problem.

Test 5-3

1. Norton is to Thevenin as ideal current source is to
   (*a*) controlled source   (*b*) ideal voltage source   (*c*) independent source
   (*d*) load-independent source .... ( )
2. Which of these belongs least?
   (*a*) load resistance   (*b*) black box   (*c*) Norton circuit
   (*d*) Thevenin circuit .... ( )
3. The following words can be rearranged to form a sentence: CIRCUIT ALWAYS
   NORTON IS BETTER. The sentence is
   (*a*) true   (*b*) false .... ( )
4. An ideal current source whose value is zero is identical to
   (*a*) a short   (*b*) an open   (*c*) zero resistance
   (*d*) infinite conductance .... ( )
5. Zero voltage source is to zero current source as zero resistance is to
   (*a*) zero conductance   (*b*) infinite resistance   (*c*) open
   (*d*) all the foregoing .... ( )

5-5. POWER

Figure 5-13 shows atoms inside a load resistance. Chemical bonds (ionic or covalent) hold these atoms together. At absolute zero temperature (−273°C) the atoms are motionless; but as the temperature rises, they spring back and forth about the positions shown in Fig. 5-13. The higher the temperature, the faster the vibrations.

   As you read more about these vibrations, hunt for the answers to

Why does current produce heat?
How is power defined?

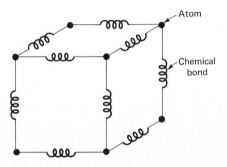

Atom

Chemical
bond

Figure 5-13. Atoms and their
bonds.

Free electrons produce heat in a resistance

When free electrons move through a load resistance, they collide with the atoms. Each collision deflects the electron and reduces its speed. At the same time, each collision makes an atom vibrate a bit faster. The effect of all atoms vibrating faster is a rise in the temperature of the load. In other words, heat is nothing more than faster vibration of the atoms.

In terms of work and energy, free electrons lose energy as they move through a load; the energy lost equals the work done on the atoms. In turn, the work done on the atoms produces heat. So what it boils down to is this: the loss of energy by the electrons equals the heat generated in the resistance.

Definition of power

Power is defined as the work done divided by the time in which it's done. As a defining formula,

$$P = \frac{W}{t} \tag{5-4}$$

where P = power
W = work
t = time

If free electrons passing through a resistor lose 5 J of energy in 2 s, the power equals

$$P = \frac{5 \text{ J}}{2 \text{ s}} = 2.5 \text{ J/s}$$

Or if 15 J of heat are produced in 3 s, the power equals

$$P = \frac{15 \text{ J}}{3 \text{ s}} = 5 \text{ J/s}$$

Power is the same as rate of work because its numerical value tells us the number of joules of work done in each second. In the first of the preceding examples, 2.5 J of work are done in each second. In the second example, 5 J of work are done during each second. The rate of work is greater in the second case than in the first.

The watt

The *watt* (abbreviated W) is the metric unit of measure for power; it's defined as

$$1 \text{ watt} = 1 \text{ joule per second}$$

or
$$1 \text{ W} = 1 \text{ J/s}$$

A watt is the value of power when 1 J of work is done during each second. With this definition, the answers in the preceding examples become

$$2.5 \text{ J/s} = 2.5 \times 1 \text{ J/s} = 2.5 \times 1 \text{ W} = 2.5 \text{ W}$$
$$5 \text{ J/s} = 5 \times 1 \text{ J/s} = 5 \times 1 \text{ W} = 5 \text{ W}$$

The watt is standard in electrical and electronics work. For this reason, you always specify power in watts rather than joules per second. Since W is always in joules and t is always in seconds, Eq. (5-4) automatically gives an answer in watts.

Incidentally, textbooks normally use italic letters (I, V, P, etc.) as abbreviations for physical quantities (current, voltage, power, etc.). On the other hand, Roman letters (A, V, W, etc.) are used as abbreviations for units of measure (ampere, volt, watt, etc.). This is why italic V represents voltage, but Roman V stands for volt; similarly, italic W represents work, but Roman W stands for watt. Watch for these differences in this and other textbooks.

EXAMPLE 5-11.
Electrons passing through a resistor lose 30,000 J in 15 s. What is the power?

SOLUTION.

$$P = \frac{W}{t} = \frac{30,000}{15} = 2000 \text{ W} = 2 \text{ kW}$$

EXAMPLE 5-12.
A house receives electric energy at a rate of 2 kW. How much energy is received in 1 h?

SOLUTION.
Multiply both sides of Eq. (5-4) by t to get

$$W = Pt$$

This says the work done (equal to the energy received) equals power multiplied by time. Since $P = 2$ kW and $t = 1$ h $= 3600$ s,

$$W = Pt = 2000 \times 3600 = 7,200,000 \text{ J} = 7.2 \text{ MJ}$$

EXAMPLE 5-13.
Power companies sell electric energy by the kilowatt-hour (abbreviated kWh). One kilowatt-hour is the energy received in 1 h when the power is 1 kW. If electric energy costs 2 cents per kilowatt-hour, what is the electrical bill for a house that has received 1500 kWh during the month?

SOLUTION.
The cost of the electric energy is

$$C = 1500 \times 0.02 = 30 \text{ dollars}$$

Test 5-4

1. Current through a conductor produces heat because
    (*a*) electrons hit atoms   (*b*) atoms hit atoms   (*c*) atoms give up energy
    (*d*) electrons gain energy with each collision .... ( )
2. The potential energy lost by electrons passing through a resistor equals the
    (*a*) speed of the electrons   (*b*) heat produced in the resistor
    (*c*) battery voltage   (*d*) rate of charge flow .... ( )
3. Which of these belongs least?
    (*a*) energy   (*b*) power   (*c*) watt   (*d*) joule per second .... ( )
4. Power most closely means
    (*a*) joules   (*b*) watts   (*c*) rate of work   (*d*) heat .... ( )
5. Watt is to power as mho is to
    (*a*) ohm   (*b*) parallel   (*c*) conductance   (*d*) rate of work .... ( )
6. $P = W/t$ does not have to be proved because it is a
    (*a*) defining formula   (*b*) experimental formula   (*c*) derived formula ... ( )

5-6. POWER THEOREMS

As you may suspect, power is related to voltage and current. Furthermore, simple rules exist for calculating power in series and parallel circuits. We can summarize these power relations and rules by stating a few theorems. Among other things, these theorems answer the following:

> *How is power related to voltage and current?*
> *Why does each load power add to the total power?*
> *How is source power related to load power?*

Relation to voltage and current

Given a resistance with voltage V and current I, the power equals the product of voltage and current. In symbols,

$$P = VI \qquad (5\text{-}5)$$

where P = power
      V = voltage
      I = current

So, if a resistance has 10 V across it and 5 A through it, the power equals

$$P = 10 \times 5 = 50 \text{ W}$$

Here is how to derive Eq. (5-5) from known formulas. First, recall that

$$V = \frac{W}{Q}$$

This rearranges into

$$W = VQ$$

Substitute this expression into the defining formula for power to get

$$P = \frac{W}{t} = \frac{VQ}{t} = VI$$

because Q/t equals I. Therefore, we have proved Eq. (5-5).

Alternative power formulas

With Eq. (5-5) we can derive two alternative power formulas. Since $V = RI$,

$$P = VI = RI \times I = RI^2$$

or as it usually is written,

$$P = I^2R \tag{5-6}$$

The derived equation says power equals the square of the current times the resistance. This formula is useful when you have the values of the current and the resistance, but not the voltage.

Another alternative formula is this. Since $I = V/R$,

$$P = VI = V\frac{V}{R}$$

or

$$P = \frac{V^2}{R} \tag{5-7}$$

This says the power equals the square of the voltage divided by the resistance; use this when you have the values of the voltage and the resistance, but not the current.

Of the three formulas, the one to remember is $P = VI$. When necessary, you can derive the other two. After a while, you automatically will remember the alternative formulas.

Power in a series circuit

In the series circuit of Fig. 5-14a, the power in each resistance is

$$P_1 = V_1I$$

and

$$P_2 = V_2I$$

These formulas tell us the rate of work done by electrons moving through each resistance. Equivalently, they tell us the rate of heat production.

In the equivalent circuit of Fig. 5-14b, the power is

$$P = VI \tag{5-8a}$$

Figure 5-14. *Power in series circuit.*

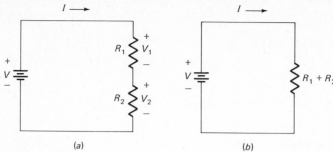

(a)　　　　　　　(b)

This formula tells us the rate of work done on the equivalent resistance, or the rate of total heat production in the circuit.

In Fig. 5-14a,

$$V = V_1 + V_2$$

Substituting this into Eq. (5-8a) gives

$$P = (V_1 + V_2)I = V_1 I + V_2 I$$

or
$$P = P_1 + P_2 \qquad (5\text{-}8b)$$

All this final result says is the total power equals the sum of the individual powers.

Equation (5-8b) makes sense. Free electrons lose potential energy passing through load resistances. In a unit of time, the total energy lost should equal the energy lost to the first resistance, plus the energy lost to the second resistance. Equivalently, the total heat produced in the circuit should equal the heat produced in the first resistance, plus the heat produced in the second resistance.

In general, no matter how many resistances are in a series circuit,

$$P = P_1 + P_2 + \cdots + P_n$$
$$\qquad (5\text{-}9)$$

To get the total power in a series circuit, you add the powers in the series resistances.

Power in parallel circuits

A similar result occurs in parallel circuits: the total power equals the sum of the individual powers. The proof for two resistances follows.

In Fig. 5-15a, the power to each resistance is

$$P = V I_1$$

and
$$P = V I_2$$

In Fig. 5-15b, the power to the equivalent resistance is

$$P = V I$$

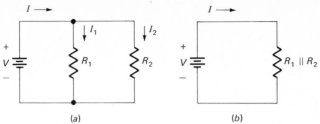

Figure 5-15. Power in parallel circuit.

Since $I = I_1 + I_2$,

$$P = V(I_1 + I_2) = VI_1 + VI_2$$

or
$$P = P_1 + P_2 \tag{5-10}$$

As we see, this derived equation is identical to the one for a series circuit. Therefore, the total power equals the sum of the individual powers. In general, no matter how many parallel resistances there are, the total power is

$$P = P_1 + P_2 + \cdots + P_n$$

Additive power theorem

Using proofs similar to the ones just given, we can prove the following theorem: the total power in any circuit equals the sum of the powers in the individual resistances. This theorem is easy to understand and believe. Power is work done per unit of time, or energy lost per unit of time. Since energy is neither created nor destroyed, the total energy lost by moving charges has to equal the energy lost to individual load resistances. This is equivalent to saying the total heat produced must equal the sum of the individual heats produced in the load resistances.

Total source power equals total load power

The next theorem may be obvious, but should be stated just in case. The sources supply energy to the load resistances. Since energy is neither created nor destroyed, all the energy received by the load resistances must come from the sources. In theorem form: the total source power equals the total load power.

As an example, the current in Fig. 5-16 is

$$I = \frac{V}{R} = \frac{10}{10,000} = 1 \text{ mA}$$

The voltage across the upper resistance is 6 V; therefore the power in this resistance is

$$P_1 = V_1 I = 6 \times 0.001 = 0.006 \text{ W} = 6 \text{ mW}$$

Figure 5-16. Total source power equals total load power.

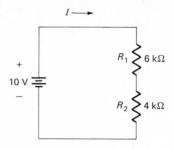

The lower resistance has 4 V across it, and a power of

$$P_2 = V_2 I = 4 \times 0.001 = 0.004 \text{ W} = 4 \text{ mW}$$

The total load power is

$$P_{\text{load}} = P_1 + P_2 = 6 \text{ mW} + 4 \text{ mW} = 10 \text{ mW}$$

And the total source power is

$$P_{\text{source}} = P_{\text{load}} = 10 \text{ mW}$$

The energy delivered to the load resistances comes from the source. Therefore, the rate of energy lost by the source equals the rate of energy absorbed by the loads. In this particular example, the source loses energy at a rate of 10 mW, and the equivalent load absorbs it at the same rate.

EXAMPLE 5-14.
A 12-V battery produces a current of 5 mA in a series circuit. What is the total load power? At what rate does the battery lose energy?

SOLUTION.
The total voltage is 12 V, and the current is 5 mA; so,

$$P = VI = 12(0.005) = 0.06 \text{ W} = 60 \text{ mW}$$

This is the total load power. Equivalently, it is the rate at which the battery loses energy.

EXAMPLE 5-15.
Twenty Christmas tree lights are in parallel across a 120-V source. If each has a resistance of 10 kΩ, what is the power in each? The total power?

SOLUTION.
Given voltage and resistance, we can use

$$P = \frac{V^2}{R}$$

Alternatively, we can calculate the current in each bulb and then use

$$P = VI$$

Either way, we get the same answers.

The power in the first bulb is

$$P_1 = \frac{V_1^2}{R_1} = \frac{120^2}{10,000} = 1.44 \text{ W}$$

The remaining 19 bulbs have the same value of power. The total power is the sum of the individual powers:

$$P = 20 \times 1.44 = 28.8 \text{ W}$$

EXAMPLE 5-16.

What is the power in the 3-Ω resistance of Fig. 5-17a?

SOLUTION.

Apply the voltage-divider theorem to get

$$V_2 = \frac{R_2}{R} V = \frac{3}{9} 18 = 6 \text{ V}$$

The current through this resistor is

$$I = \frac{V_2}{R_2} = \frac{6}{3} = 2 \text{ A}$$

And the power is

$$P_2 = V_2 I = 6(2) = 12 \text{ W}$$

EXAMPLE 5-17.

Figure 5-17b shows the schematic symbol of a *fuse*, a resistor that melts when the current is excessive. The 3-A fuse shown melts when the current exceeds 3 A; this opens the path for charge flow and protects the circuit from too much current.

If the fuse has a resistance of 0.02 Ω, what is the fuse power at the melting point? What value of R_L causes the fuse to open?

SOLUTION.

The fuse power at the melting point is

$$P = I^2 R = 3^2(0.02) = 0.18 \text{ W} = 180 \text{ mW}$$

Figure 5-17. Calculating power.

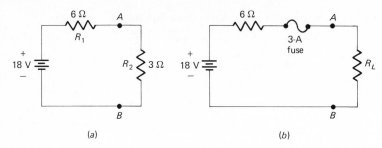

(a)   (b)

In Fig. 5-17b, the only way to get 3 A is to reduce R_L to zero. When this happens, the equivalent resistance is 6.02 Ω, and

$$I = \frac{18}{6.02} \cong 3 \text{ A}$$

So the only way to blow the fuse is by shorting the load terminals.

When R_L is greater than zero, the fuse is intact and offers only 0.02 Ω of resistance. Because of this, an intact fuse is approximated as a short and a blown fuse as an open.

Test 5-5

1. When $P = VI$ was first stated, the load was specified as
   (a) resistive   (b) inductive   (c) capacitive .... ( )
2. $P = VI$ is an example of a
   (a) defining formula   (b) experimental formula   (c) derived formula ... ( )
3. The power in each resistor of a complicated circuit adds to the total power:
   (a) never   (b) sometimes   (c) most of the time   (d) always .... ( )
4. The additive power theorem is based on
   (a) energy is neither created nor destroyed   (b) heat   (c) definition
   (d) experiment .... ( )
5. Which does not belong?
   (a) total source power   (b) total load power   (c) sometimes equal
   (d) always equal .... ( )
6. If the current through a short is 3 A, the power equals
   (a) zero   (b) 3 W   (c) 9 W   (d) infinity .... ( )

5-7. MATCHED-LOAD POWER

Figure 5-18a shows a black box driving an adjustable load resistance. A special case occurs when R_L equals the R_{TH} of the black box. The most important things to learn

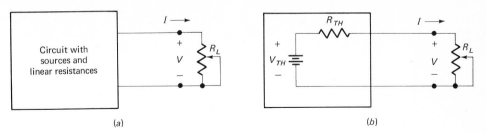

Figure 5-18. Matched load.

(a)   (b)

about this special case are

> What is the maximum load-power theorem?
> When does the theorem not apply?

The theorem

Visualize the black box of Fig. 5-18a replaced by its Thevenin circuit as shown in Fig. 5-18b. If the wiper is moved up and down, V and I vary. Because of this, the load power depends on the value of the load resistance. Either by experiment or with calculus, it's possible to prove the *maximum load-power theorem:* load power is maximum when a matched load is used. In other words, given a fixed R_{TH}, the power in the load reaches a maximum value when R_L is adjusted to equal R_{TH}.

If a black box has a Thevenin resistance of 50 Ω, a load of 50 Ω will have more power than any other value of load resistance. If a complicated circuit has a R_{TH} of 1 kΩ, a load resistance of 1 kΩ results in maximum load power. The key idea is to match a *variable* R_L to a *fixed* R_{TH}.

As a concrete example, maximum load power occurs in Fig. 5-19 when R_L is adjusted to 4 kΩ. With this matched load, the load voltage is half the Thevenin voltage:

$$V = 6 \text{ V}$$

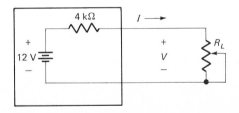

Figure 5-19. Example of maximum load power.

And the load current is

$$I = \frac{12}{4000 + 4000} = 1.5 \text{ mA}$$

The matched-load power is

$$P = VI = 6(0.0015) = 0.009 \text{ W} = 9 \text{ mW}$$

If R_L is changed to any value other than 4 kΩ, the load power is always less than 9 mW.

Misconception about the theorem

The maximum load-power theorem does not apply to the case of an adjustable R_{TH}. That is, you don't adjust R_{TH} to match the value of a fixed R_L. If R_{TH} is variable, as shown in Fig. 5-20a, the maximum load power occurs when R_{TH} is zero (Fig. 5-20b).

An adjustable R_{TH} like Fig. 5-20a is rare. In practice, you normally get a black box or circuit whose Thevenin resistance is fixed and whose load resistance is adjustable. In this case, the maximum load power occurs when R_L is adjusted to equal R_{TH}.

Applications of the theorem

When a source is weak, that is, can deliver only a small amount of energy in a unit of time, it may be necessary to match the load to the source. A TV antenna, for instance, picks up an electrical signal from a transmitting station. The antenna acts like a voltage source and a series resistance, the Thevenin circuit of Fig. 5-21a.

The schematic symbol for the voltage source of Fig. 5-21a is different from a battery because the TV signal is not a fixed voltage. Instead, this signal varies in time. Chapter 8 discusses time-varying sources; the main point at the moment is that a TV antenna has a Thevenin voltage v_{TH} and a Thevenin resistance R_{TH}.

The quality of a TV picture depends on the power to the TV receiver. In Fig. 5-21b, the receiver has a resistance of R_L; this is the load resistance connected to the antenna.

Figure 5-20. Variable Thevenin resistance.

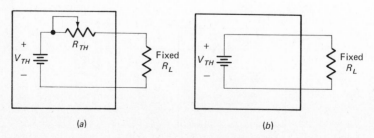

(a)   (b)

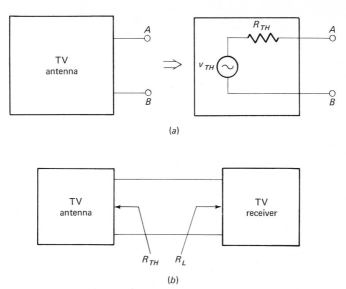

Figure 5-21. Maximum load-power theorem.

To get the best picture, therefore, we need the maximum load power. The maximum load-power theorem says make R_L equal R_{TH} to get the maximum load power. The Thevenin resistance of a TV antenna is typically 300 Ω; therefore, TV receivers are built with resistances of 300 Ω.

In telephone systems load resistances are also matched to the Thevenin resistance of the source. The matching resistance in telephone systems is 600 Ω. Likewise, in microwave systems, loads are matched to sources; in these systems, the matching resistance is 50 Ω.

You will encounter many other important applications of the maximum load-power theorem when you study electronics.

EXAMPLE 5-18.
A *signal generator* is a commercial instrument that produces a time-varying voltage. You can visualize a signal generator as a Thevenin circuit (see Fig. 5-22a). At a particular instant in time, v_{TH} equals 1 V. If R_{TH} equals 50 Ω, what is the maximum load power at the given instant?

SOLUTION.
The maximum load power occurs when the load resistance matches the *source resistance*. (Source resistance means the Thevenin resistance of the source.) In this case, R_L

Figure 5-22. Thevenin circuit of
signal generator.

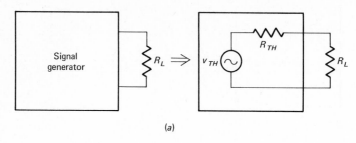

(a)

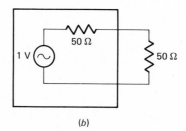

(b)

should equal 50 Ω (Fig. 5-22b). For this matching resistance, the load voltage is

$$V = \frac{v_{TH}}{2} = \frac{1}{2} = 0.5 \text{ V}$$

and the load power is

$$P = \frac{V^2}{R_L} = \frac{0.5^2}{50} = 0.005 \text{ W} = 5 \text{ mW}$$

EXAMPLE 5-19.
Prove the maximum load-power theorem for Fig. 5-23a.

SOLUTION.
For the matched case, R_L equals 1 Ω (Fig. 5-23b), and the load voltage is

$$V = \frac{v_{TH}}{2} = \frac{2}{2} = 1 \text{ V}$$

The load current is

$$I = \frac{V}{R_L} = \frac{1}{1} = 1 \text{ A}$$

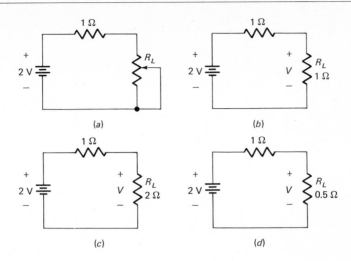

Figure 5-23.

(a)

(b)

(c)

(d)

The load power is

$$P = VI = 1 \times 1 = 1 \text{ W}$$

Suppose we move the wiper of Fig. 5-23a to get a R_L of 2 Ω (Fig. 5-23c). In this case, the voltage-divider theorem gives a load voltage of

$$V = \frac{2}{3} 2 = 1.33 \text{ V}$$

The load current equals

$$I = \frac{V}{R_L} = \frac{1.33}{2} = 0.667 \text{ A}$$

The load power is

$$P = VI = 1.33 \times 0.667 = 0.887 \text{ W}$$

which is less than 1 W. For any R_L greater than 1 Ω in Fig. 5-23a, less than 1 W of load power always results. (Try other values of R_L to convince yourself.)

On the other hand, R_L may be less than 1 Ω. Figure 5-23d shows an R_L of 0.5 Ω. In this case, the voltage-divider theorem gives a load voltage of

$$V = \frac{0.5}{1.5} 2 = 0.667 \text{ V}$$

The load current equals

$$I = \frac{V}{R_L} = \frac{0.667}{0.5} = 1.33 \text{ A}$$

The load power equals

$$P = VI = 0.667(1.33) = 0.887 \text{ W}$$

which is less than 1 W. For any load resistance less than 1 Ω, the load power in Fig. 5-23a is always less than 1 W.

Test 5-6

1. In the following series, what logically comes next?
   THEVENIN VOLTAGE, MATCHED LOAD, _____
   (a) current   (b) voltage   (c) power   (d) half of the Thevenin voltage   ( )
2. When you use the maximum load-power theorem, the Thevenin resistance must remain
   (a) in ohms   (b) fixed   (c) varying   (d) none of these   ....   ( )
3. A load resistance has a value of 10 Ω. To get the maximum load power, an adjustable Thevenin resistance should equal
   (a) 0   (b) 5 Ω (c) 10 Ω (d) 20 Ω ....   ( )
4. A microwave antenna has a Thevenin resistance of 50 Ω. To get the maximum load power, the load resistance should equal
   (a) 0   (b) 25 Ω (c) 50 Ω (d) 100 Ω ....   ( )
5. An oscillator is an electronic circuit you will learn about in a later course. If an oscillator has a Thevenin resistance of 1 kΩ, we can get the maximum load power by using a load resistance equal to
   (a) zero   (b) half of the Thevenin resistance   (c) 1000 Ω (d) infinity   ( )

SUMMARY OF FORMULAS

DEFINING

$$G = \frac{I}{V} \tag{5-1a}$$

$$P = \frac{W}{t} \tag{5-4}$$

DERIVED

$$G = \frac{1}{R} \tag{5-1b}$$

$$I = VG$$

$$G = G_1 + G_2 + \cdots + G_n \quad \text{(parallel)} \qquad (5\text{-}3)$$

$$P = VI \qquad (5\text{-}5)$$

$$P = I^2 R \qquad (5\text{-}6)$$

$$P = \frac{V^2}{R} \qquad (5\text{-}7)$$

$$P = P_1 + P_2 + \cdots + P_n \quad \text{(all circuits)}$$

Problems

5-1.   A load has 2 V across it and 5 mA through it. What is its conductance?

5-2.   Calculate the conductance for each of these: a 10-Ω resistor, a 1-kΩ resistor, and a 50-kΩ resistor.

5-3.   What is the conductance of 500 ft of AWG-22 copper wire?

5-4.   A load has a conductance of 5000 μmhos. What does the load current equal when the load voltage is 10 V?

5-5.   A mhometer reads conductances of 100 μmhos, 300 μmhos, and 600 μmhos. If these are connected in parallel, what is the equivalent conductance? If the measured conductances are connected in series, what is the equivalent conductance?

5-6.   What is the voltage across the 10-kΩ resistance of Fig. 5-24a?

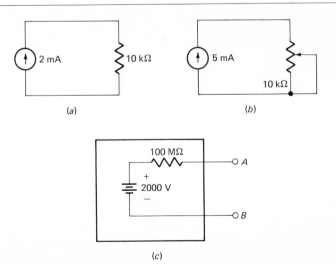

Figure 5-24.

(a)

(b)

(c)

5-7. Figure 5-24b shows a potentiometer with a total resistance of 10 kΩ. What is the load voltage when the wiper is at the top? At the bottom? In the middle?

5-8. Suppose you have a stock of resistors with values from 10 Ω to 1 MΩ. If you connect one of these between the AB terminals of Fig. 5-24c, what is the minimum possible current? The maximum possible current? (Three-digit accuracy is adequate for the answers.) To a close approximation, what value does the load current have for any load resistance between 10 Ω and 1 MΩ?

5-9. What is the Thevenin voltage between the AB terminals of Fig. 5-25a? The Thevenin resistance?

5-10. Thevenize Fig. 5-25b left of the AB terminals.

5-11. What is the V_{TH} between the AB terminals of Fig. 5-25c? The R_{TH}?

5-12. Thevenize Fig. 5-25d left of the CD terminals. Next, what is the Thevenin voltage between the AB terminals? The Thevenin resistance between AB?

5-13. The current source of Fig. 5-25e has a value of I_{CBO}. If I_{CBO} equals 1 μA, what is the Thevenin voltage between the AB terminals? The Thevenin resistance?

5-14. What is the Norton circuit left of the AB terminals in Fig. 5-26a?

5-15. Work out the Norton circuit left of the AB terminals in Fig. 5-26b.

5-16. In Fig. 5-26c, Thevenize the circuit left of the AB terminals. What is the Norton circuit left of the AB terminals?

5-17. Work out the Norton circuit to the left of the AB terminals in Fig. 5-26d.

5-18. A car battery loses 24,000 J of energy during 10 s while starting an engine. How many kilowatts is this?

5-19. A *horsepower* is a British unit of measure for power. The relation between this British unit and the metric unit is

$$1 \text{ hp} = 746 \text{ W}$$

Figure 5-25.

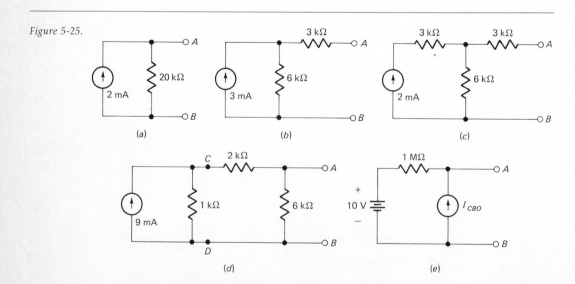

(a)   (b)   (c)

(d)   (e)

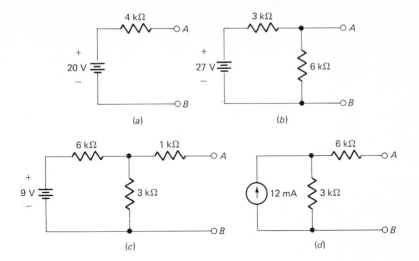

Figure 5-26.

If an electric drill does 1865 J of work in 5 s what is the horsepower of the drill?

5-20. The battery in a transistor radio loses 5.4 J per minute when the radio is on. What is the power from the battery?

5-21. A light bulb consumes 180,000 J during an hour. What is the power?

5-22. The battery voltage in a transistor radio equals 9 V. If the current out of the battery is 10 mA, what is the power from the battery?

5-23. A light bulb has 120 V across it. If its resistance is 144 Ω, what is the power in the light bulb?

5-24. An AWG-22 copper wire 500 ft in length has a current of 0.5 A through it. What is the power in the wire?

5-25. How much power is there in the 5-kΩ resistance of Fig. 5-27a? The 15-kΩ resistance? The total load power?

5-26. Calculate the power in each resistance of Fig. 5-27b. What is the total source power?

5-27. Work out the power in each resistance of Fig. 5-27c. What is the total load power? The total source power?

5-28. Ten circuits are in parallel across a 9-V source. Five of the circuits each have 1 mA of current. Three circuits have 2 mA each. And two circuits have 4 mA each. What is the total load power? Source power?

5-29. A *power supply* is an instrument or circuit that produces a steady voltage similar to a battery. A good power supply produces an almost constant voltage and has almost zero Thevenin resistance.

   In the power supply of Fig. 5-28a there is an adjustment for changing the output voltage from 10 to 30 V. This being the case, what is the minimum load power with a 100-Ω load? The maximum load power?

Figure 5-27.

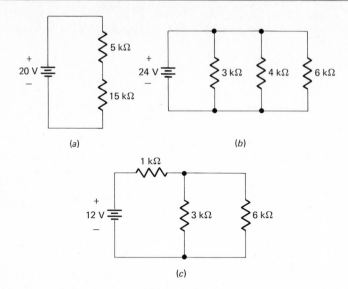

(a) (b)

(c)

5-30. What current rating should the fuse of Fig. 5-28*b* have if it is to blow when
   a. A short is between the *AB* terminals.
   b. A 2-Ω load is between the *AB* terminals.
   c. A 75-Ω load is between the *AB* terminals.

5-31. To get the maximum load power in Fig. 5-28*c*, what value should R_L have? What is the resulting load power?

Figure 5-28.

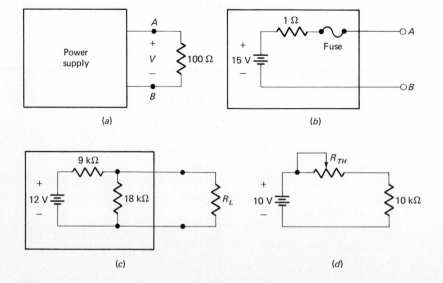

(a) (b)

(c) (d)

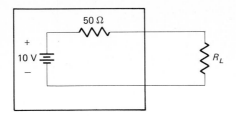

Figure 5-29.

5-32. A radar antenna has a Thevenin voltage of 100 μV and a Thevenin resistance of 50 Ω. What value should a load resistance have to get the maximum power? What is the maximum load power?

5-33. We want to get the maximum load power in the 10-kΩ resistance of Fig. 5-28d. What value should R_{TH} be adjusted to? What is the maximum power in the 10-kΩ resistance?

5-34. Calculate the power in the 10-kΩ resistance of Fig. 5-28d when R_{TH} is adjusted to equal 10 kΩ. Also, work out the load power when R_{TH} is adjusted to zero.

5-35. A signal generator has a Thevenin voltage of 10 V and a Thevenin resistance of 600 Ω. What is the maximum load power you can get out of this signal generator?

5-36. A TV antenna has a Thevenin voltage of 20 mV and a Thevenin resistance of 300 Ω. When the receiver is matched to the antenna, what is the power to the receiver?

5-37. Figure 5-29 shows a black box with a Thevenin voltage of 10 V and a Thevenin resistance of 50 Ω. Convince yourself that the maximum load power occurs when R_L is 50 Ω by calculating the load power for each of these:
   a. $R_L = 0$
   b. $R_L = 25$ Ω
   c. $R_L = 50$ Ω
   d. $R_L = 100$ Ω
   e. $R_L = \infty$ (same as an open circuit).
   (If necessary, select more values of R_L until you believe the maximum load-power theorem.)

ANSWERS TO TESTS

5-1. *d, d, c, b, c*
5-2. *d, b, b, c*
5-3. *b, a, b, b, d*
5-4. *a, b, a, c, c, a*
5-5. *a, c, d, a, c, a*
5-6. *d, b, a, c, c*

6. Kirchhoff Methods

Technicians and engineers solve most problems with Ohm's law, Thevenin's theorem, and other methods discussed earlier. But there are some circuits for which earlier methods either won't work or are too difficult to apply.

This chapter is about methods that always work, at least in theory. By applying Kirchhoff's laws to a circuit, we can get two or more equations involving unknown currents or voltages. Solving these equations gives the unknown values.

6-1. SIMULTANEOUS EQUATIONS

Simultaneous equations are equations that exist in the same experiment, problem, or discussion. For instance, if x is twice as old as y, and also 5 years older, the simultaneous equations are

$$x = 2y$$

$$x = y + 5$$

Solving these equations gives $x = 10$ and $y = 5$. These values of x and y are called the *simultaneous solution*.

In the review that follows, recall the answers to these questions:

> *What are the three methods of simultaneous solution?*
> *When are simultaneous equations independent?*

Addition-and-subtraction method

One of the methods learned in algebra is the *addition-and-subtraction method*. Here is an example. Suppose we have these two simultaneous equations:

$$x + 3y = 14 \qquad \text{(first)}$$

$$2x - y = 7 \qquad \text{(second)}$$

Multiply each term of the first equation by 2, and the simultaneous equations become

$$2x + 6y = 28 \qquad \text{(first)}$$

$$2x - y = 7 \qquad \text{(second)}$$

Next, subtract the second equation from the first to get

$$7y = 21$$

Solving for y gives

$$y = 3$$

This is one of the unknown values. To get the other, substitute the y value into either of the original equations and solve for x. Substituting $y = 3$ into the first original equation,

$$x + 3y = x + 3(3) = 14$$

$$x + 9 = 14$$

or $$x = 5$$

So, the simultaneous solution of the two original equations is $x = 5$ and $y = 3$.

In general, here are the steps in the addition-and-subtraction method for two equations in two unknowns:

1. Multiply the terms of either equation (or both) by whatever factor (or factors) is needed to make the coefficients of x or the coefficients of y equal in both equations. (In our example, we multiplied the first equation by 2 to make the coefficients of x equal in both equations.)
2. Add or subtract the equations to eliminate one of the unknowns. (In the example, we subtracted the second equation from the first to eliminate x.)
3. Solve for the remaining unknown. (We solved for y.)
4. Substitute the newly found value into either of the original equations and solve for the other unknown. (Finally, we solved for x.)

Substitution method

Here's how to solve the same equations by the *substitution method*. The equations are

$$x + 3y = 14 \quad \text{(first)}$$

$$2x - y = 7 \quad \text{(second)}$$

Solving the first equation for x in terms of y gives

$$x = 14 - 3y \quad \text{(derived)}$$

Next, substitute the right side of this derived equation into the second equation to get

$$2(14 - 3y) - y = 7$$

$$28 - 6y - y = 7$$

or

$$y = 3$$

This is one of the unknowns. To get the other, substitute the y value into the derived equation and solve for x as follows:

$$x = 14 - 3y = 14 - 3(3) = 14 - 9$$

or

$$x = 5$$

So, the simultaneous solution by the substitution method is $x = 5$ and $y = 3$.

To summarize the steps in the substitution method:

1. Solve either equation for one unknown in terms of the other. The resulting equation is a derived equation. (In our example, we solved for x in terms of y.)
2. Substitute the right side of the derived equation into the other original equation. (In our example, we substituted into the second equation.)
3. Solve for one of the unknowns. (We found the y value.)
4. Substitute the newly found value into the derived equation and solve for the other unknown. (We substituted the y value and solved for the x value.)

Determinant method

After solving enough simultaneous equations, you begin to see shortcuts. The *determinant method* is a shortcut derived from the addition-and-subtraction method. There are too many details in the determinant method to permit an adequate discussion in this book. (The right place to learn the method is in an algebra course.) For the rest of this book all you need to know is that the determinant method is normally used to solve three or more simultaneous equations.

Even with the determinant method, a simultaneous solution can be tedious work. When many equations are involved, a digital computer is often used to solve for the unknowns. In other words, solving more than three simultaneous equations is work for an expert dealing with problems not encountered in everyday electronics.

Our main interest in simultaneous equations, therefore, is not how to solve them, but how to set them up. Being able to do this gives much insight into how circuits work, and may suggest new ways to a solution. More is said about this later.

Independent equations

Suppose we have

$$x + 2y = 4 \quad \text{(original)}$$

Multiplying each term by 3 gives

$$3x + 6y = 12 \quad \text{(derived)}$$

The derived equation is not fundamentally different from the original, because the relation between x and y is unchanged. In other words, the derived equation contains no new information about x and y, because it has the same graph as the original equation. Any equation that can be derived from another is called a *dependent equation*.

Here is another example of a dependent equation. Given

$$3x + 2y = 4 \quad \text{(first)}$$

$$12x + 8y = 16 \quad \text{(second)}$$

notice each term in the second equation differs by a factor of 4 from the corresponding term in the first equation. The second equation is therefore dependent and contains no new information about the relation between x and y.

Independent equations are different. Each independent equation adds new information about x and y. The way to recognize independent equations is this: if the x terms differ by one factor and the y terms by another, the two equations are independent. For instance, given

$$x + y = 5$$

$$2x + 3y = 4$$

notice the x terms differ by a factor of 2, but the y terms by a factor of 3. Because of this, the two equations are independent. If plotted, independent equations always produce distinct graphs.

In a circuit problem, you can throw away any equation that is not independent, because only independent equations add useful information. Without exception, if a problem has n unknowns, you have to write and solve n independent equations involving these unknowns. In other words, a complete circuit analysis takes two steps:

1. Writing independent equations based on Kirchhoff's current and voltage laws.
2. Solving these simultaneous equations for the unknown currents and voltages.

As mentioned earlier, our main interest is in step 1, because being able to set up equations often gives great insight into circuit action. The rest of this chapter, therefore, is mostly about setting up independent equations.

EXAMPLE 6-1.
Given

$$x - y = 3$$

$$2x + 2y = 3$$

Are these equations independent?

SOLUTION.
Equations are always independent when x terms differ by one factor and y terms by another. In this case, the x terms differ by 2 and the y terms by -2. These factors are *different* because of the minus sign; therefore, the equations are independent.

EXAMPLE 6-2.
Are the following equations independent?

$$x + y + 2z = 8 \qquad \text{(first)}$$

$$3x + 3y - 6z = 12 \qquad \text{(second)}$$

SOLUTION.
The x terms differ by a factor of 3, the y terms by a factor of 3, and the z terms by a factor of -3. Even one variation in the factor is enough to make the equations independent. So the foregoing equations are independent.

Incidentally, when testing for independence, forget about the constants; they have nothing to do with independence. In other words, the right sides of the first and second equations contain 8 and 12; the ratio 1.5 is not used in testing for independence. Only the unknown terms are involved in testing for independence.

Test 6-1

1. You multiply to make coefficients equal in which of these methods?
   (*a*) addition and subtraction   (*b*) substitution   (*c*) determinants .... ( )
2. One of the original equations is solved for one unknown in terms of another in which of these methods?
   (*a*) addition and subtraction   (*b*) substitution   (*c*) determinants .... ( )
3. Normally, when there are three or more simultaneous equations, which of the following is used?
   (*a*) addition and subtraction   (*b*) substitution   (*c*) determinants .... ( )
4. Which of these does not belong?
   (*a*) no new information   (*b*) independent   (*c*) derived
   (*d*) dependent .... ( )
5. Two equations in two unknowns are independent if the x terms differ by a factor

and the y terms differ by

(*a*) the same factor   (*b*) another factor   (*c*) the reciprocal only

(*d*) none of these .... ( )

6. What comes next in the following series?

            DEPENDENT, SAME GRAPH, INDEPENDENT, ____

(*a*) similar graph   (*b*) same graph   (*c*) derived graph

(*d*) different graph .... ( )

6-2. BRANCH METHOD

One way to find all the voltages and currents in a circuit is with the *branch method*. To understand this approach, learn the answers to

What is a branch?

What are branch equations?

Definition of a branch

Branches are the series paths that make up a circuit; each path with the same current is a branch. For example, the bridge of Fig. 6-1*a* has these branches:

Branch *A-D*   (contains the source)
Branch *A-B*   (contains R_1)
Branch *B-D*   (contains R_2)
Branch *A-C*   (contains R_3)
Branch *C-D*   (contains R_4)
Branch *B-C*   (contains R_5)

Each branch contains a single *circuit element* (a source, resistance, or other device).

Branches may contain more than one circuit element. Figure 6-1*b* shows a circuit with seven equipotential points, *A* through *G*. Notice the circuit has one-element branches like branch *A-B*, branch *B-C*, etc. But is also has branches with more than one element like

Branch *B-E-G*   (contains R_4 and R_5)
Branch *C-D-F-G*   (contains R_3, R_7, and R_8)

An *equivalent branch* is the branch you get when you lump all resistances in a branch into a single equivalent resistance. For instance, R_4 and R_5 in branch *B-E-G* of Fig. 6-1*b* can be lumped into a single resistance $R_4 + R_5$, as shown in Fig. 6-1*c*. Likewise, branch *C-D-F-G* can be replaced by the equivalent branch of Fig. 6-1*c*.

Figure 6-1. Branches of a circuit.

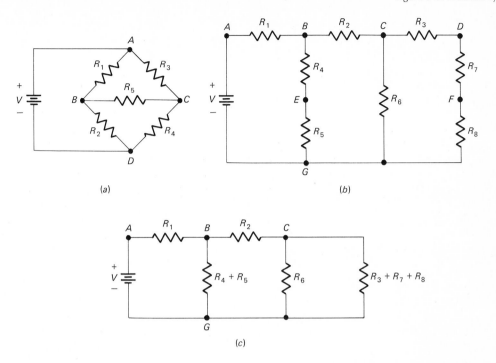

(a)

(b)

(c)

Setting up equations

Suppose we want the total solution for Fig. 6-2*a*. This means we want the voltage and current for every resistance in the circuit. The first step in the branch method is to reduce each branch with more than one resistance to an equivalent branch. The 2-Ω and 1-Ω resistances of branch *A-C-D* can be lumped into a 3-Ω resistance; the two 1-Ω resistances of branch *A-B-D* combine into a 2-Ω resistance. Figure 6-2*b* shows the circuit with equivalent branches.

Each resistance in Fig. 6-2*b* has an unknown voltage and current. Since the circuit has three resistances, there are six unknowns to start with. But Ohm's law eliminates three of these unknowns as follows. Because

$$V = RI$$

once we have a branch current, we can calculate the corresponding voltage. In other words, there are ultimately only three unknown branch currents; the unknown voltages are automatically known once the branch currents are found.

Figure 6-2*c* shows the circuit with the three unknown branch currents. Since there are three unknowns, you need to write three independent equations involving these

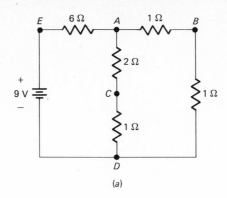

Figure 6-2. Using the branch method.

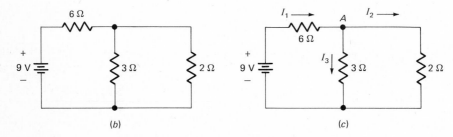

(b)   (c)

unknowns. You can get these three equations by applying Kirchhoff's laws as follows. At point A, the current law gives

$$I_1 = I_2 + I_3 \tag{6-1}$$

Next, sum voltages around the left loop to get

$$6I_1 + 3I_3 - 9 = 0 \tag{6-2}$$

and around the right loop to get

$$2I_2 - 3I_3 = 0 \tag{6-3}$$

To compare like terms, rearrange Eqs. (6-1) through (6-3) in the following form:

$$I_1 - I_2 - I_3 = 0$$
$$6I_1 \qquad + 3I_3 = 9 \tag{6-4}$$
$$2I_2 - 3I_3 = 0$$

We call this orderly arrangement *standard form;* all it means is vertically lining up the corresponding unknowns and transposing the constants to the right side. With equations in standard form, it is easy to see each equation is independent of the other two.

In other words, simultaneous equations (6-4) are three independent equations in three unknowns.

The grinding out of a simultaneous solution is an algebra problem. Any of the three methods described in Sec. 6-1 leads to the same solution:

$$I_1 = 1.25 \text{ A}$$

$$I_2 = 0.75 \text{ A}$$

$$I_3 = 0.5 \text{ A}$$

With these branch currents, you can return to the original circuit (Fig. 6-2a) and use Ohm's law to calculate the voltage across each resistance. In this way, you get the total solution for the circuit.

Summary

The general idea of the branch method is this. The ultimate unknowns are the branch currents, because Ohm's law automatically gives the unknown voltages once the branch currents are found. If a circuit has B branches, there are B unknown currents. To get a solution, we have to write B independent equations based on Kirchhoff's current and voltage laws.

The branch method is straightforward, but not used as much as the *loop* and *node* methods discussed later. For this reason, all you have to remember about the branch method are the main ideas:

1. The branches of a circuit are the series paths.
2. A circuit with B branches has B unknown branch currents.
3. A total solution requires B independent equations based on Kirchhoff's current and voltage laws.

Test 6-2

1. A branch contains how many resistances?
   (a) 1   (b) 2   (c) any number .... ( )
2. Which of the following belongs least?
   (a) one resistance   (b) branch   (c) series path   (d) same current ... ( )
3. Complicated branch is to equivalent branch as many resistances are to
   (a) no resistances   (b) one resistance   (c) any number of resistances ... ( )
4. In conjunction with Ohm's law, branch currents determine the values of all
   (a) currents   (b) resistances   (c) voltages .... ( )
5. If a circuit has seven branches, the branch method requires how many independent equations?
   (a) 14   (b) 7   (c) none of these .... ( )

6-3. LOOP METHOD

A circuit with B branches has B unknown branch currents. Because of this, you have to write B independent equations involving branch currents. With some circuits this leads to many equations.

The *loop method* is different. It automatically eliminates some of the unknowns before you start writing equations. Because of this, fewer independent equations are needed with the loop method than with the branch method. To understand why, you need to know

What are loop currents?
What are windowpanes in a circuit?
How do you write loop equations?

Loop currents

In Fig. 6-3a, the current law applied to point A gives

$$I_1 = I_2 + I_3$$

Solving for I_3 gives

$$I_3 = I_1 - I_2$$

This says the current through the middle resistance equals the difference between I_1 and I_2. To emphasize this relation, we can relabel the branch currents as shown in Fig. 6-3b.

Figure 6-3c shows an equivalent way to visualize the current through the middle resistor; this current is the algebraic sum of two components, I_1 and I_2. That is, you can think of the total current as the superposition of a downward component I_1 and an upward component I_2; the algebraic sum equals $I_1 - I_2$, the same as the original current in Fig. 6-3b.

I_1 flows through the loop on the left, and I_2 through the loop on the right. To emphasize this, we can label the currents through each circuit element as shown in Fig. 6-3d. This stresses the idea that I_1 is the current in the left loop, and I_2 the current in the right loop.

There's no need to draw four separate current arrows for each loop. Since the four currents in the left loop are the same, we can draw a single *loop arrow* as shown in the left loop of Fig. 6-3e. Similarly, the currents through the right loop are the same, so we draw a loop arrow in the right loop as shown.

We call the currents of Fig. 6-3e *loop currents,* because each flows through a loop or closed path. Since the two loops have a common branch (the middle resistance), the net current in this branch is the algebraic sum of the loop currents.

Compare the loop currents of Fig. 6-3e to the branch currents of Fig. 6-3a. Notice there are two unknown loop currents, but three unknown branch currents. The reason

Figure 6-3. Deriving loop currents.

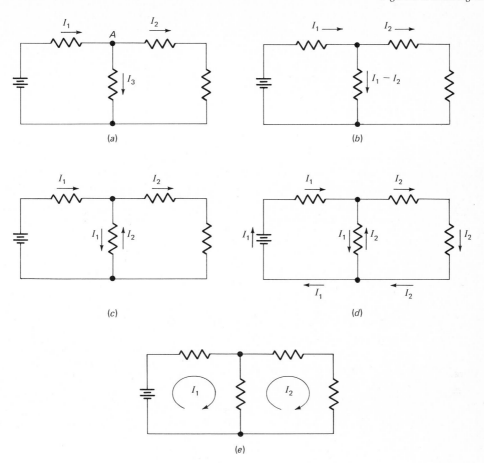

is simple. To get the loop currents, we applied Kirchhoff's current law at point A; this eliminated one of the unknown currents.

Loop currents always reduce the number of unknown currents, because loop currents are derived through the application of Kirchhoff's current law. Because there are always fewer loop currents than branch currents, we need fewer independent equations with loop currents than with branch currents.

Windowpanes

We're not about to apply Kirchhoff's current law every time we want loop currents. This would be too much work and really unnecessary, because there is a shortcut for finding loop currents. We call this shortcut the *windowpane technique.* Here's how it works.

In Fig. 6-4*a*, temporarily forget about the sources and resistances; think of the whole circuit as a large window with four panes, as shown in Fig. 6-4*b*. For convenience, number the windowpanes the way you read, from left to right and down.

Since each pane corresponds to a loop, draw a loop current in each pane as shown in Fig. 6-4*c*. To avoid confusion, always draw loop currents in a *clockwise* direction. If the true direction of current is counterclockwise, a negative sign will turn up. For instance, if you work out a problem and find I_4 equals -5 mA, this means the true direction of I_4 is counterclockwise in Fig. 6-4*c*.

Figure 6-4*d* shows the original circuit with the loop currents. Since there are four loop currents, we need to write four independent equations involving these loop currents. This is much easier than using the branch method. Figure 6-4*a* has eight branches, or eight unknown branch currents; therefore, a branch-current solution

Figure 6-4. Windowpanes of a circuit.

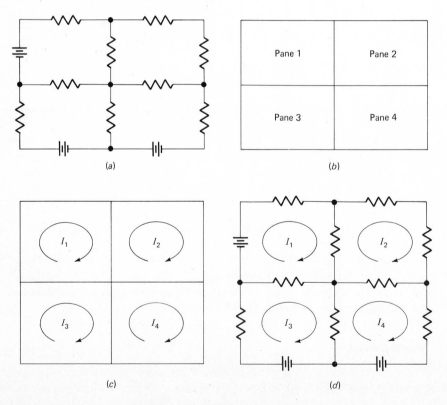

(a) (b) (c) (d)

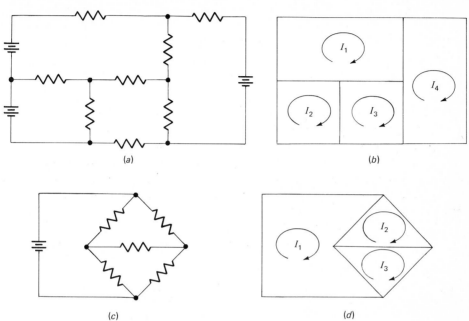

Figure 6-5. More examples of windowpanes.

would require eight independent equations involving branch currents. Now you can see why the loop-current method is preferred.

Figure 6-5a shows another circuit. If you look carefully, you will find nine branches; this means nine unknown branch currents. A branch-current solution requires nine independent equations based on Kirchhoff's current and voltage laws. But the windowpane technique automatically eliminates some of the unknown currents, because it is based on Kirchhoff's current law. Visualizing the circuit in terms of its windowpanes, we come up with the four loop currents shown in Fig. 6-5b.

Similarly, a branch-current solution of the Wheatstone bridge (Fig. 6-5c) requires six independent equations, because there are six branches. But a loop-current solution needs only three independent equations, because there are only three windowpanes (Fig. 6-5d).

Writing loop equations

After you have drawn loop currents for a circuit, the next step is to apply Kirchhoff's voltage law to each loop. Since each loop gives one voltage equation, a circuit with L loops gives L voltage equations. Because each loop is different from the others,

Figure 6-6. Writing loop equations.

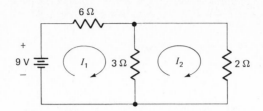

the voltage equations are independent. In other words, the loop method guarantees a complete solution, because you get as many independent equations as there are unknowns.

A *unique resistance* is one with only a single loop current. The 6-Ω resistance of Fig. 6-6 is unique to the left loop, because I_1 is the only current through this resistance. Similarly, the 2-Ω resistance is unique to the right loop, because only I_2 flows through it.

A *common resistance* has two loop currents through it. For instance, the 3-Ω resistance of Fig. 6-6 is common to both loops, because I_1 and I_2 flow through it. No matter how complicated a circuit, all resistances are either unique (one loop current) or common (two loop currents).[1]

To have an orderly method of summing voltages, we always will sum in a clockwise direction. Because of this, each *unique resistance* always contributes a *positive voltage* to the total loop sum. For instance, in Fig. 6-6, when summing around the right loop, we write

$$+2I_2$$

for the contribution of the 2-Ω resistance.

To make things easy, each common resistance also contributes a positive voltage, if we take the positive direction of current the same as the *local loop current* (the loop current in the loop being summed). For instance, summing voltages in the left loop, we take the direction of I_1 as the positive direction. Because of this, the algebraic sum of currents through the common resistance is

$$I_1 - I_2$$

and the voltage contribution of the common resistance is

$$+3(I_1 - I_2)$$

On the other hand, when summing voltages around the right loop, we take the direction of I_2 as the positive direction. As a result, the algebraic sum of currents

---

[1] This statement is valid for all two-dimensional circuits, the ones you can draw on paper without crossing branches. Practical circuits are almost always two-dimensional.

through the common resistance is

$$I_2 - I_1$$

and the voltage contribution of the common resistance is

$$+3(I_2 - I_1)$$

So you see, each loop is treated separately; that is, the positive direction of current is always taken as the direction of the local loop current.

Here's how it works when we put all the foregoing ideas together. In Fig. 6-6, summing around the left loop gives

$$6I_1 + 3(I_1 - I_2) - 9 = 0$$

Summing around the right loop gives

$$2I_2 + 3(I_2 - I_1) = 0$$

Rearranging the two loop equations gives the standard form:

$$9I_1 - 3I_2 = 9 \tag{6-5}$$

$$-3I_1 + 5I_2 = 0$$

Carrying out the rest of the algebra gives this simultaneous solution:

$$I_1 = 1.25 \text{ A}$$

$$I_2 = 0.75 \text{ A}$$

More examples follow. As you read them, look for these basic ideas about the loop method:

1. In each loop, sum the voltages in the direction of the loop current (clockwise).
2. A unique resistance always contributes a positive voltage equal to the loop current times the unique resistance.
3. A common resistance always contributes a positive voltage equal to the difference between two loop currents times the common resistance. (Remember to subtract the foreign loop current from the local loop current.)

EXAMPLE 6-3.
Write the loop equations for Fig. 6-7.

SOLUTION.
The left loop gives

$$2I_1 + 4(I_1 - I_2) - 10 = 0$$

or

$$6I_1 - 4I_2 = 10$$

Figure 6-7. Example of loop equations.

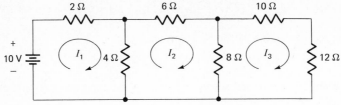

The middle loop gives

$$6I_2 + 8(I_2 - I_3) + 4(I_2 - I_1) = 0$$

or

$$-4I_1 + 18I_2 - 8I_3 = 0$$

The right loop gives

$$10I_3 + 12I_3 + 8(I_3 - I_2) = 0$$

or

$$-8I_2 + 30I_3 = 0$$

In standard form the three loop equations are

$$6I_1 - 4I_2 \qquad\quad = 10$$
$$-4I_1 + 18I_2 - 8I_3 = 0 \qquad (6\text{-}6)$$
$$-8I_2 + 30I_3 = 0$$

EXAMPLE 6-4.

Write the voltage equations for Fig. 6-8 and collect in standard form.

SOLUTION.
The first loop gives

$$3(I_1 - I_2) + 6(I_1 - I_3) - 9 = 0$$

or

$$9I_1 - 3I_2 - 6I_3 = 9$$

Figure 6-8. Writing loop equations for a Wheatstone bridge.

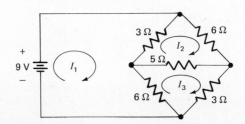

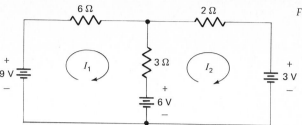

Figure 6-9. More loop equations.

The second loop leads to

$$6I_2 + 5(I_2 - I_3) + 3(I_2 - I_1) = 0$$

or

$$-3I_1 + 14I_2 - 5I_3 = 0$$

The third loop produces

$$3I_3 + 6(I_3 - I_1) + 5(I_3 - I_2) = 0$$

or

$$-6I_1 - 5I_2 + 14I_3 = 0$$

Collecting the three loop equations in standard form,

$$9I_1 - 3I_2 - 6I_3 = 9$$

$$-3I_1 + 14I_2 - 5I_3 = 0 \qquad (6\text{-}7)$$

$$-6I_1 - 5I_2 + 14I_3 = 0$$

EXAMPLE 6-5.

Write the two loop equations for Fig. 6-9.

SOLUTION.

The left loop produces

$$6I_1 + 3(I_1 - I_2) + 6 - 9 = 0$$

or

$$9I_1 - 3I_2 = 3$$

The right loop gives

$$2I_2 + 3 - 6 + 3(I_2 - I_1) = 0$$

or

$$-3I_1 + 5I_2 = 3$$

Test 6-3

1. Which of these always flows on the outside boundary of a pane?
    (*a*) branch current (*b*) algebraic sum (*c*) unique current
    (*d*) loop current .... ( )

2. Unique resistance is to one loop current as common resistance is to
    (*a*) one loop current   (*b*) two loop currents   (*c*) three loop currents ... ( )
3. Which of the following does not belong?
    (*a*) unique resistance   (*b*) common resistance   (*c*) algebraic sum
    (*d*) two loop currents .... ( )
4. The following words can be rearranged into a sentence: THERE'S PER PANE BRANCH ONE CURRENT. The sentence is
    (*a*) true   (*b*) false .... ( )
5. The number of panes does not indicate which of the following?
    (*a*) the number of unknown loop currents
    (*b*) the number of independent voltage equations
    (*c*) the number of independent loop equations
    (*d*) the number of branches .... ( )
6. Because we sum voltages in a clockwise direction and subtract the foreign loop current from the local loop current, the sign on each voltage contribution is algebraically
    (*a*) minus   (*b*) plus   (*c*) either   (*d*) neither .... ( )

6-4. NODE METHOD

A *node* is another name for an equipotential point. In Fig. 6-10*a*, points *A*, *B*, *C*, and *D* are nodes. The *node method* is a useful alternative to the loop method.

To appreciate the node method, learn the answers to

What is a node voltage?
How do you write node equations?

Node voltage

The first step in the node method is to select a *reference node*, a point from which all other voltages are measured. We can use any node (*A* through *D*) of Fig. 6-10*a* for the reference node. For a reason given later, we often choose the bottom node, in this case, node *D*. Figure 6-10*b* shows the circuit with *D* relabeled as the reference node.

A *node voltage* is the voltage between a node and the reference node. In Fig. 6-10*b*, *A* has a node voltage of 9 V, *B* has an unknown node voltage V_B, and *C* has a node voltage of 1 V. Figure 6-10*c* summarizes these node voltages.

An *independent node* is defined as a node whose voltage is unknown. Node *B* is the only independent node in Fig. 6-10*c*. Independent nodes are important for this reason: each independent node leads to one independent equation. A circuit with *N* independent nodes results in *N* independent equations.

Figure 6-10. Node voltages.

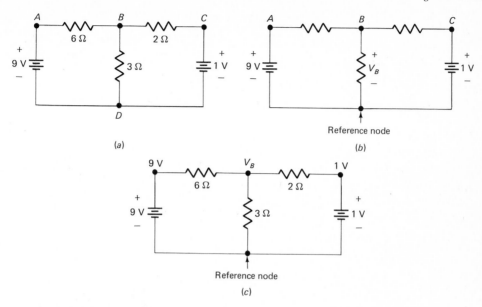

(a)

(b)

(c)

Reference node

Voltages across resistors

The original circuit of Fig. 6-10a can be redrawn as shown in Fig. 6-11. A has a node voltage of 9 V, B has an unknown node voltage V_B, and C has a node voltage of 1 V. The reference node always has a node voltage of 0 V.

The voltage across a resistor equals the difference between the node voltages at each end of the resistor. In Fig. 6-11, the node voltage at the left end of the 6-Ω resistor is 9 V, and the node voltage at the right end is V_B; the voltage across the resistor therefore equals

$$V_1 = 9 - V_B$$

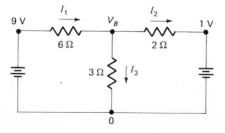

Figure 6-11. Writing node equations.

The 2-Ω resistor has a node voltage of V_B on its left end, and a node voltage of 1 V on its right end; therefore, the voltage across this resistor is

$$V_2 = V_B - 1$$

Finally, the 3-Ω resistor has V_B on one end, and 0 V on the other end; so the voltage across it equals

$$V_3 = V_B - 0 = V_B$$

Direction of currents

The direction of assumed current is important; it determines which node voltage is subtracted from which. For instance, in Fig. 6-11 if we draw I_2 the opposite way, the voltage across the 2-Ω resistor will be

$$V_2 = 1 - V_B$$

As mentioned several times already, if the true current in a circuit is opposite the direction assumed while setting up equations, a minus sign turns up for the wrongly assumed current.

Since it makes no difference what direction we assume for the branch currents, we will always draw currents to the right and down. For instance, I_1 and I_2 point to the right in Fig. 6-11; I_3 points down. This means the voltages across resistors always are obtained by taking

Left node voltage $-$ right node voltage

or   Upper node voltage $-$ lower node voltage

Writing node equations

In the node method, you apply Kirchhoff's current law to each independent node. Figure 6-11 has only one independent node. Applying the current law to node B gives

$$I_1 = I_2 + I_3$$

or in terms of voltages and resistances,

$$\frac{9 - V_B}{6} = \frac{V_B - 1}{2} + \frac{V_B}{3}$$

This equation has only one unknown. By clearing the equation of fractions and solving for V_B,

$$V_B = 2 \text{ V}$$

Now, all node voltages are known in Fig. 6-11. Therefore, we can calculate the branch currents as follows:

$$I_1 = \frac{9-2}{6} = 1.167 \text{ A}$$

$$I_2 = \frac{2-1}{2} = 0.5 \text{ A}$$

$$I_3 = \frac{2}{3} = 0.667 \text{ A}$$

Generalization

Given any circuit, select a reference node and then figure out which nodes are dependent (known voltage) and which are independent (unknown voltage). Apply Kirchhoff's current law to each independent node to get one independent equation. A circuit with N independent nodes therefore gives N independent equations involving the unknown node voltages.

To eliminate unnecessary work, any branch with more than one resistance should be replaced by its equivalent resistance. A circuit like Fig. 6-12a has two unknown node voltages. You can apply the current law to the independent nodes to get two independent equations. But this adds unnecessary difficulty to the solution. It's more sensible to reduce the branch with the 5 Ω and 1 Ω to an equivalent 6-Ω branch as shown in Fig. 6-12b. Then, there's only one independent node to worry about. After you find the value of I_1 (1.167 A found earlier), you can calculate V_A in the original circuit (Fig. 6-12a).

In summary, here are the main ideas of the node method:

1. Replace complicated branches by equivalent branches.
2. Label each node with its node voltage. The reference node has zero voltage, and all other nodes have either known or unknown voltages.

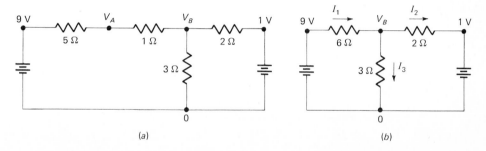

Figure 6-12. Simplifying branches before writing equations.

(a)

(b)

3. Apply Kirchhoff's current law to each independent node. Express each current as the difference between two node voltages divided by a resistance.

4. If desired, solve the N simultaneous equations for N unknown voltages.

EXAMPLE 6-6.
Solve for all voltages and currents in Fig. 6-13a.

SOLUTION.
Here's how it's done with the node method. First, label the node voltages as shown in Fig. 6-13b, and show unknown currents to the right and down. The circuit has only one independent node. At this node,

$$I_1 = I_2 + I_3$$

or

$$\frac{6 - V_B}{6} = \frac{V_B - 12}{2} + \frac{V_B}{3}$$

Multiplying through by 6 gives

$$6 - V_B = 3V_B - 36 + 2V_B$$

Solving for V_B,

$$V_B = 7 \text{ V}$$

Figure 6-13. Applying the node method.

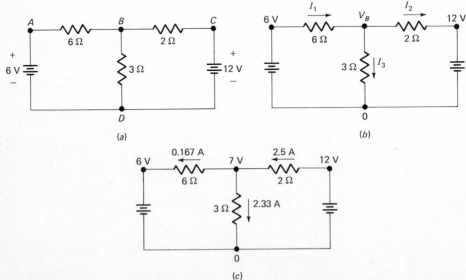

(a)

(b)

(c)

Next, the currents in Fig. 6-13b are

$$I_1 = \frac{6-7}{6} = -0.167 \text{ A}$$

$$I_2 = \frac{7-12}{2} = -2.5 \text{ A}$$

$$I_3 = \frac{7}{3} = 2.33 \text{ A}$$

In Fig. 6-13b, I_3 equals 2.33 A in the direction shown; currents I_1 and I_2 equal 0.167 A and 2.5 A, in the opposite directions.

Figure 6-13c summarizes the solution with all the node voltages and currents in the true directions.

EXAMPLE 6-7.
Write the node equations for Fig. 6-14a.

SOLUTION.
The circuit has five nodes, A through E. When we label nodes as shown in Fig. 6-14b, we see only one independent node; therefore, we need to write only one independent equation. Applying the current law to node B gives

$$I_1 = I_2 + I_3$$

$$\frac{10 - V_B}{4000} = \frac{V_B + 2}{6000} + \frac{V_B - 5}{8000}$$

Notice the voltage across the 6-kΩ resistance is

$$V_B - (-2) = V_B + 2$$

Figure 6-14. More node equations.

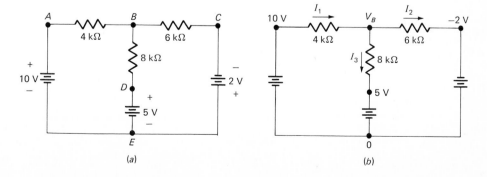

(a)   (b)

The rest of the solution is straightforward. You can solve for V_B, and then calculate the currents. If you do, you will find the true direction of each current is the same as shown in Fig. 6-14b.

EXAMPLE 6-8.
Earlier, we applied the loop method to the bridge of Fig. 6-15a. This resulted in three loop equations involving three loop currents.
Apply the node method to Fig. 6-15a to get the node equations.

SOLUTION.
There are four nodes in Fig. 6-15a. After labeling all nodes as shown in Fig. 6-15b, there are only two independent nodes, B and C. Therefore, we need to write two node equations.
Applying Kirchhoff's current law to node B gives

$$I_1 = I_3 + I_4$$

or
$$\frac{9 - V_B}{3} = \frac{V_B - V_C}{5} + \frac{V_B}{6} \qquad (6\text{-}8a)$$

Applying the current law to node C produces

$$I_2 + I_3 = I_5$$

or
$$\frac{9 - V_C}{6} + \frac{V_B - V_C}{5} = \frac{V_C}{3} \qquad (6\text{-}8b)$$

By straightforward algebra, Eqs. (6-8a) and (6-8b) rearrange into this standard form:

$$21V_B - 6V_C = 90$$
$$\qquad (6\text{-}9)$$
$$6V_B - 21V_C = -45$$

Figure 6-15. Applying node method to Wheatstone bridge.

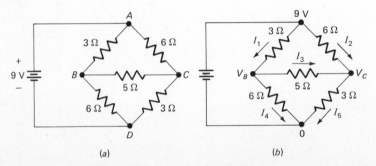

(a)   (b)

These are two independent equations in two unknowns; therefore, we can solve them by any of the three methods described in Sec. 6-1.

In Example 6-4, we wrote the loop equations for the same Wheatstone bridge. Because of the three loop currents, three simultaneous equations resulted. In the case of a Wheatstone bridge, therefore, it's easier to solve the circuit by the node method.

In general, use whichever method gives fewer equations. You can tell which method gives fewer equations by counting windowpanes and independent nodes. When a circuit has fewer panes than independent nodes, use the loop method. If the circuit has fewer independent nodes than panes, use the node method. If the number of panes equals the number of independent nodes, use either method.

Test 6-4

1. The reference node is always
   (*a*) the bottom node   (*b*) a node other voltages are measured from
   (*c*) a point where more than two elements meet   (*d*) none of these .... ( )
2. A node always has
   (*a*) two or more elements connected to it   (*b*) a node voltage
   (*c*) a voltage with respect to the reference node   (*d*) all of these .... ( )
3. Voltage law is to loop as current law is to
   (*a*) windowpane   (*b*) node voltage   (*c*) reference node
   (*d*) independent node .... ( )
4. The voltage across a resistor always equals the difference between which of these?
   (*a*) left-node voltage and right-node voltage
   (*b*) upper-node voltage and lower-node voltage
   (*c*) node voltage where current goes in and node voltage where current leaves
   (*d*) none of these .... ( )
5. The kind of elements connected to a node determine whether it's independent or not. A node must be independent if the elements connected to it are
   (*a*) sources only   (*b*) sources and resistances   (*c*) resistances only
   (*d*) resistances and sources .... ( )
6. A circuit has B branches, L loops, N nodes, and three independent nodes. How many times do you have to apply Kirchhoff's current law using the node method?
   (*a*) B (*b*) L (*c*) N (*d*) 3 .... ( )
7. A circuit has B branches, L windowpanes, and N independent nodes. If B is greater than L, but L is less than N, you write the fewest number of equations with the
   (*a*) branch method   (*b*) loop method   (*c*) node method .... ( )

6-5. THE CHASSIS

Electronic components are often mounted on a metal base called a *chassis*. As shown in Example 2-16, a chassis is a good conductor. Because of this, we can use the chassis as part of a circuit. For instance, in Fig. 6-16*a* the two sources and the 3-Ω resistor connect

Figure 6-16. Chassis symbol.

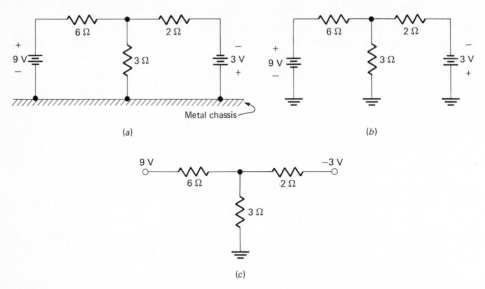

(a)

Metal chassis

(b)

(c)

to the chassis. Because of its low resistance, the chassis provides a conducting path, the same as a piece of wire. In fact, because the chassis resistance is so low, the usual practice is to visualize the entire chassis as an equipotential point.

Chassis symbol

Figure 6-16b shows the schematic symbol for the chassis. When you see this symbol, visualize a metal chassis completing the path between components.

We can simplify schematic diagrams by deleting all source symbols, and showing only the source voltages driving the nodes. For instance, Fig. 6-16c shows 9 V driving the left node, and −3 V driving the right node. These node voltages produce exactly the same currents as the sources of Fig. 6-16b.

Ground

Often, the power plug of electronics equipment has a third prong on it. When connected to an ac outlet, this third prong is *grounded* (put in contact with the earth) back at the electric power company. When this kind of plug is used, the third prong places the chassis in contact with the earth; this is why the chassis is often referred to as *ground*.

Even if the power plug does not have a third prong, it's still customary to refer to the chassis as ground. In describing the circuit of Fig. 6-16b, we'd say the negative terminal of the 9-V battery is grounded, the bottom of the 3-Ω resistor is grounded, and the positive terminal of the 3-V battery is grounded.

Because the chassis is an excellent conductor, all ground points represent the same equipotential point. In other words, in all preliminary analysis the voltage between any pair of ground points is approximated as zero.

EXAMPLE 6-9.
Calculate node voltage V_A and current I in Fig. 6-17a.

SOLUTION.
With the voltage-divider theorem,

$$V_A = \frac{R_2}{R} V = \frac{3}{9} 18 = 6 \text{ V}$$

The current equals

$$I = \frac{18 - V_A}{6} = \frac{18 - 6}{6} = 2 \text{ A}$$

EXAMPLE 6-10.
The black box of Fig. 6-17b contains a single resistance R. If we measure V_A equal to 3.75 V, what is the value of R?

SOLUTION.
Since we are given V_A equal to 3.75 V, we have the node voltages on each end of the 5-kΩ resistance. Therefore, the current is

$$I = \frac{10 - 3.75}{5000} = 1.25 \text{ mA}$$

The second step is to realize this current goes through the unknown resistance R.

Figure 6-17. Examples of using node voltage.

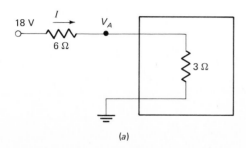

(a)

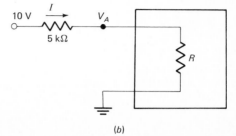

(b)

Since the voltage across R equals V_A,

$$R = \frac{V_A}{I} = \frac{3.75}{0.00125} = 3 \text{ k}\Omega$$

Therefore, the black box contains a 3-kΩ resistance.

Simple as it is, the foregoing method is used a great deal in finding the *input resistance* of a transistor circuit. The input resistance of the transistor acts like the R of Fig. 6-17b. You will learn more about this in electronics.

6-6. LADDER METHOD

If the windowpanes of a circuit can be laid out end to end as shown in Fig. 6-18a, the circuit is called a *ladder*. Figure 6-18b is an example of a three-loop ladder. If a circuit does not look like a ladder but can be redrawn as a ladder, it's still called a ladder. For instance, Fig. 6-18c does not look like a ladder, but we can redraw it as shown in Fig. 6-18d; therefore, Fig. 6-18c is a ladder.

Ladders are used a great deal in practice. Because of this, you should know the shortcut solution to a ladder circuit. In what follows, find the answers to

What is a new-source circuit?
How do you scale new-source answers?

Figure 6-18. Ladder circuits.

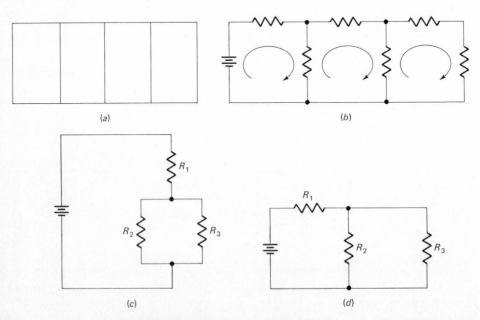

(a)   (b)

(c)   (d)

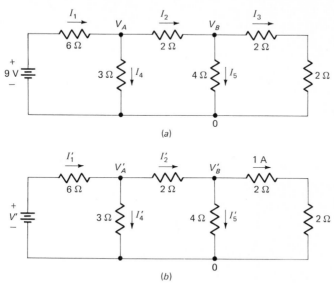

Figure 6-19. Converting original circuit to new-source circuit.

(a)

(b)

The new-source circuit

Here's the shortcut for solving a ladder. In Fig. 6-19a, the 9-V source sets up different currents and voltages throughout the ladder. The final current I_3 is unknown. The first step in the ladder method is to visualize a new circuit as shown in Fig. 6-19b. This new circuit has the *same resistances* as the original circuit. But notice the differences:

1. We deliberately make the final current equal to 1 A.
2. Except for a coincidence, I_3 in the original circuit will not equal 1 A. To account for this difference, we have to use an unknown source voltage V' as shown in Fig. 6-19b.

In other words, replace the original source of Fig. 6-19a by an unknown source voltage to allow the final current to equal 1 A, as shown in Fig. 6-19b. The new circuit is called the *new-source circuit*. To keep all new currents and voltages distinct from the original ones, use prime marks on each current and voltage (I_1', V_A', etc.).

Solving the new-source circuit

It's easy to find the currents and voltages in the new-source circuit. Starting at the load end, we can work back toward the source, finding currents and voltages as we go. For instance, in Fig. 6-19b the 1 A flows through an equivalent branch resistance of

$$R = 2 + 2 = 4 \ \Omega$$

With Ohm's law,

$$V'_B = RI = 4(1) = 4 \ \text{V}$$

Now use this value to get

$$I'_5 = \frac{V'_B}{R_5} = \frac{4}{4} = 1 \ \text{A}$$

Applying the current law to node B gives

$$I'_2 = I'_5 + 1 = 1 + 1 = 2 \ \text{A}$$

This 2 A flows through the 2-Ω resistance between nodes A and B; therefore,

$$V'_A = R_2 I'_2 + V'_B = 2(2) + 4 = 8 \ \text{V}$$

Since we now have the voltage of node A,

$$I'_4 = \frac{V'_A}{R_4} = \frac{8}{3} = 2.67 \ \text{A}$$

Applying the current law to node A gives

$$I'_1 = I'_2 + I'_4 = 2 + 2.67 = 4.67 \ \text{A}$$

This 4.67 A flows through a 6-Ω resistance; therefore, the new-source voltage must equal

$$V' = R_1 I'_1 + V'_A = 6(4.67) + 8 = 36 \ \text{V}$$

We have calculated every current and voltage in the circuit. We did it by finding the currents and voltages one after another instead of simultaneously. Each calculation is a straightforward application of Ohm's or Kirchhoff's law using previously calculated quantities. The routine calculations remind us of falling dominoes, because each newly calculated quantity allows us to find the next unknown.

In summary, here are the currents and voltages in Fig. 6-19b:

$$I'_1 = 4.67 \ \text{A}$$

$$I'_2 = 2 \ \text{A}$$

$$I'_3 = 1 \ \text{A}$$

$$I'_4 = 2.67 \ \text{A}$$

$$I'_5 = 1 \ \text{A}$$

$$V' = 36 \ \text{V}$$

$$V'_A = 8 \ \text{V}$$

$$V'_B = 4 \ \text{V}$$

Scaling new-source answers

The currents and voltages just found are for the new-source circuit. Because the resistances of the original circuit have the same values as those of the new-source circuit, the original currents and voltages are *proportional* to the new-source currents and voltages. In other words, there is a factor F that relates the new-source answers to the currents and voltages of the original circuit.

The *scaling factor F* is defined as

$$F = \frac{\text{Original source}}{\text{New source}} \qquad (6\text{-}10)$$

In Fig. 6-19a, the original source equals 9 V. Analysis of the new-source circuit gave a new-source voltage V' equal to 36 V; therefore, the scaling factor is

$$F = \frac{9}{36} = 0.25$$

Scaling the new-source answers means multiplying them by F. Here's how it's done:

$$I_1 = FI_1' = 0.25(4.67) = 1.17 \text{ A}$$

$$I_2 = FI_2' = 0.25(2) = 0.5 \text{ A}$$

$$I_3 = FI_3' = 0.25(1) = 0.25 \text{ A}$$

$$I_4 = FI_4' = 0.25(2.67) = 0.667 \text{ A}$$

$$I_5 = FI_5' = 0.25(1) = 0.25 \text{ A}$$

$$V_A = FV_A' = 0.25(8) = 2 \text{ V}$$

$$V_B = FV_B' = 0.25(4) = 1 \text{ V}$$

These are the currents and voltages in the original ladder.

Conclusion

After working out a few ladders, you will realize how easy the foregoing method really is. Like other shortcuts (equivalent resistance, Thevenin's theorem, the super-position theorem, etc.) the *ladder method* temporarily disregards the original problem while an equivalent or similar problem is being solved; then, by interpreting the results properly, you can get the original currents and voltages.

To summarize the key ideas,

1. Visualize the circuit with a new source producing 1 A through the final load resistor.

2. Work from right to left, using Ohm's and Kirchhoff's laws as needed to get all the currents and voltages.
3. Scale the new-source answers by the ratio of the original source to the new source.

Test 6-5

1. The new-source circuit and the original circuit always have the same
   (*a*) source voltage   (*b*) final load current   (*c*) resistance
   (*d*) intermediate answers .... ( )
2. A new-source circuit seldom has
   (*a*) windowpanes   (*b*) the same resistances as the original circuit
   (*c*) a final load current of 1 A
   (*d*) the original currents and voltages .... ( )
3. In the ladder method the unknowns are eliminated
   (*a*) simultaneously   (*b*) one after another
   (*c*) by direct attack on the original circuit
   (*d*) by either the loop or node method .... ( )
4. New-source answers don't apply to the original circuit until they
   (*a*) are expressed in amperes and volts
   (*b*) are multiplied by the ratio of the new source to the original source
   (*c*) all have been found   (*d*) have been properly scaled .... ( )
5. Which of these is the least useful method discussed in this chapter?
   (*a*) branch method   (*b*) loop method   (*c*) node method
   (*d*) ladder method .... ( )

6-7. SUMMARY

In everyday electronics, you seldom have to solve three or more simultaneous equations, because you almost always can find easier ways to get currents and voltages. As a rule, only a specialist works with circuits where a simultaneous solution is the best solution.

Being able to write circuit equations is useful in itself, because it gives you insight into how circuits work. Furthermore, knowing about branches, loops, and nodes makes it easier to discuss the action of electronic circuits.

The number of windowpanes in a circuit tells you how many unknown loop currents there are. In turn, this indicates the number of loop equations needed for a total solution by the loop method.

The number of independent nodes tells us how many node equations are required for a total solution by the node method.

The idea of a reference node and node voltages has practical applications. In measuring voltages, the chassis is almost always used as a reference node, and other voltages are measured with respect to it. Voltage-measuring instruments (such as volt-meters, oscilloscopes, etc.) normally have one lead on the chassis and the other lead on the node whose voltage is being measured.

The ladder method is a shortcut solution for all ladder circuits. Normally, you use this method for all ladders with more than two windowpanes.

SUMMARY OF FORMULAS

DEFINING

$$F = \frac{\text{Original source}}{\text{New source}}$$

Problems

6-1. Solve for I_1 and I_2 in these two equations:

$$4I_1 + 2I_2 = 10$$

$$3I_1 - 4I_2 = 2$$

6-2. Solve these equations simultaneously:

$$3V_1 + 2V_2 = 18$$

$$2V_1 - 4V_2 = -4$$

6-3. Suppose we have a pair of simultaneous equations

$$R_{11}I_1 + R_{12}I_2 = K_1$$

$$R_{21}I_1 + R_{22}I_2 = K_2$$

where R_{11}, R_{12}, R_{21}, R_{22}, K_1, and K_2 are numbers. Are the equations independent when

a. $R_{11} = 2$, $R_{12} = 4$, $R_{21} = 6$, and $R_{22} = 8$
b. $R_{11} = 2$, $R_{12} = 4$, $R_{21} = 6$, and $R_{22} = 12$
c. $R_{11} = 1$, $R_{12} = 2$, $R_{21} = 3$, and $R_{22} = -6$
d. $R_{11} = 4$, $R_{12} = -4$, $R_{21} = -4$, and $R_{22} = 4$

6-4. How many branches does Fig. 6-20a have? Figure 6-20b?

6-5. The circuit of Fig. 6-20c has how many branches? Of Fig. 6-20d?

Figure 6-20.

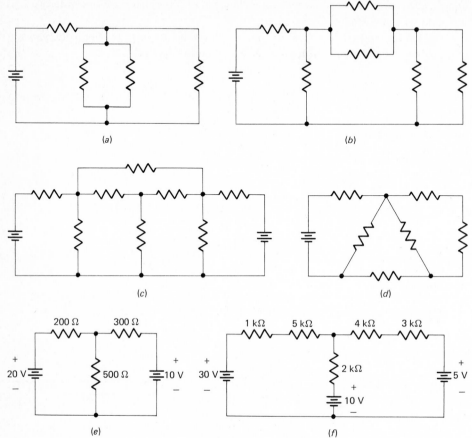

(a)

(b)

(c)

(d)

200 Ω   300 Ω

1 kΩ   5 kΩ   4 kΩ   3 kΩ

2 kΩ

+
20 V

500 Ω

+
10 V

+
30 V

+
10 V

+
5 V

(e)

(f)

6-6. If the branch method is used, how many equations do you need to write for each of these circuits:
    a. Fig. 6-20*a*
    b. Fig. 6-20*b*
    c. Fig. 6-20*c*
    d. Fig. 6-20*d*

6-7. How many windowpanes are there in Fig. 6-20*a*? In Fig. 6-20*b*?

6-8. Figure 6-20*c* has how many windowpanes? Figure 6-20*d*?

6-9. Write the loop equations for Fig. 6-20*e*.

6-10. Express in standard form the loop equations for Fig. 6-20*f*.

6-11. How many windowpanes does Fig. 6-21*a* have? How many independent loop equations can you write?

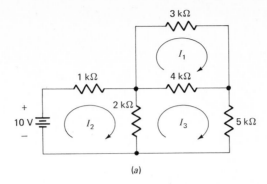

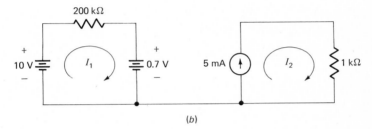

(a)

(b)

Figure 6-21.

6-12. Write the loop equations for Fig. 6-21a.

6-13. In Fig. 6-21b, write the loop equation for the left loop. What value does the current have in the left loop? What is the loop current in the right loop? The voltage across the 1-kΩ resistance?

6-14. Write the three loop equations for the Wheatstone bridge of Fig. 6-22.

6-15. How many nodes are there in
   a. Fig. 6-20a
   b. Fig. 6-20b
   c. Fig. 6-20c
   d. Fig. 6-20d

6-16. How many independent nodes does Fig. 6-20e have? Figure 6-20f?

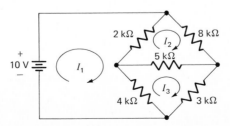

Figure 6-22.

Figure 6-23.

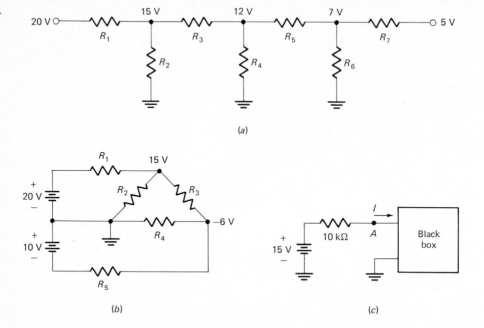

(a)

(b)   (c)

6-17.   The circuit of Fig. 6-21a has how many independent nodes? The left loop of Fig. 6-21b?

6-18.   Write the node equations for the Wheatstone bridge of Fig. 6-22. (Label unknown voltages and currents the same as in Fig. 6-15b.)

6-19.   What is the voltage across each resistance of Fig. 6-23a?

6-20.   Work out the voltage across each resistance in Fig. 6-23b.

6-21.   The black box of Fig. 6-23c acts like a resistance. If you measure a node voltage of 6 V from A to ground, what is the resistance of the black box?

6-22.   A transistor circuit is inside the black box of Fig. 6-23c. This transistor circuit has the same effect between node A and ground as a resistance R_5. If you measure a node voltage V_A equal to 8 V, what is the value of R?

6-23.   Use the ladder method to find all currents and voltages in Fig. 6-24a.

6-24.   Find all branch currents and node voltages in the ladder of Fig. 6-24b.

6-25.   Figure 6-24c is not a ladder. Nevertheless, it is possible to find all currents and voltages by using a method similar to the ladder method. That is, you make one of the currents equal 1 A, and use a new source voltage. By working out intermediate answers and scaling them, you can get the currents and voltages in Fig. 6-24c.

      Find all currents and voltages in the foregoing way by making I_2' equal to 1 A.

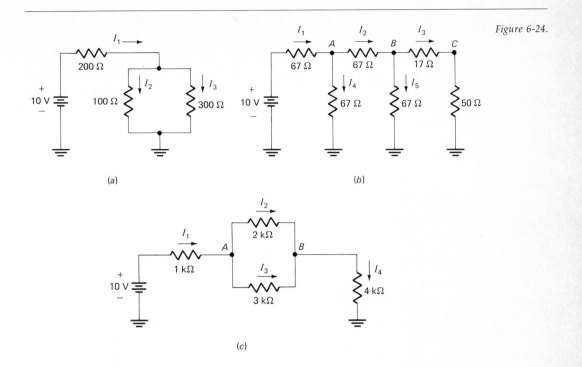

Figure 6-24.

(a)

(b)

(c)

ANSWERS TO TESTS

6-1. *a, b, c, b, b, d*
6-2. *c, a, b, c, b*
6-3. *d, b, a, b, d, b*
6-4. *b, d, d, c, c, d, b*
6-5. *c, d, b, d, a*

7. Basic Measurements

Measuring current, voltage, and resistance is important. To begin with, they are the most basic electrical quantities. Furthermore, many nonelectrical quantities such as temperature, light, pressure, etc. can be converted into current, voltage, or resistance. So being able to measure the three basic electrical quantities means we can indirectly measure many other physical quantities.

7-1. THE MOVING-COIL METER

A *magnet* can attract pieces of iron, steel, nickel, and other magnetic materials. Because of this, a magnetic field exists around a magnet (discussed in Chap. 10). An ammeter uses two kinds of magnets. As you read the following material, get the answers to

What are the two kinds of magnets used?
How does an ammeter work?
Why is an ammeter connected in series?

Two kinds of magnets

A *permanent magnet* is made of hard iron, steel, or certain other materials. It's called a permanent magnet because it retains its magnetic field indefinitely. Even though used repeatedly, this kind of magnet has a field whose strength does not change with time.

An *electromagnet* is different. It's a coil of wire wound on a soft-iron core (see Fig. 7-1a). When charges flow through the coil of wire, a magnetic field appears around the electromagnet. In other words, it acts like a permanent magnet in that it can attract magnetic materials. The strength of the field, however, depends on the amount of current. Doubling the current doubles the magnetic field, tripling the current triples the field, and so on. And if you turn the current off, the field disappears.

How an ammeter works

Figure 7-1b shows a *moving-coil meter*, the common type used in electronics. A permanent magnet surrounds an electromagnet. With no current in the electromagnet, no force exists between the two magnets, and the needle points to zero as shown. But when there is current, the electromagnet and the permanent magnet repel; being movable, the electromagnet rotates clockwise and pushes against a spring (not shown). The amount of current determines how far the electromagnet rotates. If the current is 5 mA, the needle stops at *half scale* (middle reading). When the current equals 10 mA, the needle shows *full scale* (maximum reading).

Increasing the number of turns on the electromagnet increases the *sensitivity* of the ammeter. More turns mean a stronger magnetic field (discussed in Chap. 10). With enough turns, we can get full-scale readings in the microampere region. For instance, adding enough turns results in a meter with a full-scale reading of 50 μA (a common value in measuring instruments).

Connecting an ammeter

Figure 7-2a shows the schematic symbol of an ammeter. The manufacturer puts a plus sign on one end, and a minus sign on the other, to indicate the polarity of the voltage across the ammeter when it is properly connected in a circuit. In other words,

---

Figure 7-1. (a) *Electromagnet.*
(b) *Moving-coil meter.*

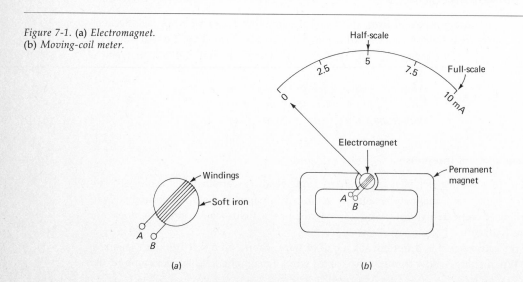

(a)

(b)

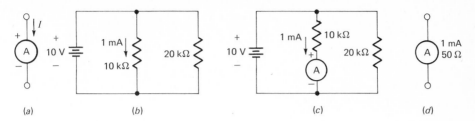

Figure 7-2. Schematic symbol for
ammeter.

(a)   (b)   (c)   (d)

to get upscale readings, you have to connect the ammeter so that conventional flow enters the plus end.

An ammeter is intended to measure branch current. For this reason, you have to make the ammeter part of a branch, that is, connect the ammeter *in series* with the elements of a branch. For instance, the 10-kΩ resistor of Fig. 7-2b has 1 mA through it. To measure this branch current, connect the ammeter in series with the 10-kΩ resistor as shown in Fig. 7-2c. In this way, the current is the same through the ammeter as through the 10-kΩ resistor.

The general rule is this: always connect an ammeter in series, never in parallel.

Resistance of ammeter

If an *ideal ammeter* could be built, it would have zero resistance; when connected in a branch, it would not change the resistance of the branch. Although manufacturers try to keep the resistance of ammeters as low as possible, all ammeters have some resistance. Therefore, whenever you add an ammeter to a branch, you are adding some resistance to the branch, and this *decreases* the branch current slightly.

The resistance of an ammeter comes from its electromagnet, in particular, the coil of wire. This resistance and the full-scale current of the ammeter are two of the most important ammeter specifications. On schematic diagrams, we will indicate both of these quantities when necessary. For instance, Fig. 7-2d shows an ammeter with a full-scale current of 1 mA and a resistance of 50 Ω.

EXAMPLE 7-1.
What is the current through the 2-kΩ resistor of Fig. 7-3a? If an ammeter is connected as shown in Fig. 7-3b, what is the current in the branch?

SOLUTION.
The current in the 2-kΩ resistor of Fig. 7-3a is

$$I = \frac{V}{R} = \frac{2}{2000} = 1 \text{ mA} \qquad \text{(original)}$$

Figure 7-3. Current through
ammeter.

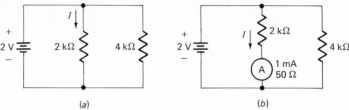

(a)   (b)

When the ammeter is connected, the equivalent branch resistance equals

$$R = 2000 + 50 = 2050 \ \Omega$$

The current then decreases to

$$I = \frac{V}{R} = \frac{2}{2050} = 0.976 \ \text{mA} \qquad (\text{measured})$$

The measured current is slightly less than the original current; this always happens; that is, adding an ammeter to a branch always lowers the branch current.

EXAMPLE 7-2.
Manufacturers specify the accuracy of an ammeter in terms of full-scale current. A meter with an accuracy of ±2 percent means *any reading* can be off by as much as ±2 percent of the *full-scale* value.
    If a 1-mA ammeter has an accuracy of ±2 percent, how far off can the readings be at midscale? At quarter scale?

SOLUTION.
The full-scale value is 1 mA; therefore, any reading can be off as much as

$$\pm 2 \ \text{percent} \times 1 \ \text{mA} = \pm 0.02 \ \text{mA}$$

    At midscale, the ammeter reads 0.5 mA. This reading may be off as much as ±0.02 mA. In other words, the ammeter current may actually be as low as

$$0.5 \ \text{mA} - 0.02 \ \text{mA} = 0.48 \ \text{mA}$$

or as high as

$$0.5 \ \text{mA} + 0.02 \ \text{mA} = 0.52 \ \text{mA}$$

    At quarter scale, the ammeter reads 0.25 mA. Therefore, the actual current through the ammeter may be as low as

$$0.25 \ \text{mA} - 0.02 \ \text{mA} = 0.23 \ \text{mA}$$

or as high as

$$0.25 \text{ mA} + 0.02 \text{ mA} = 0.27 \text{ mA}$$

As you see, the readings are less accurate downscale. This is the reason you should always try to get upscale readings. The closer to full scale, the better.

The kind of error discussed in this example is known as *calibration error*; it's caused by friction and other effects that depend on ammeter construction. The error discussed in the preceding example is called *loading error*; it's caused by the effect of ammeter resistance on branch current. These two different types of error are analyzed elsewhere.[1]

Test 7-1

1. Which of these does not belong?
   (*a*) permanent magnet   (*b*) field gets weaker
   (*c*) field strength does not change   (*d*) hard iron .... ( )
2. The quantity that controls the field around an electromagnet is the
   (*a*) current   (*b*) coil resistance   (*c*) soft-iron core
   (*d*) size of the electromagnet   .... ( )
3. The electromagnet of an ammeter is moved by a
   (*a*) spring   (*b*) iron core   (*c*) permanent magnet
   (*d*) force between the magnets .... ( )
4. If the current through the electromagnet doubles, which of these does not double in an ammeter?
   (*a*) the needle deflection   (*b*) the field around the electromagnet
   (*c*) the force between magnets
   (*d*) the field around the permanent magnet .... ( )
5. Which of these belongs least?
   (*a*) more sensitive ammeter   (*b*) less resistance   (*c*) more turns of wire
   (*d*) stronger field around an electromagnet .... ( )
6. An ammeter is never connected
   (*a*) in series with branch elements
   (*b*) in series with the element whose current you want to measure
   (*c*) in such a way that conventional current enters the plus end
   (*d*) in parallel with a branch whose current you want to measure .... ( )
7. The following words can be rearranged into a sentence: CONNECT ALWAYS SERIES IN AN AMMETER. This advice is
   (*a*) good   (*b*) bad   .... ( )

[1] See Malvino, A. P.: *Electronic Instrumentation Fundamentals*, McGraw-Hill Book Company, New York, 1967, pp. 33–53.

7-2. AMMETER RANGES

Commercial ammeters often have several ranges or choices of full-scale current to suit the particular measurement. This section answers

What does shunting an ammeter mean?
Why is a make-before-break switch sometimes used?
What is an Ayrton shunt?

Direct method

Figure 7-4 shows one way to get different ranges; all we are doing is switching in different meters. In the position shown, the maximum input current is 100 μA. In the middle position, maximum input current is 300 μA, and in the right position it's 1 mA. A setup like Fig. 7-4 is not often used because of cost; each range requires one meter.

Shunt method

Figure 7-5 shows the usual way of getting several ranges. Only a single meter is used; the extra ranges are a result of *shunt resistors* (shunt means parallel). When the switch is in the left position, the input current passes through the ammeter; therefore, we can measure input currents up to

$$I_{in} = 100 \ \mu A \qquad \text{(left position)}$$

When the switch is in the middle, a 500-Ω resistance is across the meter. The ammeter and resistor have the same voltage, but the 500-Ω resistor has twice as much current as the 1-kΩ meter. When the ammeter has 100 μA through it, the shunt resistor has 200 μA and the maximum input current is

$$I_{in} = I_m + I_{sh} = 100 \ \mu A + 200 \ \mu A$$
$$= 300 \ \mu A \qquad \text{(middle position)}$$

Figure 7-4. Three-range ammeter.

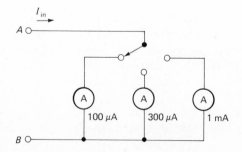

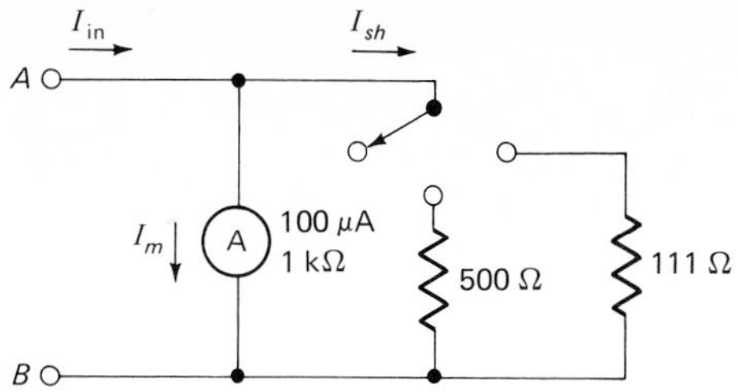

Figure 7-5. Shunted ammeter.

When the switch of Fig. 7-5 is in the right position, most of the input current is through the 111-Ω resistor. As learned in Chap. 3, the ratio of the currents in a parallel connection equals the inverse ratio of the load resistances. Therefore,

$$\frac{I_{sh}}{I_m} = \frac{R_m}{R_{sh}} = \frac{1000}{111} = 9.01 \cong 9$$

Solving for I_{sh} gives

$$I_{sh} = 9I_m$$

This says the shunt current is nine times the meter current when the switch is in the right position. Therefore, when the meter current is 100 μA, the shunt current is 900 μA and the total input current is

$$I_{in} = I_m + I_{sh} = 100 \ \mu A + 900 \ \mu A$$
$$= 1 \ mA \quad \text{(right position)}$$

The key equations for any shunted ammeter are

$$\frac{I_{sh}}{I_m} = \frac{R_m}{R_{sh}} \tag{7-1a}$$

and

$$I_{in} = I_m + I_{sh} \tag{7-1b}$$

With these, you can analyze or design a shunted ammeter.

Make-before-break switch

In Fig. 7-6a, the total input current is 200 μA. With the switch in the right position, the shunt resistor gets nine times as much current as the meter. This works out at 180 μA through the shunt resistor and 20 μA through the 100-μA meter.

Suppose the switch is moved to the middle position. During the act of switching, there is an instant when the meter is temporarily *unshunted* (see Fig. 7-6b). Because of this, the entire 200 μA of input current passes through the 100-μA meter as shown. This excessive current *pegs* the meter (drives the needle hard into the full-scale stop).

To avoid needle damage while changing ranges, a *make-before-break* switch is

Figure 7-6. Make-before-break
switch.

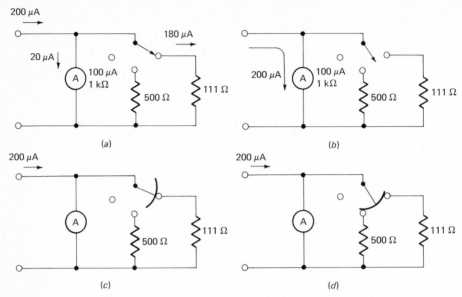

(a)　　　　(b)

(c)　　　　(d)

sometimes used (Fig. 7-6c). The wiper on this kind of switch is broad enough that it makes contact with the middle position before breaking contact with the right position. Because of this, the meter remains shunted between switch positions as shown in Fig. 7-6d. This is one way to eliminate meter pegging and needle damage when switching ranges.

Ayrton shunt

Another way to prevent meter pegging is with an *Ayrton* shunt (Fig. 7-7). With this kind of arrangement, the meter is always shunted, even though an *ordinary switch* is used. In fact, between switch positions, there's no current in the meter.

By analyzing Fig. 7-7, we can come up with these maximum input currents:

$$I_{max} = 300 \ \mu A \qquad \text{(position 1)}$$

$$I_{max} = 1 \ mA \qquad \text{(position 2)}$$

$$I_{max} = 3 \ mA \qquad \text{(position 3)}$$

Test 7-2

1. Parallel is to series as shunt is to
    (a) series　　(b) parallel　　(c) series-parallel　　(d) parallel-series　　......... ( )

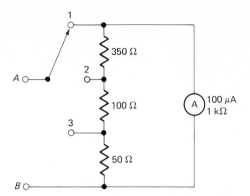

Figure 7-7. Ayrton shunt.

2. Compared with the meter current, the shunt current is always
    (*a*) smaller   (*b*) greater   (*c*) of a different value   (*d*) none of these ... ( )
3. Between switch positions, a make-before-break switch does not
    (*a*) prevent pegging the meter   (*b*) keep the meter shunted
    (*c*) decrease the meter current to zero
    (*d*) parallel two of the shunt resistors .... ( )
4. Which of these is not related to an Ayrton shunt?
    (*a*) meter always shunted   (*b*) current in all resistances
    (*c*) meter branch often contains resistor
    (*d*) meter is sometimes unshunted .... ( )

7-3. SIMPLE VOLTMETERS

An *X-to-Y transducer* is a device or circuit that converts an input quantity *X* to an output quantity *Y*. For instance, a solar cell is a sunlight-to-voltage transducer, a loudspeaker is a current-to-sound transducer, etc. In the following discussion, look for the answers to

> *How does a simple voltmeter work?*
> *How do you connect a voltmeter?*
> *What is the input resistance of a voltmeter?*

Basic action

Simple voltmeters are nothing more than *voltage-to-current* transducers that include a moving-coil meter. The input voltage produces a current through the meter; therefore, the meter makes an indirect measurement of the voltage.

Figure 7-8*a* shows the way to build a simple voltmeter. In this case, the transducer equation is given by Ohm's law:

Figure 7-8. Voltmeter.

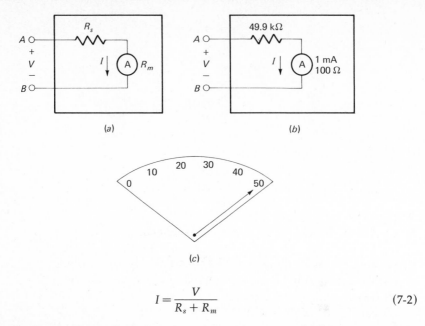

(a)

(b)

(c)

$$I = \frac{V}{R_s + R_m} \tag{7-2}$$

This says the input voltage divided by the total resistance equals the output current. Because current and voltage are directly proportional, we can mark the meter face with voltage values instead of current values.

As an example, Fig. 7-8*b* shows a simple voltmeter that can measure input voltages from 0 to 50 V. Here is how it works. The equivalent resistance between the *AB* terminals is 50 kΩ; therefore, the transducer equation becomes

$$I = \frac{V}{50,000} \tag{7-2a}$$

With this equation, we can calculate the output current produced by each input voltage.

The meter of Fig. 7-8*b* has a full-scale current of 1 mA. To find the corresponding full-scale voltage V_{fs}, substitute into Eq. (7-2a):

$$0.001 = \frac{V_{fs}}{50,000}$$

or

$$V_{fs} = 50 \text{ V}$$

So, when the input voltage equals 50 V in Fig. 7-8*b*, the needle indicates full-scale. This is why we mark 50 at the full-scale point on the meter face (see Fig. 7-8*c*).

The other points on the meter face are found by direct proportion. Since *V* and *I* are directly proportional in Eq. (7-2a), we can get these corresponding pairs of current and voltage:

I	V
0.2 mA	10 V
0.4 mA	20 V
0.6 mA	30 V
0.8 mA	40 V

This explains why the meter face of Fig. 7-8c applies to the voltmeter shown in Fig. 7-8b.

Key formula

That's all there is to a simple voltmeter. You put a resistance R_s in series with a moving-coil meter whose full-scale current is I_{fs} and whose resistance is R_m. Because I_{fs} and R_m are usually known, it's convenient to rewrite Eq. (7-2) in terms of full-scale values:

$$I_{fs} = \frac{V_{fs}}{R_s + R_m} \tag{7-3}$$

This is the key equation for analyzing or designing simple voltmeters; all you have to do is substitute the known quantities and solve for the unknown quantity (see Examples 7-3 through 7-5).

Connecting a voltmeter

Figure 7-9a shows the schematic symbol for any kind of voltmeter. If the voltmeter is like the simple type just discussed, it contains a series resistance and a moving-coil meter as shown in Fig. 7-9b. The manufacturer stamps a plus sign on one end of the voltmeter and a minus sign on the other. This indicates the polarity of input voltage that produces upscale readings.

A voltmeter is intended to measure the voltage between two nodes. Often, one of the nodes is ground, so the voltmeter measures node voltages. Especially important, you always connect a voltmeter between a pair of nodes. In other words, a voltmeter does not become part of an existing branch like an ammeter; rather, the voltmeter *adds a new branch* to the circuit. Stated another way, you always connect a voltmeter in parallel with the voltage you are trying to measure.

If you want to measure the voltage across the 3-kΩ resistor of Fig. 7-9c, connect the voltmeter in parallel with this resistor as shown in Fig. 7-9d. This is the only way to connect the voltmeter to get the voltage between the *AB* terminals.

In general, always connect a voltmeter in parallel, never in series.

Figure 7-9. Connecting a
voltmeter.

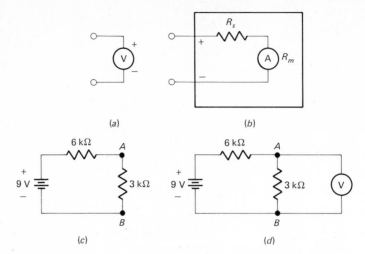

(a)   (b)

(c)   (d)

Resistance of voltmeter

If an *ideal voltmeter* could be built, it would have infinite resistance; when connected in a circuit, it would not provide a new path for current. Manufacturers try to
build voltmeters with very high resistances to minimize *voltmeter loading error* (see Example 7-6). Nevertheless, whenever you connect a voltmeter in a circuit, you are adding a new branch to the circuit; this tends to *lower* the voltage slightly.

The input resistance of a simple voltmeter like Fig. 7-10a equals

$$R_{in} = R_s + R_m \tag{7-4}$$

where R_s is the resistance in series with the meter, and R_m is the meter resistance. The
input resistance of the voltmeter shown in Fig. 7-10b therefore equals

$$R_{in} = 198{,}000 + 2000 = 200 \text{ k}\Omega$$

Figure 7-10. Input resistance of
voltmeter.

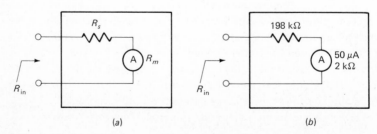

(a)   (b)

A voltmeter like this shunts 200 kΩ across the circuit being tested. This may produce a serious loading error.

Electronic voltmeters use transistors, vacuum tubes, and other devices to increase the input resistance. Typical electronic voltmeters have an input resistance of more than 10 MΩ. With this kind of voltmeter, loading error is usually small enough to neglect.

EXAMPLE 7.3.

What is the maximum voltage the voltmeter of Fig. 7-10*b* can measure?

SOLUTION.

Equation (7-3) says

$$I_{fs} = \frac{V_{fs}}{R_s + R_m}$$

Substituting the values given in Fig. 7-10*b*,

$$50(10^{-6}) = \frac{V_{fs}}{200(10^3)}$$

Solving for V_{fs},

$$V_{fs} = 50(10^{-6})200(10^3) = 10 \text{ V}$$

Therefore, the voltmeter can measure up to 10 V.

EXAMPLE 7-4.

A moving-coil meter has these specifications:

$$I_{fs} = 50 \ \mu\text{A}$$
$$R_m = 2 \text{ k}\Omega$$

If this moving-coil meter is used in Fig. 7-10*a*, what value should R_s be for the full-scale voltage to equal 50 V?

SOLUTION.

Again, substitute the given values into Eq. (7-3):

$$I_{fs} = \frac{V_{fs}}{R_s + R_m}$$

$$50(10^{-6}) = \frac{50}{R_s + 2000}$$

Solve for R_s to get

$$R_s = \frac{50}{50(10^{-6})} - 2000 = 998 \text{ k}\Omega$$

Put this in series with the given meter, and you have a simple voltmeter that can measure up to 50 V.

EXAMPLE 7-5.
Prove the full-scale voltages of Fig. 7-11 are 2.5 V, 10 V, and 50 V. Also, what is the input resistance on each range?

SOLUTION.
Equation (7-3) rearranges into this form:

$$V_{fs} = (R_s + R_m)I_{fs}$$

When the switch is in the left position,

$$V_{fs} = (50{,}000)50(10^{-6}) = 2.5 \text{ V}$$

With the switch in the middle position,

$$V_{fs} = (200{,}000)50(10^{-6}) = 10 \text{ V}$$

And in the right position,

$$V_{fs} = (10^{6})50(10^{-6}) = 50 \text{ V}$$

The input resistance equals the sum of the series resistance and the meter resistances. From left to right, the input resistances are

$$R_{in} = 50 \text{ k}\Omega \qquad \text{(left)}$$

$$R_{in} = 200 \text{ k}\Omega \qquad \text{(middle)}$$

$$R_{in} = 1 \text{ M}\Omega \qquad \text{(right)}$$

Figure 7-11. Multirange voltmeter.

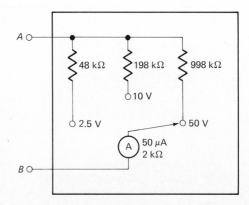

Notice the input resistance increases with the full-scale voltage; this means less loading error on higher voltage ranges.

Incidentally, the series resistors in a voltmeter like Fig. 7-11 are sometimes called *multiplier resistances;* they give the voltmeter its different ranges.

EXAMPLE 7-6.
If the voltmeter of Fig. 7-12a were ideal, what voltage would it indicate? What does the voltmeter actually read?

SOLUTION.
If ideal, the voltmeter would have infinite resistance, equivalent to an open. Therefore, the voltage being measured would equal

$$V_2 = \frac{R_2}{R} V = \frac{100,000}{200,000} 40 = 20 \text{ V}$$

The voltmeter actually has an input resistance of 1 MΩ. This resistance is in parallel with the 100-kΩ resistor, so that the actual voltage is slightly lower than 20 V. A neat way to calculate the actual voltage is to Thevenize the circuit left of the *AB* terminals. This results in Fig. 7-12b. Now, the voltage across the voltmeter equals

$$V_2 = \frac{R_2}{R} V = \frac{1,000,000}{1,050,000} 20 = 19 \text{ V}$$

So you see, there's a loading error. The voltage between the *AB* terminals without the voltmeter is 20 V. But with the voltmeter it drops to 19 V, a 5 percent error. Looking at Fig. 7-12b, it's clear that the higher the input resistance of the voltmeter, the less the loading error.

As a guide, you get less than 1 percent loading error if the voltmeter input resistance is 100 times greater than the Thevenin resistance facing it; the loading error is less than 5 percent if the voltmeter input resistance is more than 20 times the Thevenin resistance.

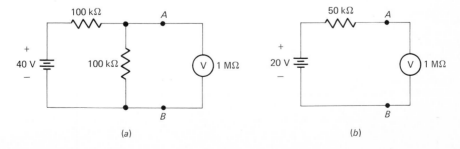

Figure 7-12. Loading error.

(a)

(b)

Test 7-3

1. A simple voltmeter is a transducer that converts
   (*a*) current to voltage   (*b*) current to current
   (*c*) voltage to current   (*d*) voltage to voltage .... ( )
2. The moving-coil meter of a simple voltmeter responds to current, but the meter face
   is marked in
   (*a*) amperes   (*b*) volts   (*c*) ohms   (*d*) none of these .... ( )
3. Which of these is not true of a simple voltmeter?
   (*a*) the moving-coil meter responds to current
   (*b*) the voltage scale is linear
   (*c*) current and voltage are inversely proportional
   (*d*) the needle indicates voltage   (*e*) none of these .... ( )
4. You never connect a voltmeter
   (*a*) in parallel   (*b*) between two nodes   (*c*) across a resistance
   (*d*) in series with a branch .... ( )
5. What comes next in this series:
   AMMETER, LOW RESISTANCE, VOLTMETER, _____
   (*a*) high resistance   (*b*) high current   (*c*) high voltage
   (*d*) low resistance .... ( )
6. The input resistance of a simple voltmeter is not affected by
   (*a*) multiplier resistance   (*b*) voltage being measured
   (*c*) meter resistance   (*d*) full-scale voltage .... ( )

7-4. METER SENSITIVITY

The smaller the full-scale current, the more sensitive the meter is. A 50-μA meter, for instance, is more sensitive than a 1-mA meter. In the discussion that follows, we answer two important questions:

What is the defining formula for sensitivity?
How is input resistance related to sensitivity?

Defining formula

Meter sensitivity S is defined as the reciprocal of full-scale current. As a defining formula,

$$S = \frac{1}{I_{fs}} \tag{7-5}$$

where S = sensitivity
  I_{fs} = full-scale current

The smaller the full-scale current, the more sensitive the meter. If $I_{fs} = 1$ mA,

$$S = \frac{1}{0.001} = 1000$$

Or if I_{fs} is 50 μA,

$$S = \frac{1}{50(10^{-6})} = 20,000$$

This says a 50-μA meter is 20 times more sensitive than a 1-mA meter.

We left out the units in the preceding calculations. The reciprocal of amperes is ohms per volt, evident from the following:

$$\frac{1}{I} = \frac{1}{V/R} = \frac{R}{V}$$

or

$$\frac{1}{\text{amperes}} = \frac{\text{ohms}}{\text{volt}} = \Omega/V$$

When we add these units to the earlier calculations, we find that a 1-mA meter has a sensitivity of 1000 Ω/V, and a 50-μA meter has a sensitivity of 20,000 Ω/V.

The sensitivity, sometimes called the *ohms-per-volt rating*, is normally printed on the meter face of nonelectronic voltmeters. If a 50-μA meter is used in a voltmeter, the meter face has 20,000 Ω/V on it.

Relation to input resistance

Equation (7-3) says the full-scale current in a voltmeter equals

$$I_{fs} = \frac{V_{fs}}{R_s + R_m}$$

which rearranges into

$$R_s + R_m = \frac{V_{fs}}{I_{fs}}$$

Since the input resistance of a simple voltmeter equals $R_s + R_m$, the foregoing equation becomes

$$R_{in} = \frac{V_{fs}}{I_{fs}}$$

With Eq. (7-5), the input resistance equals

$$R_{in} = SV_{fs} \tag{7-6}$$

This is an important result. When you use a nonelectronic voltmeter, you will know the value of S (it's printed on the meter face) and the value of V_{fs} (the range you select). Because of this, you can quickly calculate the input resistance of the voltmeter.

EXAMPLE 7-7.

A voltmeter has ranges of 2.5 V, 10 V, 50 V, 250 V, and 1000 V. If the sensitivity is 20,000 Ω/V, what is the input resistance on each range?

SOLUTION.

Multiply 20,000 Ω/V by the full-scale voltage of each range as follows:

$$R_{in} = SV_{fs} = 20{,}000(2.5) = 50 \text{ k}\Omega \qquad (2.5\text{-V range})$$

$$R_{in} = 20{,}000(10) = 200 \text{ k}\Omega \qquad (10\text{-V range})$$

$$R_{in} = 20{,}000(50) = 1 \text{ M}\Omega \qquad (50\text{-V range})$$

$$R_{in} = 20{,}000(250) = 5 \text{ M}\Omega \qquad (250\text{-V range})$$

$$R_{in} = 20{,}000(1000) = 20 \text{ M}\Omega \qquad (1000\text{-V range})$$

Test 7-4

1. Meter sensitivity is an example of a
   (a) defining formula   (b) experimental formula   (c) derived formula
   (d) none of these .... ( )
2. The sensitivity of a moving-coil meter decreases when
   (a) full-scale current decreases
   (b) more turns are used on the electromagnet
   (c) fewer turns are used on the moving coil
   (d) meter resistance increases .... ( )
3. You are most likely to find which of these on the meter face of a simple voltmeter?
   (a) full-scale current   (b) input resistance   (c) meter resistance
   (d) sensitivity .... ( )
4. Input resistance will decrease if
   (a) sensitivity increases   (b) full-scale current decreases
   (c) full-scale voltage increases   (d) sensitivity decreases .... ( )
5. Which of these does not belong?
   (a) low sensitivity   (b) high input resistance
   (c) low full-scale current   (d) high meter resistance .... ( )

7-5.   THE OHMMETER

The following discussion answers these questions:

> *How does a simple ohmmeter work?*
> *Why does R_{TH} occur at half-scale?*

Basic idea

Think of a simple ohmmeter as a *resistance-to-current* transducer with a moving-coil meter. The unknown resistance being measured is the input quantity; the current through the meter is the output quantity.

Figure 7-13a shows the way to build a simple ohmmeter. For this circuit, the transducer equation is

$$I = \frac{V}{R_z + R_m + R} \tag{7-7}$$

where I = current through meter
V = internal battery voltage
R_z = resistance of zero-adjust
R_m = resistance of meter
R = resistance being measured

This equation indicates that the current through the meter depends on the resistance being measured. Because each value of R produces a different value of I, we can mark off the meter face with values of R rather than with values of I.

As an example, Fig. 7-13b shows a simple ohmmeter. The equivalent resistance right of the AB terminals is

$$R_z + R_m = 28,000 + 2000 = 30 \text{ k}\Omega$$

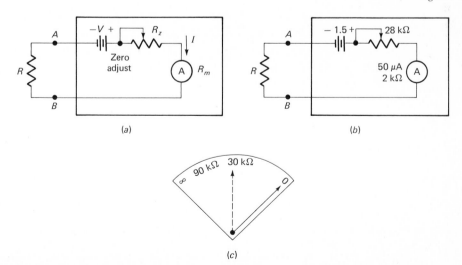

Figure 7-13. Ohmmeter.

Since the battery voltage equals 1.5 V, Eq. (7-7) becomes

$$I = \frac{1.5}{30{,}000 + R} \tag{7-8}$$

With this equation, we can calculate the current produced by each value of resistance.
When the resistance is zero, the current is maximum and equals

$$I = \frac{1.5}{30{,}000} = 50 \ \mu A \qquad (R = 0)$$

This is enough current to produce full-scale deflection, as shown by the solid needle of Fig. 7-13c.
If the resistance equals 30 kΩ, Eq. (7-8) gives

$$I = \frac{1.5}{30{,}000 + 30{,}000}$$

$$= 25 \ \mu A \qquad (R = 30 \ k\Omega)$$

This represents half-scale current, shown by the dashed needle of Fig. 7-13c.
When the resistance is 90 kΩ, the current drops to

$$I = \frac{1.5}{30{,}000 + 90{,}000}$$

$$= 12.5 \ \mu A \qquad (R = 90 \ k\Omega)$$

This is quarter-scale current, and explains why 90 kΩ is marked at the quarter-scale point in Fig. 7-13c.

With Eq. (7-8) we can locate as many other points on the meter face as desired. Because of the inverse relation between I and R, small current values mean large resistances. In other words, infinite resistance (open) is the extreme left point, whereas zero resistance (short) is the extreme right point on the meter face of Fig. 7-13c.

Key formula

That's the basic idea behind simple ohmmeters. To have different ranges, different internal resistances are switched into the circuit. For any ohmmeter design, we can Thevenize the circuit inside the ohmmeter and come up with a single battery and resistance, as shown in Fig. 7-14a. Resistance R_{TH} includes the meter resistance. As usual, V_{TH} is the voltage between the AB terminals when the load is open (R removed from the circuit).

In general, therefore, the current I in Fig. 7-14a equals

$$I = \frac{V_{TH}}{R_{TH} + R} \tag{7-9}$$

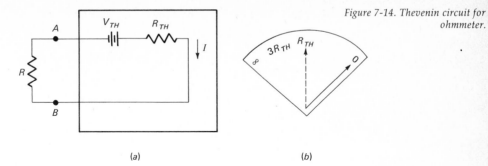

Figure 7-14. Thevenin circuit for ohmmeter.

(a)   (b)

For many ohmmeter designs, this current or a fixed fraction of it will pass through the moving-coil meter (see Problem 7-20).

In Eq. (7-9), the current is maximum when the measured resistance R is zero. This is why the meter face of Fig. 7-14b has zero marked at full-scale deflection (solid arrow). When the measured resistance R equals the Thevenin resistance R_{TH} of the ohmmeter, Eq. (7-9) gives the half-scale current (dashed arrow in Fig. 7-14b). Similarly, an R of $3R_{TH}$ results in quarter-scale deflection.

The zero-adjust (Fig. 7-13a) is needed to take care of variations in battery voltage, internal resistors, etc. As the battery ages, its voltage decreases; the zero-adjust can compensate for this. Also, when an ohmmeter has several ranges, different resistors are switched into the circuit; the zero-adjust can compensate for the tolerance in these resistors.

Test 7-5

1. A simple ohmmeter is a transducer that converts
    (a) current to voltage   (b) voltage to current   (c) current to current
    (d) resistance to current .... ( )
2. Which of these is not true of a simple ohmmeter?
    (a) the moving-coil meter responds to current
    (b) the ohms scale is nonlinear
    (c) resistance and current are directly proportional
    (d) I decreases when R increases .... ( )
3. What comes next in this series:
                    OHMMETER, THEVENIN RESISTANCE, _____
    (a) quarter scale   (b) half scale   (c) full scale   (d) none of these ...... ( )
4. The Thevenin resistance of an ohmmeter does not depend on
    (a) the ohmmeter range   (b) the half-scale mark
    (c) the unknown resistance   (d) the zero-adjust   .... ( )

5. Which of these belongs least?
   (*a*) resistance being measured   (*b*) zero-adjust   (*c*) battery aging
   (*d*) changes in internal resistors .... ( )

7-6. THE VOLT-OHM-MILLIAMMETER

With a meter, a switch, resistors, and a zero-adjust, we can build a *volt-ohm-milliam-meter* (VOM), also known as a *multimeter*. This instrument combines a voltmeter, an ohmmeter, and a milliammeter. The simplest VOMs are nonelectronic (no transistors, vacuum tubes, integrated circuits, and so on); these nonelectronic VOMs use the circuits discussed in this chapter.

The main disadvantage of nonelectronic VOMs is the loading error they produce when measuring voltage or current. Often, the Thevenin resistance of the circuit being tested is very high, and excessive voltmeter loading error takes place; or the Thevenin resistance may be very low, and excessive ammeter loading error occurs.[2]

In closing this chapter, you should take away with you the idea of using a transducer to convert one quantity into another. This chapter showed how to transduce voltage to current (voltmeter), and resistance to current (ohmmeter). Since many other physical quantities are transducible to electrical quantities, a whole new world is now opening up for you.

SUMMARY OF FORMULAS

DEFINING

$$S = \frac{1}{I_{fs}}$$ (7-5)

DERIVED

$$\frac{I_{sh}}{I_m} = \frac{R_m}{R_{sh}}$$ (7-1*a*)

$$I_{in} = I_m + I_{sh}$$ (7-1*b*)

$$I_{fs} = \frac{V_{fs}}{R_s + R_m}$$ (7-3)

$$R_{in} = SV_{fs}$$ (7-6)

$$I = \frac{V_{TH}}{R_{TH} + R}$$ (7-9)

---

[2] If you are interested in electronic voltmeters that measure millivolts with more than 10 MΩ, or in electronic ammeters that measure microamperes with less than 1 Ω, see Malvino, A. P.: *Electronic Principles*, McGraw-Hill Book Company, New York, 1973, pp. 534–540.

Problems

7-1. Calculate the value of I in Fig. 7-15a for an ideal ammeter. For an ammeter with an R_m of 1 kΩ.

7-2. What are the approximate full-scale currents for each range of Fig. 7-15b?

7-3. Calculate the approximate full-scale current for each range of Fig. 7-15c.

7-4. An ammeter has ranges of 10 μA, 100 μA, 1 mA, 10 mA, 100 mA, and 1 A. If the moving-coil meter has an I_{fs} of 10 μA and an R_m of 10 kΩ, what are the shunt resistances on each range?

7-5. An ammeter has a meter accuracy of $\pm$3 percent and a full-scale current of 5 mA. If the meter reads 2.5 mA, what is the lowest possible value of ammeter current? The highest?

7-6. What is the current in the 1-kΩ resistor of Fig. 7-16a? If an ammeter with an R_m of 200 Ω is connected in series with the 1-kΩ resistor, what will the new value of current be?

7-7. How much current is there in the 500-Ω resistor of Fig. 7-16b? If an ammeter with an R_m of 100 Ω is used to measure the current through the 500-Ω resistor, what will the new value of current be?

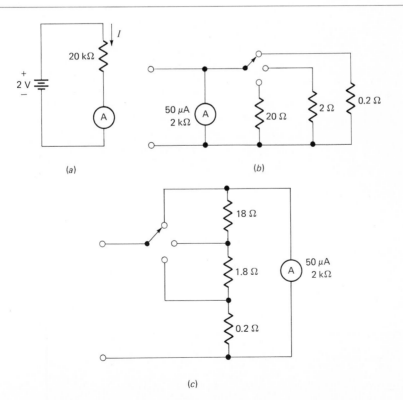

Figure 7-15.

(a)

(b)

(c)

Figure 7-16.

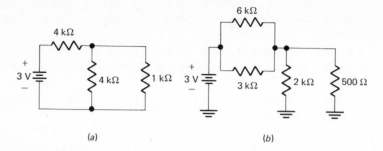

(a)   (b)

7-8. Figure 7-17a shows a voltmeter with three ranges. Calculate the full-scale voltage on each range.

7-9. What is the input resistance on each range of the voltmeter shown in Fig. 7-17a?

7-10. Calculate the values of multiplier resistances for each range of Fig. 7-17b.

7-11. A meter has a full-scale current of 20 μA. What is its sensitivity?

7-12. What is the ohms-per-volt rating of the meter shown in Fig. 7-17b?

7-13. A voltmeter has an S of 5000 Ω/V. What is its input resistance on these ranges: 3 V, 10 V, 30 V, and 100 V?

7-14. A voltmeter has a meter accuracy of ±2.5 percent of full-scale. If it reads 20 V on the 50-V range, what is the lowest possible voltage across the voltmeter? The highest?

7-15. What is the voltage across the 30-kΩ resistor of Fig. 7-18a? If a voltmeter with an input resistance of 200 kΩ is connected across this resistor, what will the new voltage be?

7-16. A voltmeter is to be connected across the 3-kΩ resistor of Fig. 7-18b. If the loading error is to be less than 5 percent, what input resistance should the voltmeter have?

7-17. If the voltmeter of Fig. 7-18c were ideal, what would be the voltage between the AB terminals? If the voltmeter has an input resistance of 100 kΩ, what is the voltage between these terminals?

7-18. An ohmmeter like Fig. 7-19a is called a *series type* because the moving-coil meter

Figure 7-17.

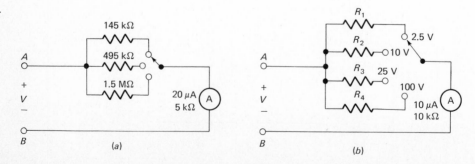

(a)   (b)

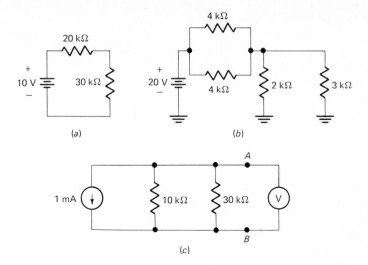

Figure 7-18.

is in series with the resistance being measured. When the *AB* terminals are shorted, the zero-adjust can be varied to get full-scale deflection. What is the value of R_z that zeroes the ohmmeter?

7-19. In Fig. 7-19*a*, what resistance does half-scale deflection represent? Quarter-scale deflection?

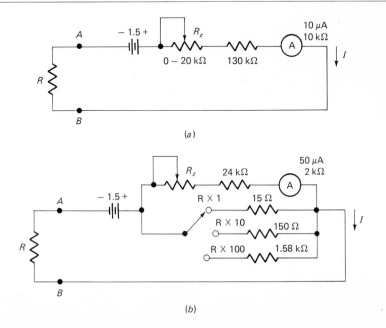

Figure 7-19.

7-20. When measured resistances are small, a *shunt-type* ohmmeter like Fig. 7-19*b* is often used.

    a. What value does R_z have when the ohmmeter is zeroed?

    b. With the *AB* terminals shorted, how much current is there in the 15-Ω resistor?

    c. If the resistance being measured equals 15 Ω, how much current is there in the moving-coil meter?

7-21. In Fig. 7-19*b*, the circuit right of the *AB* terminals can be Thevenized to get a V_{TH} of 1.5 V, and an R_{TH} that depends on the switch position.

    a. On the $R \times 1$ range, what does half-scale deflection represent?

    b. On the $R \times 10$ range, what does quarter-scale deflection represent?

    c. On the $R \times 100$ range, what resistance is being measured if the needle indicates 5 μA.

ANSWERS TO TESTS

7-1. *b, a, d, d, b, d, a*

7-2. *a, d, c, d*

7-3. *c, b, c, d, a, b*

7-4. *a, c, d, d, a*

7-5. *d, c, b, c, a*

8. Time

In many circuits, *time* is just as important as current, voltage, and resistance. This chapter tells how to extend all you've learned to *time-varying signals,* currents or voltages that change from one instant to the next.

8-1. BASIC IDEAS

To understand how to apply Ohm's law, Kirchhoff's laws, Thevenin's theorem, etc., to time-varying signals, you have to know more about time. In what follows, your job is to learn

> *What is time?*
> *How do you visualize it?*
> *What is a waveform?*

Nature of time

To begin with, time is a *fundamental quantity,* along with length, mass, temperature, and charge. These five quantities are the foundations of the physical universe; all other physical quantities can be defined in terms of the five fundamental quantities.

Because it is a fundamental quantity, time cannot be defined mathematically. That is, there is no defining formula for t in terms of other fundamental quantities. But we can measure time by defining a unit of measure. The metric unit of time is the *second,* defined as

$$1 \text{ second} = \frac{\text{Average solar day}}{86,400}$$

An average solar day is the length of a day averaged over many years. With the foregoing value, we can build clocks and other instruments to measure time.[1]

Visualizing time

Time is so elusive we need a visual aid to pin it down. The usual way to visualize
time is with a horizontal axis like Fig. 8-1a. Each point along this axis stands for a point
in time called an *instant*.

In Fig. 8-1a, the instant where $t = 0$ is known as the *origin* or *reference instant,*
because all other points in time are measured from it. Instants left of the origin occur
earlier in time; those right of it take place later.

When setting up problems, we are free to locate the reference instant anywhere.
It's like choosing a reference node in a circuit; we can pick any node as the reference
node, and measure other voltages from it. Likewise with the reference instant. We can
let $t = 0$ mark the instant of any event, and measure other instants from it.

Usually, we let $t = 0$ coincide with something important, such as the blast-off of a
rocket, the start of a race, the closing of a switch, etc. For example, after the switch of
Fig. 8-1b is closed, the current equals 2 mA. In discussing a circuit like this, it helps to
let the closing of the switch coincide with $t = 0$. Then, negative values of t represent the
circuit before the switch is closed, and positive values represent the circuit after the
switch is closed. A point in time like $t = 2$ s represents the circuit 2 s after the switch is
closed.

At a particular instant, time stops and all action ceases. It's analogous to taking a
photograph of a moving object. When we refer to a particular instant, therefore, everything stops while we examine a circuit at the particular point in time.

Waveforms

Signals carry information. Radio and TV signals are good examples; they have variations proportional to the sound and pictures being transmitted. Most currents and

---

[1] In 1967, the 13th General Conference of Weights and Measures met in Paris and redefined the
second in terms of a cesium standard (an atomic clock).

Figure 8-1. Visualizing time.

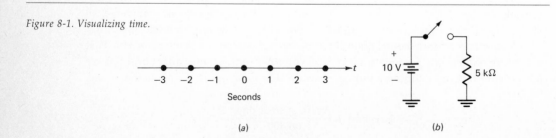

(a)   (b)

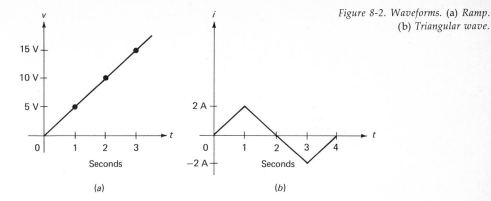

Figure 8-2. Waveforms. (a) Ramp.
(b) Triangular wave.

voltages in electronics are time-varying. For this reason, we have to work with *instantaneous values,* the values of current and voltage at each instant in time.

To help us visualize signals in time, we use *waveforms;* these are graphs of instantaneous current or voltage versus time. For instance, Fig. 8-2a shows a time-varying voltage. Right of the origin, we read these instantaneous values:

$$v = \ 5 \text{ V} \qquad \text{at} \qquad t = 1 \text{ s}$$
$$v = 10 \text{ V} \qquad \text{at} \qquad t = 2 \text{ s}$$
$$v = 15 \text{ V} \qquad \text{at} \qquad t = 3 \text{ s}$$

and so on. So the voltage increases 5 V during each second; this is why the graph of Fig. 8-2a is linear. Any waveform that increases linearly with time is called a *ramp.*

Figure 8-2b shows a *triangular* waveform, another example of a time-varying signal. Notice it's made up of different ramps: an increasing ramp between $t = 0$ and $t = 1$, a decreasing ramp from $t = 1$ to $t = 3$, and an increasing ramp between $t = 3$ and $t = 4$.

Think of a waveform as a sequence of values. When you look at Fig. 8-2a, for instance, take the waveform apart in your mind and see Fig. 8-3a, then Fig. 8-3b, then Fig. 8-3c, and finally Fig. 8-3d. Taking a signal apart like this is essential in circuit analysis.

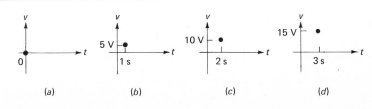

Figure 8-3. Succession of
instantaneous values.

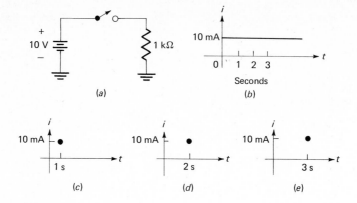

Figure 8-4. Example of instantaneous current.

(a)

(b)

(c)

(d)

(e)

The oscilloscope

An *oscilloscope* is a *voltage-to-waveform* transducer, an electronic instrument that measures time-varying voltage and displays the waveform on a screen. An oscilloscope works on a point-by-point basis, because it shows the input voltage at the instant $t = 0$, then it shows the input voltage at the next instant, and so on until all instantaneous voltages have been displayed. The spacing between the displayed points is so small the final picture looks continuous.

EXAMPLE 8-1.
Figure 8-4a shows a 10-V battery, a switch, and a 1-kΩ load. If the switch closes at $t = 0$, what does the current waveform look like?

SOLUTION.
Before $t = 0$, the current is zero. After $t = 0$, the current equals 10 mA. Therefore, we can draw the waveform shown in Fig. 8-4b. Don't forget to take the waveform apart in your mind; see it as a succession of current values at different points in time like Figs. 8-4c, d, and e. Each point on the waveform gives the current value at a particular instant.

EXAMPLE 8-2.
Figure 8-5a shows a circuit with three switches. Switch A closes at $t = 0$, switch B closes at $t = 2$, and switch C closes at $t = 4$. Draw the waveform of current i.

SOLUTION.
Before $t = 0$, all switches are open and current i equals zero. Between $t = 0$ and $t = 2$, switch A is closed, but B and C are open. Therefore, the circuit is identical to Fig. 8-5b, and the current equals

$$i = \frac{9}{6000} = 1.5 \text{ mA}$$

Between $t = 2$ and $t = 4$, switches A and B are closed, but C is open. So, the active part of the circuit is that shown in Fig. 8-5c. The parallel connection of the 6-kΩ resistances results in an equivalent resistance of 3 kΩ and a current of

$$i = \frac{9}{3000} = 3 \text{ mA}$$

After $t = 4$, all switches are closed and the circuit looks like Fig. 8-5d. The equivalent resistance is 2 kΩ, and the current is

$$i = \frac{9}{2000} = 4.5 \text{ mA}$$

Figure 8-5e shows the current waveform. Notice how nicely this graph summarizes the circuit action; it tells the whole story. Notice also how the current *steps* at $t = 0$, $t = 2$, and $t = 4$. You get a step when each switch is closed.

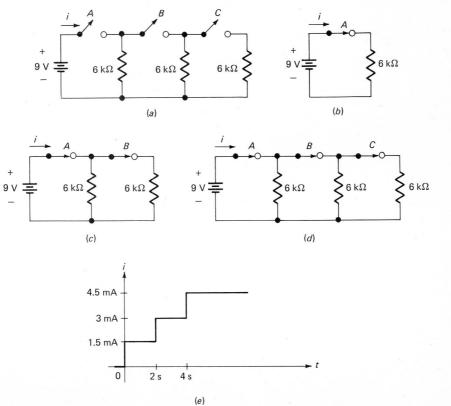

Figure 8-5. Example of step current.

Test 8-1

1. Time is not
   (*a*) measured in seconds   (*b*) a fundamental quantity
   (*c*) defined in terms of fundamental quantities   (*d*) measurable .... ( )
2. Which of these belongs least?
   (*a*) origin   (*b*) any point in time   (*c*) any instant
   (*d*) any point on a time axis .... ( )
3. Reference node is to any node as origin is to
   (*a*) second   (*b*) waveform   (*c*) reference instant
   (*d*) any instant .... ( )
4. Waveform most closely means
   (*a*) ramp   (*b*) triangular signal   (*c*) signal
   (*d*) graph of current or voltage versus time .... ( )
5. A ramp is not
   (*a*) a waveform   (*b*) nonlinear   (*c*) related to time
   (*d*) a uniform increase or decrease .... ( )
6. Which of these does not apply to an oscilloscope?
   (*a*) shows each successive voltage value
   (*b*) displays voltage waveforms   (*c*) works on a point-by-point basis
   (*d*) vertical axis represents time .... ( )

8-2. THE TIME THEOREM

Almost everything you learned earlier applies to circuits with time-varying sources, provided certain conditions are satisfied. To take the big step from fixed sources to time-varying sources, you must know

> *What are resistive circuits?*
> *What is the time theorem?*
> *What are three basic types of current?*

Resistive circuits

All loads have properties called *resistance, capacitance,* and *inductance.* Earlier chapters dealt with resistance; later chapters discuss capacitance and inductance. All three properties are present in a circuit; therefore, an exact analysis includes all three. Fortunately, most loads are designed to emphasize one of the properties. Because of this, we usually can neglect two of the properties in preliminary analysis. For instance, when a carbon resistor is properly used, only its resistance matters; its capacitance and inductance are negligible.

A *resistive circuit* is one in which all capacitances and inductances are negligible; only the resistances have significant effects on current and voltage. Many electronic circuits act like resistive circuits when operated correctly. In fact, resistive circuits are by far the most common.

The theorem

Here's a big step. With two exceptions, all formulas and theorems learned earlier *apply at each instant in a resistive circuit.* We call this important statement the *time theorem.* Given a resistive circuit with time-varying sources, we can analyze it at each instant by applying Ohm's law, Kirchhoff's laws, Thevenin's theorem, etc.

As an example, Fig. 8-6*a* shows a resistive circuit with a time-varying source. The current through the 10-kΩ resistor will be time-varying, because the source voltage is a ramp. To find the current, apply Ohm's law at each instant. At $t = 0$, $v = 0$, and

$$i = \frac{v}{R} = \frac{0}{10,000} = 0$$

At $t = 1$, $v = 5$, and

$$i = \frac{v}{R} = \frac{5}{10,000} = 0.5 \text{ mA}$$

At $t = 2$, $v = 10$, and

$$i = \frac{v}{R} = \frac{10}{10,000} = 1 \text{ mA}$$

At $t = 3$, $v = 15$, and

$$i = \frac{v}{R} = \frac{15}{10,000} = 1.5 \text{ mA}$$

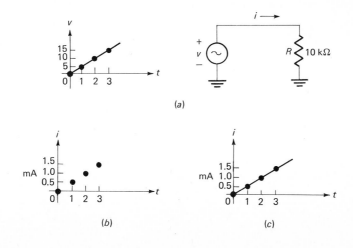

Figure 8-6. Time theorem.

Figure 8-6*b* shows each value of the current at the corresponding instant. If we calculate intermediate current values, we will find they lie along the ramp shown in Fig. 8-6*c*. In other words, in a resistive circuit a ramp of source voltage produces a ramp of current.

The analysis of a resistive circuit requires analyzing the circuit at successive instants. Instead of having one problem, we have several problems (one for each instant). Usually, calculations at three or four instants are enough to indicate how the circuit acts at all other instants. That is, after calculating the first few points like Fig. 8-6*b*, we deduce the current waveform is a ramp like Fig. 8-6*c*. (Whenever in doubt about intermediate points, calculate more values.)

Types of current

If the sources driving a resistive circuit do not vary in time, the currents will be constant. For instance, a battery puts out a fixed or constant voltage. If it is the only source in a resistive circuit, all currents will be constant in time; their values will not change from one instant to the next.

Direct current (dc) is another name for constant current; the value does not change in time. Figure 8-7*a* shows the waveform of direct current. No matter what instant we select, the current has the same value.

Alternating current (ac) is a special kind of time-varying current. Figure 8-7*b* shows what it looks like; this unique waveform is called a *sine wave*. As shown, the current reaches a maximum positive value at $t = t_1$, and a maximum negative value at $t = t_3$.

The positive and negative values of current in Fig. 8-7*b* indicate the true direction of current in a circuit. If the waveform of Fig. 8-7*b* represents the current in Fig. 8-7*c*, then the true direction agrees with the arrow for all instants between $t = 0$ and $t = t_2$; on the other hand, the true direction of current is opposite the arrow for all instants between $t = t_2$ and $t = t_4$. Therefore, the current alternates its direction.

Time-varying current (tc) is the general name for any current that varies in time. Alternating current (Fig. 8-7*b*) is a special case of time-varying current. Likewise, the ramp of Fig. 8-6*c*, the staircase of Fig. 8-5*e*, and the triangular wave of Fig. 8-2*b* are special cases of time-varying current.

Instantaneous current is the value of current at an instant or point in time.

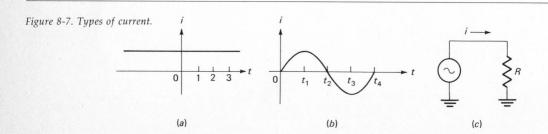

Figure 8-7. Types of current.

(a)

(b)

(c)

Therefore, direct current is a succession of instantaneous currents of the same value. Alternating current is a succession of instantaneous currents that increase to a maximum first in one direction then in the other; the waveshape is that of a sine wave. Time-varying current is any succession of instantaneous currents that change with time.

Capital and lowercase letters

Earlier chapters analyzed *dc circuits,* those in which all currents are constant in time. Whenever currents or voltages are constant in time, use capital I and V to represent them. In this way, you can tell at a glance when a formula involves constant currents and voltages. For instance, when applying Ohm's law to dc circuits, write

$$I = \frac{V}{R}$$

On the other hand, when currents and voltages are time-varying, use lowercase i and v. When applying Ohm's law to a *tc circuit,* write

$$i = \frac{v}{R}$$

This notation is useful, especially when dc and tc sources are in the same circuit. (A dc or tc source is one that produces direct or time-varying current in a resistive circuit.)

Exceptions to the time theorem

The time theorem has two exceptions, two formulas we cannot use if the currents and voltage are time-varying. The formulas are

$$I = \frac{Q}{t} \qquad \text{(dc only)} \tag{8-1}$$

and

$$P = \frac{W}{t} \qquad \text{(dc only)} \tag{8-2}$$

Both of these are dc formulas; they apply only to dc circuits, ones in which all currents and voltages are constant in time. For tc circuits, the definitions of current and power are more complicated.[2]

We don't need Eqs. (8-1) and (8-2) for tc-circuit analysis because we can calculate current and power using alternative formulas like

$$i = \frac{v}{R}$$

---

[2] If you have studied calculus, you can understand these defining formulas: $i = dQ/dt$ and $p = dW/dt$.

and

$$p = vi$$

where i, v, and p are the values of current, voltage, and power at each instant in a resistive circuit.

Therefore, after discarding Eqs. (8-1) and (8-2), everything you learned in Chaps. 1 through 7 applies to resistive circuits with any kind of source (dc, ac, or tc).

EXAMPLE 8-3.

Draw the waveform of current through the 2-kΩ resistor of Fig. 8-8*a*.

SOLUTION.

The easy way to do this is by Thevenizing the circuit left of the *AB* terminals. Temporarily remove the 2-kΩ resistor. With the voltage-divider theorem, the Thevenin

Figure 8-8. Thevenin's theorem for instantaneous quantities.

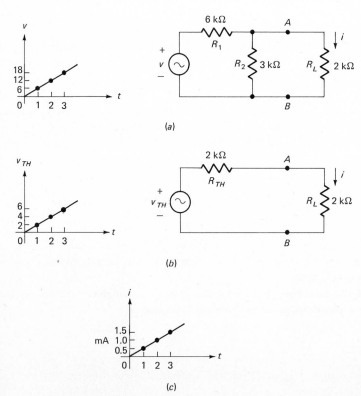

voltage is

$$v_{TH} = \frac{R_2}{R} v = \frac{3000}{9000} v = \frac{v}{3}$$

The Thevenin resistance is

$$R_{TH} = R_1 \| R_2 = 6000 \| 3000 = 2 \text{ k}\Omega$$

Notice v_{TH} is one-third of v. Because of this, v_{TH} is a ramp of voltage whose values are one-third the values of the source ramp.

Figure 8-8b shows the Thevenized circuit. The equivalent series resistance equals 4 kΩ. At the instant $t = 1$, $v_{TH} = 2$ and

$$i = \frac{v_{TH}}{R_{TH} + R_L} = \frac{2}{4000} = 0.5 \text{ mA}$$

At $t = 2$, $v_{TH} = 4$ and

$$i = \frac{4}{4000} = 1 \text{ mA}$$

At $t = 3$, $v_{TH} = 6$ and

$$i = \frac{6}{4000} = 1.5 \text{ mA}$$

Figure 8-8c shows the current waveform. So a ramp of source voltage results in a ramp of load current.

EXAMPLE 8-4.

Draw the waveform of current through the 10-kΩ resistor of Fig. 8-9a.

SOLUTION.

With the 10-kΩ load disconnected, no charges flow through the 3-kΩ resistor. Because of this, the voltage between the AB terminals is the same as the voltage between the CD terminals. With the voltage-divider theorem,

$$v_{TH} = \frac{R_2}{R} v = \frac{4000}{8000} v = \frac{v}{2}$$

The Thevenin resistance equals

$$R_{TH} = R_3 + R_1 \| R_2 = 3000 + 4000 \| 4000 = 5 \text{ k}\Omega$$

Figure 8-9b shows the Thevenized circuit. The equivalent resistance equals

$$R_{TH} + R_L = 5000 + 10{,}000 = 15 \text{ k}\Omega$$

At $t = 1$, $v_{TH} = 5$ and

$$i = \frac{v_{TH}}{R_{TH} + R_L} = \frac{5}{15{,}000} = 0.333 \text{ mA}$$

Figure 8-9. Example of
instantaneous analysis.

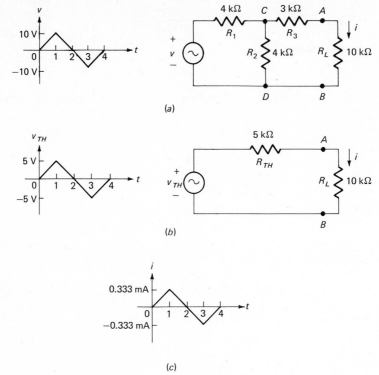

Because of Ohm's law, all other instantaneous currents are proportional to the instantaneous source voltage; this is why the current waveform is the triangular wave of Fig.
8-9c.

Negative values of current represent a reversal of direction. Between $t = 2$ and
$t = 4$, the true direction of i is opposite that shown in Fig. 8-9a.

EXAMPLE 8-5.
The source waveform of Fig. 8-10a is called a *square wave*. What is the waveform of
voltage across the 20-kΩ resistor?

SOLUTION.
Between $t = 0$ and $t = 1$, the current source produces conventional flow in the direction
shown by the arrow. The voltage at each instant therefore equals

$$v = Ri = 20{,}000(0.001)$$
$$= 20 \text{ V}$$

with the polarity shown.

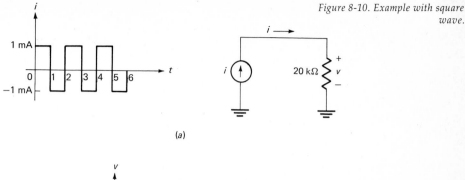

Figure 8-10. Example with square wave.

Between $t = 1$ and $t = 2$, the current equals -1 mA; this means the true direction of current is opposite the arrow. Because of this, the voltage across the load is

$$v = Ri = 20{,}000(-0.001)$$
$$= -20 \text{ V}$$

A negative voltage means the polarity is opposite that shown in Fig. 8-10a.

Figure 8-10b shows the voltage waveform. This means a square wave of source current produces a square wave of load voltage.

Test 8-2

1. Any circuit you build has
   (a) resistance   (b) capacitance   (c) inductance   (d) all three .... ( )
2. Which of the following is not true of the capacitance and inductance of a resistive circuit?
   (a) they are negligible
   (b) they have insignificant effect on currents and voltages
   (c) they are included in the analysis
   (d) they drop out of the analysis .... ( )
3. The time theorem extends almost all earlier theory to
   (a) any circuit   (b) resistive circuits   (c) dc circuits
   (d) capacitive circuits .... ( )

4. Which of these is not related to the time theorem?
   (*a*) each instant treated as separate problem
   (*b*) formulas and theorems apply at each instant
   (*c*) circuit must be resistive   (*d*) works for any circuit .... ( )
5. Which of these belongs least?
   (*a*) flow is in one direction but varies   (*b*) current is constant
   (*c*) direct current   (*d*) current does not change in value .... ( )
6. Alternating current has a waveform whose shape is
   (*a*) similar to constant current   (*b*) non-time-varying
   (*c*) a sine wave   (*d*) has only positive values .... ( )
7. A ramp is an example of which kind of current?
   (*a*) time-varying   (*b*) direct   (*c*) constant   (*d*) alternating .... ( )

8-3. THE SIMILARITY THEOREM

What is the similarity theorem?

In earlier examples, a ramp of source voltage produced a ramp of load current, a triangular wave of source voltage resulted in a triangular wave of load current, and so on. We can summarize these results with the *similarity theorem*. It says all current and voltage waveforms have the same shape as the source waveform, provided the circuit is *resistive, linear,* and has only *one source.*

Here's the proof. Because of Ohm's law, current and voltage are proportional in a linear resistance; therefore, the current waveform and the voltage waveform have the same shape. If the current through a resistor is a ramp, the voltage must be a ramp; if the current is a sine wave, the voltage must be a sine wave, and so on. The only way to satisfy Kirchhoff's current and voltage laws everywhere in the circuit is for all currents and voltages to have the same shape as the source waveform.

As an example, Fig. 8-11*a* shows a triangular wave of source voltage. The similarity

---

Figure 8-11. Similarity theorem.

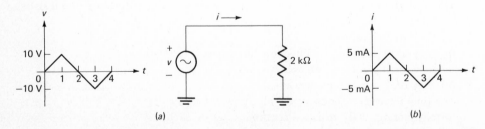

(*a*)   (*b*)

theorem says the load current is also a triangular wave. When the source reaches its *positive peak value* (the highest point on the source waveform), $t = 1$, $v = 10$, and

$$i = \frac{v}{R} = \frac{10}{2000} = 5 \text{ mA}$$

Figure 8-11*b* shows this current. Because all other current values are proportional, the *negative peak* of source voltage (maximum negative value) produces -5 mA, as shown at $t = 3$.

EXAMPLE 8-6.

Figure 8-12*a* shows a source waveform called a *sawtooth*. Draw all current and voltage waveforms in the circuit.

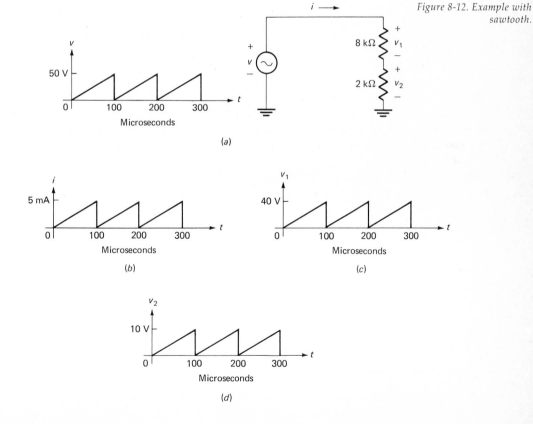

Figure 8-12. Example with sawtooth.

(a)

(b)

(c)

(d)

Figure 8-13. Sawtooth driving ladder.

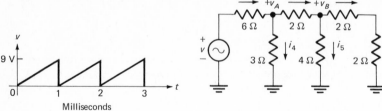

Milliseconds

SOLUTION.

Because of the similarity theorem, the sawtooth of source voltage produces a sawtooth of load current. At $t = 100$ μs, $v = 50$ V and

$$i = \frac{50}{10,000} = 5 \text{ mA} \qquad \text{(peak current)}$$

Figure 8-12*b* shows the current waveform.

The voltage waveforms v_1 and v_2 are sawtooths with peak values of

$$v_1 = R_1 i = 8000(0.005) = 40 \text{ V}$$

and

$$v_2 = R_2 i = 2000(0.005) = 10 \text{ V}$$

Figure 8-12*c* and *d* show these voltage waveforms.

A sawtooth is a common waveform in TV receivers, oscilloscopes, and other visual-display instruments.

EXAMPLE 8-7.

Describe the current and voltage waveforms for the ladder of Fig. 8-13.

SOLUTION.

The similarity theorem tells us a sawtooth source produces a sawtooth current through each resistance, and a sawtooth voltage at each node. The peak values of these saw-tooths occur at the same instants as the source peak values, that is, at $t = 1$ ms, $t = 2$ ms, $t = 3$ ms, and so on.

The source has a positive peak of 9 V. If we apply the ladder method at the instant the source equals 9 V, we get

$$i_1 = 1.17 \text{ A}$$

$$i_2 = 0.5 \text{ A}$$

$$i_3 = 0.25 \text{ A}$$

$$i_4 = 0.667 \text{ A}$$

$$i_5 = 0.25 \text{ A}$$

$$v_A = 2 \text{ V}$$

$$v_B = 1 \text{ V}$$

(These values were worked out in Sec. 6-6.)

Test 8-3

1. You cannot apply the similarity theorem if
   (*a*) resistances are linear   (*b*) a current source is used
   (*c*) one resistance is nonlinear   (*d*) a tc source is involved .... ( )
2. A sine-wave source drives a circuit with 100 linear resistances. The current in each resistance must have the shape of a
   (*a*) dc signal   (*b*) sine wave   (*c*) square wave   (*d*) ramp .... ( )
3. When the similarity theorem applies, we have to analyze the circuit at
   (*a*) one instant   (*b*) two instants   (*c*) at least three or four instants
   (*d*) many instants   .... ( )

8-4. THE SINE WAVE

The sine wave is the most fundamental of all waveforms, because it is the natural output of many circuits. Furthermore, transduced physical quantities often result in sine waves of current or voltage. As a result, the sine wave is the most important waveform studied in this book.

The discussion that follows captures the basic ideas behind the sine wave. In particular, look for the answers to

What is the period of a sine wave?
How is frequency defined?
How is frequency related to period?

Period

Look at the sine wave of Fig. 8-14*a*. Here is what you should see. At $t = 0$, $v = 0$; this is the starting point of the sine wave. At $t = 1$ ms, $v = 10$ V; this is the positive peak point; it has the largest voltage in the positive direction. At $t = 2$ ms, $v = 0$; this is the crossover point where the voltage reverses its polarity. At $t = 3$ ms, $v = -10$ V; this is the negative peak point; the voltage here is maximum and of opposite polarity. At $t = 4$ ms, $v = 0$; at this point the sine wave starts over and repeats all voltage values.

A *cycle* is the smallest part of a waveform that establishes the size and shape of the waveform. In Fig. 8-14*a*, the waveform between $t = 0$ and $t = 4$ ms is a cycle, the part between $t = 4$ ms and $t = 8$ ms is another cycle, the section from $t = 8$ ms to $t = 12$ ms

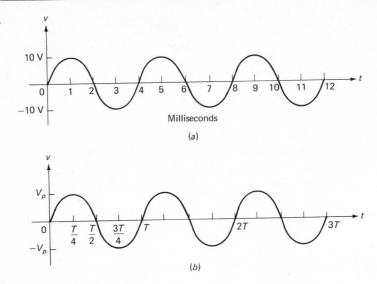

Figure 8-14. Sine wave.

is a third cycle, and so on. Because the sine wave is repetitive, the first cycle tells the whole story; all other cycles are duplications of the first cycle.

The *period* of a waveform equals the duration of a cycle, the time required to go through the cycle. In Fig. 8-14a, the period equals 4 ms, because this is how long it takes to get from the beginning to the end of the cycle. We will use T for the period of a waveform.

Figure 8-14b shows the graph of any sine wave. We use V_p for the positive peak voltage and $-V_p$ for the negative peak voltage. Notice how the first cycle begins at $t = 0$ and ends at $t = T$; this agrees with what we just said about period. So if someone tells you a sine wave has a period of $T = 10$ μs, you can substitute this into the values shown along the time axis of Fig. 8-14b to get

$$t = \frac{T}{4} = \frac{10 \text{ μs}}{4} = 2.5 \text{ μs} \qquad \text{(instant of first positive peak)}$$

$$t = \frac{T}{2} = \frac{10 \text{ μs}}{2} = 5 \text{ μs} \qquad \text{(crossover point)}$$

$$t = \frac{3T}{4} = \frac{30 \text{ μs}}{4} = 7.5 \text{ μs} \qquad \text{(instant of first negative peak)}$$

$$t = T = 10 \text{ μs} \qquad \text{(end of first cycle)}$$

$$t = 2T = 20 \text{ μs} \qquad \text{(end of second cycle)}$$

$$t = 3T = 30 \text{ μs} \qquad \text{(end of third cycle)}$$

and so on.

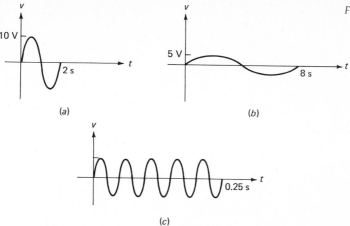

Figure 8-15. *Peak value and period.*

V_p and T are the most important values of a sine wave; once you know what they are, you have the particular sine wave pinned down. For instance, Fig. 8-15*a* shows a sine wave with a V_p of 10 V and a T of 2 s; automatically, the negative peak voltage is −10 V. Also, the positive peak occurs at $t = 0.5$ s, the crossover point at $t = 1$ s, the negative peak at $t = 1.5$ s, and so on.

As another example, Fig. 8-15*b* shows a sine wave with a V_p of 5 V and a T of 8 s. Automatically, the negative peak voltage is −5 V. Furthermore, the positive peak occurs at $t = 2$ s, the crossover point at $t = 4$ s, the negative peak at $t = 6$ s, and so on.

Frequency

The *frequency* of a sine wave is defined as the number of cycles divided by the time in which they occur. The defining formula for frequency is

$$f = \frac{n}{t} \qquad (8\text{-}3)$$

where f = frequency
   n = number of cycles
   t = time

If 12 cycles (hereafter abbreviated c) occur in 4 s, the frequency of the sine wave is

$$f = \frac{n}{t} = \frac{12 \text{ c}}{4 \text{ s}} = 3 \text{ c/s}$$

As another example, Fig. 8-15*c* shows 5 c occurring in 0.25 s; this means the sine wave has a frequency of

$$f = \frac{n}{t} = \frac{5 \text{ c}}{0.25 \text{ s}} = 20 \text{ c/s}$$

These two examples indicate the nature of frequency; it's the same as rate of repetition, because it tells us how many cycles occur in 1 s. In the first of the preceding examples, 3 c occur during each second; in the second example, 20 c take place in each second. Therefore, the second sine wave has the higher repetition rate.

Hertz

The *hertz* (abbreviated Hz) is the metric unit of measure for frequency; it's defined as

$$1 \text{ hertz} = 1 \text{ cycle per second}$$

or

$$1 \text{ Hz} = 1 \text{ c/s}$$

With this definition, the answers in the preceding examples are written as 3 Hz and 20 Hz. Since the hertz is the standard unit for frequency, answers are specified in hertz rather than cycles per second.

Frequency meters and *electronic counters* are instruments that measure frequency. Put a sine wave into either, and it indicates the frequency in hertz.

Relation of frequency to period

Frequency is defined as

$$f = \frac{n}{t}$$

where n is the number of cycles in a time t. From the discussion of period, we know 1 c occurs in time T (Fig. 8-16a). Since $n = 1$ and $t = T$,

$$f = \frac{1}{T} \tag{8-4}$$

Figure 8-16. Frequency and period.

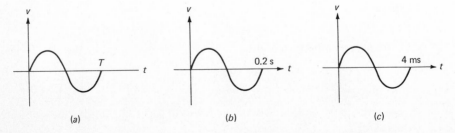

(a)   (b)   (c)

Equation (8-4) is widely used. For one thing, you can measure the period of a sine wave with an oscilloscope; then you can calculate the frequency using Eq. (8-4).

EXAMPLE 8-8.
Calculate the frequencies of the sine waves shown in Figs. 8-16b and c.

SOLUTION.
The sine wave of Fig. 8-16b has a period of 0.2 s; therefore, its frequency is

$$f = \frac{1}{T} = \frac{1}{0.2} = 5 \text{ Hz}$$

The sine wave of Fig. 8-16c has a period of 4 ms, so its frequency is

$$f = \frac{1}{T} = \frac{1}{0.004} = 250 \text{ Hz}$$

EXAMPLE 8-9.
An oscilloscope displays the sine wave of Fig. 8-17a when connected across a 1-kΩ resistor. What is the frequency? What does the current through this resistor look like?

SOLUTION.
Figure 8-17a shows a period of 10 μs; therefore,

$$f = \frac{1}{10(10^{-6})} = 100 \text{ kHz}$$

The current through the resistor is a sine wave with the same frequency as the voltage. The peak current equals

$$I_p = \frac{V_p}{R} = \frac{5}{1000} = 5 \text{ mA}$$

EXAMPLE 8-10.
If the sine wave of Fig. 8-17b is across an 8-kΩ resistance, what is the frequency and peak value of the current through the resistor?

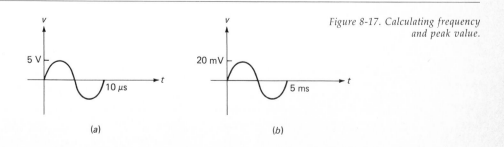

Figure 8-17. Calculating frequency and peak value.

(a)   (b)

SOLUTION.

The peak current is

$$I_p = \frac{V_p}{R} = \frac{0.02}{8000} = 2.5 \ \mu\text{A}$$

and the frequency is

$$f = \frac{1}{T} = \frac{1}{0.005} = 200 \text{ Hz}$$

EXAMPLE 8-11.

The voltage out of a typical ac outlet in a home is a sine wave with a peak of approximately 165 V and a frequency of 60 Hz. Calculate the period.

SOLUTION.

We can rearrange Eq. (8-4) to get

$$T = \frac{1}{f}$$

Substituting the given frequency,

$$T = \frac{1}{60} = 16.7 \text{ ms}$$

EXAMPLE 8-12.

What are the current and voltage waveforms of Fig. 8-18.

SOLUTION.

We first analyzed this ladder for a dc source of 9 V (Sec. 6-6). Recently, we analyzed this ladder for a sawtooth source with a peak of 9 V (Example 8-7). Now we are going to analyze it for a sine-wave source with a peak of 9 V.

The first step is realizing the similarity theorem applies. Because of this, every current and voltage waveform throughout the ladder has the shape of a sine wave.

The second step is realizing the peak values of the currents and voltages are identical to those worked out earlier. Therefore, the current and voltage sine waves in Fig. 8-18 have these peak values:

$$I_1 = 1.17 \text{ A}$$

$$I_2 = 0.5 \text{ A}$$

$$I_3 = 0.25 \text{ A}$$

$$I_4 = 0.667 \text{ A}$$

$$I_5 = 0.25 \text{ A}$$

$$V_A = 2 \text{ V}$$

$$V_B = 1 \text{ V}$$

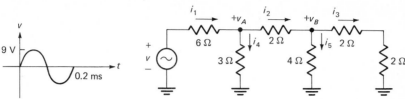

Figure 8-18. Sine wave driving ladder.

(Lowercase letters are used with time-varying currents and voltages, not with fixed values. The peak values of a sine wave do not change with time; this is why we use capital I's and V's in the foregoing list.)

The final step is to work out the frequency. Because of the similarity theorem, all sine waves have the same frequency as the source. Therefore, they all have a frequency of

$$f = \frac{1}{T} = \frac{1}{0.2(10^{-3})} = 5 \text{ kHz}$$

Test 8-4

1. To establish the size and shape of a sine wave, it is sufficient to look at
   (*a*) one cycle   (*b*) a couple of cycles   (*c*) many cycles .... ( )
2. Which of these belongs least?
   (*a*) $\frac{1}{10}$ of a cycle   (*b*) duration of a cycle   (*c*) period
   (*d*) time from beginning to end of cycle .... ( )
3. Frequency always equals
   (*a*) number of cycles   (*b*) hertz   (*c*) period
   (*d*) number of cycles divided by corresponding time .... ( )
4. Charge is to current as number of cycles is to
   (*a*) time   (*b*) period   (*c*) frequency   (*d*) amperes .... ( )
5. Which of these belongs least?
   (*a*) cycle   (*b*) frequency   (*c*) $1/T$ (*d*) repetition rate .... ( )
6. Conductance is to resistance as frequency is to
   (*a*) mho   (*b*) period   (*c*) second   (*d*) hertz .... ( )
7. The following words can be rearranged to form a sentence: RECIPROCAL FREQUENCY IS PERIOD OF THE. The sentence is
   (*a*) true   (*b*) false .... ( )

8-5. PREVIEW OF PART 2

The first half of this book has emphasized resistive circuits, by far the most important in electronics. But the two other properties of loads can be very important in some applications. Therefore, our theory will not be complete until we discuss capacitance and inductance in Part 2. Here is a glimpse of what is coming.

Capacitance

The second property of a load is its *capacitance*, measured in a unit called the *farad* (abbreviated F). When a sine-wave source drives a circuit with capacitances, each capacitance opposes the flow of charge. The opposition to ac flow is called *capacitive reactance*. The larger this capacitive reactance, the smaller the sine wave of the current.

As derived in Part 2, capacitive reactance equals

$$X_C = \frac{1}{2\pi fC} \tag{8-5}$$

where X_C = capacitive reactance, ohms
$\quad f$ = frequency of sine wave, hertz
$\quad C$ = capacitance, farads

Therefore, given the frequency f and the capacitance C, we can calculate the capacitive reactance X_C. The greater X_C, the smaller the current.

Figure 8-19a shows the schematic symbol for capacitance. On a schematic diagram, the value of capacitance is given in farads. Figure 8-19b shows a capacitance of

$$C = 5 \ \mu F$$

which is equivalent to

$$C = 5(10^{-6}) \ F$$

If a sine wave with a frequency of 60 Hz drives this capacitor, Eq. (8-5) gives

$$X_C = \frac{1}{2\pi fC} = \frac{1}{2\pi(60)5(10^{-6})}$$
$$= 530 \ \Omega$$

Figure 8-19. Capacitance and inductance.

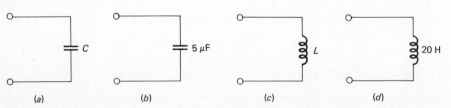

(a) $\qquad\qquad$ (b) $\qquad\qquad$ (c) $\qquad\qquad$ (d)

Inductance

The third property of all loads is *inductance*, measured in a unit called the *henry* (abbreviated H). When a sine-wave source drives a circuit with inductances, each inductance opposes the flow of charge. This opposition to flow is known as *inductive reactance*. The larger the inductive reactance, the smaller the current.

Part 2 derives this formula for inductive reactance:

$$X_L = 2\pi f L \qquad (8\text{-}6)$$

where X_L = inductive reactance, ohms
f = frequency of sine wave, hertz
L = inductance, henrys

Given the values of f and L, we can calculate X_L; this inductive reactance is the opposition to ac flow.

Figure 8-19c shows the schematic symbol of inductance. On a diagram, the value is given in henrys. The inductance of Fig. 8-19d, for instance, has a value of 20 H. If a sine wave with a frequency of 60 Hz drives this inductance, Eq. (8-6) gives

$$X_L = 2\pi f L = 2\pi(60)20$$
$$= 7.54 \text{ k}\Omega$$

SUMMARY OF FORMULAS

DEFINING

$$f = \frac{n}{t} \qquad (8\text{-}3)$$

DERIVED

$$f = \frac{1}{T} \qquad (8\text{-}4)$$

Problems

8-1. For the ramp of Fig. 8-20a, what does the voltage equal when $t = 1$ s? When $t = 2$ s?

Figure 8-20.

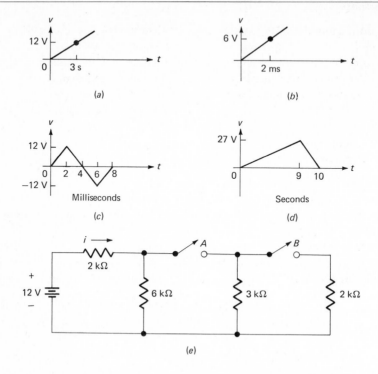

(a)

(b)

(c)

(d)

(e)

8-2. In Fig. 8-20b, what does v equal when $t = 1$ ms? When $t = 10$ ms?

8-3. The equation for any straight line passing through the origin is

$$y = mx$$

For a voltage ramp, this becomes

$$v = mt$$

   a. What is the equation for the ramp of Fig. 8-20a?
   b. For the ramp of Fig. 8-20b?

8-4. For the triangular wave of Fig. 8-20c, what is the voltage for each of these instants: $t = 2$ ms; $t = 4$ ms; $t = 6$ ms; and $t = 8$ ms?

8-5. In Fig. 8-20d, read off the voltage at each of these instants: $t = 9$ s; $t = 1$ s; $t = 2$ s; and $t = 9.5$ s.

8-6. Switch A of Fig. 8-20e closes at $t = 0$. Switch B closes at $t = 5$ ms. Draw the waveform of current i through the upper 2-kΩ resistance.

8-7. Draw the waveform of current i through the 1-kΩ resistance of Fig. 8-21a.

8-8. In Fig. 8-21a, draw the waveform of node voltage A.

8-9. Show the waveform of current through the 4-kΩ resistance of Fig. 8-21b. Also, draw the voltage waveform across this resistance.

8-10. Draw the waveform of node voltage A in Fig. 8-21b.

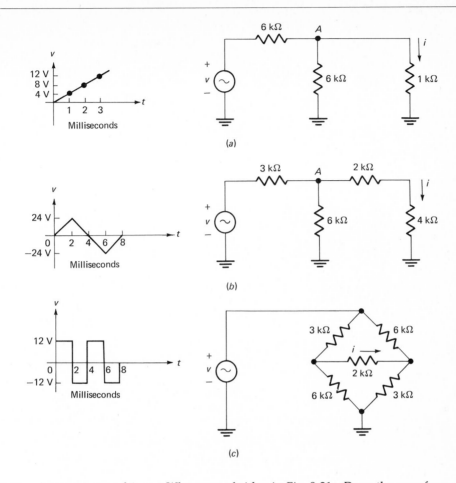

Figure 8-21.

(a)

(b)

(c)

8-11.   A square wave drives a Wheatstone bridge in Fig. 8-21c. Draw the waveform of current through the 2-kΩ resistance. What does the voltage waveform across this resistance look like?

8-12.   A sawtooth source with a peak of 20 V drives eight identical resistances in series. Describe the voltage waveform across each resistance. What will the current waveform look like if each resistance equals 1 kΩ?

8-13.   What is the output waveform for each switch position of Fig. 8-22a?

8-14.   A triangular wave drives the 100-Ω resistance of Fig. 8-22b. What does the waveform of current i through this resistance look like? The current source has a value of 50i. What does its waveform of current look like? What does the voltage waveform across the 10-kΩ load look like?

8-15.   If we change the 100-Ω resistance to a 200-Ω resistance in Fig. 8-22b, what does the new current waveform i look like? What does the 50i waveform look like? The voltage waveform across the 10-kΩ resistance?

Figure 8-22.

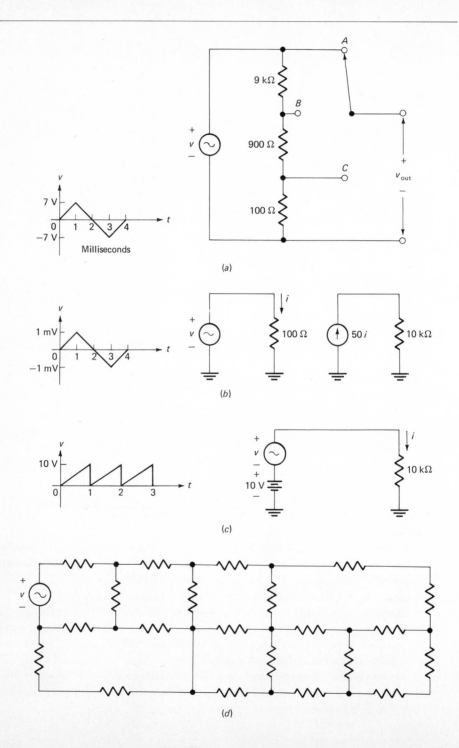

(a)

(b)

(c)

(d)

8-16. A dc source and a tc source are in series in Fig. 8-22c. By applying the super-position theorem, we can reduce this two-source circuit to two separate one-source circuits; then, the similarity theorem applies. Knowing this, you can answer these questions:

    a. When the tc source is reduced to zero, what does the waveform of current through the 10-kΩ resistance look like?

    b. When the dc source is reduced to zero, what does the waveform of current look like?

8-17. In Fig. 8-22d, suppose the tc source is a triangular wave. What does the waveform of every current and voltage in the circuit look like? If the tc source is a sine wave, what does every waveform look like?

8-18. Because of the time theorem, $p = vi$ applies at each instant in a resistive circuit. In Fig. 8-22a, calculate the total load power at each of these instants: $t = 0$; $t = 1$ ms; and $t = 3$ ms.

8-19. A sawtooth drives the series circuit of Fig. 8-23a. If the moving-coil meter looks like a pure resistance of 100 Ω, what does the waveform of current through the meter look like? Because of the inertia of moving parts, the needle of the moving-coil meter cannot follow the current waveform through it. Instead the needle points to the average current through the meter. What value of current does the needle indicate?

8-20. In Fig. 8-23b, switch A automatically closes and opens according to this schedule:

    a. It closes at $t = 0$.

    b. It opens at $t = 2$ s.

    c. It closes at $t = 4$ s.

    d. It opens at $t = 6$ s.

    Draw the waveform of current i between $t = 0$ and $t = 8$ s.

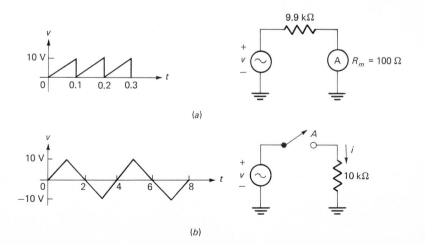

Figure 8-23.

Figure 8-24.

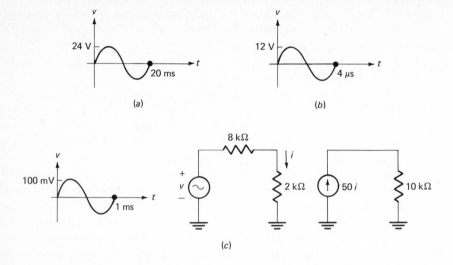

(a)

(b)

(c)

8-21. Given the sine wave of Fig. 8-24a, what is the voltage at each of these instants: $t = 0$; $t = 5$ ms; $t = 10$ ms; $t = 15$ ms; and $t = 20$ ms?

8-22. What is the period of the sine wave in Fig. 8-24a? What does $T/2$ equal? $T/4$?

8-23. Seventy-five cycles of a sine wave occur in 3 s. What is the frequency of the sine wave? The period?

8-24. If an oscilloscope shows 3 c of a sine wave occurring in 15 μs, what is the frequency of the sine wave? The period?

8-25. In Fig. 8-24b, what is the period of the sine wave? The frequency?

8-26. In Europe, the frequency of the sine-wave voltage out of an ac outlet is 50 Hz. What is the period of the sine wave?

8-27. Channel 4 of a TV receiver operates at a frequency of 69 MHz. If a sine wave has this frequency, what is its period?

8-28. In Fig. 8-24c, what is the frequency of the sine wave? What is the peak value of i? What is the peak value of $50i$? The peak voltage across the 10-kΩ resistance?

ANSWERS TO TESTS

8-1. c, a, d, d, b, d
8-2. d, c, b, d, a, c, a
8-3. c, b, a
8-4. a, a, d, c, a, b, a

PART 2.
REACTIVE
CIRCUITS

9. Capacitance

All loads have three basic properties: resistance, capacitance, and inductance. A *capacitor* is a load optimized for its capacitance.

9-1. THE BASIC IDEA

The first things to learn about capacitance are

> *What is transferred charge?*
> *How is capacitance defined?*
> *What is a farad?*

Transferred charge

A capacitor can store charge. Figure 9-1a shows an example of a capacitor: a pair of metal plates separated by air. Before the switch is closed, the number of cores in each plate equals the number of free electrons; therefore, each plate has a net charge of zero. This means there's no electric field between the plates, equivalent to zero voltage between the plates.

When the switch is closed, free electrons leave the upper plate and flow to the positive battery terminal (Fig. 9-1b). At the same time, free electrons leave the negative battery terminal and flow to the lower plate. Therefore, the upper plate becomes positively charged, and the lower plate negatively charged. The displacement of negative charge from one plate to the other produces an electric field between the plates as shown.

Where there's electric field, there's voltage. This is why voltage v appears between the plates after the switch is closed. This voltage increases while the electrons are

Figure 9-1. Charging a capacitor.

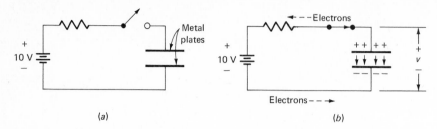

(a) (b)

flowing. When voltage v has increased to the value of source voltage, the electrons will stop flowing. In this state, the capacitor is *fully charged.*

Since Fig. 9-1b is a series circuit, each electron leaving the upper plate means one electron arrives at the lower plate. For this reason, the total negative charge that leaves the upper plate must equal the total negative charge that arrives at the lower plate. This is why the net charges on the plates are equal in magnitude but opposite in sign.

Transferred charge is the magnitude of negative charge transferred from one plate to the other. If 3 C move from the upper to the lower plate, the transferred charge is 3 C; in this case, the upper plate has a charge of +3 C and the lower plate a charge of −3 C.

Defining formula

Capacitance is defined as the transferred charge divided by the voltage between the plates. In symbols,

$$C = \frac{Q}{V} \tag{9-1}$$

where C = capacitance
  Q = transferred charge
  V = voltage

This defining formula says capacitance is directly proportional to the transferred charge and inversely proportional to the voltage between the plates. The more charge transferred per volt, the greater the capacitance.

Here are examples. If we transfer 12 C from one plate to another and the resulting voltage is 2 V (see Fig. 9-2a), the capacitance is

$$C = \frac{Q}{V} = \frac{12 \text{ C}}{2 \text{ V}} = 6 \text{ C/V}$$

If another pair of plates has a transferred charge of 5 C and a voltage of 2 V (Fig. 9-2b),

$$C = \frac{Q}{V} = \frac{5 \text{ C}}{2 \text{ V}} = 2.5 \text{ C/V}$$

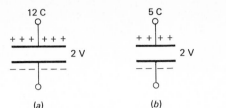

Figure 9-2. Examples of calculating capacitance.

If a capacitor is charged and then disconnected from the source, the charges are trapped on each plate, as shown in Figs. 9-2*a* and *b*. Therefore, the capacitor is a charge-storing device; it stores positive charge on one plate and negative charge on the other. Since the transferred charge equals the magnitude of charge stored on either plate, capacitance indicates how much charge is stored per volt. In the preceding examples, the first capacitor has more capacitance because it stores 6 C for each volt across it, whereas the second capacitor only stores 2.5 C for each volt.

The farad

The *farad* (abbreviated F) is the metric unit of measure for capacitance. It's defined as

$$1 \text{ farad} = 1 \text{ coulomb per volt}$$

or

$$1 \text{ F} = 1 \text{ C/V}$$

A farad is the amount of capacitance when a coulomb is transferred and a volt appears between the plates.

The answers to the preceding examples may be rewritten in terms of farads:

$$6 \text{ C/V} = 6 \times 1 \text{ C/V} = 6 \times 1 \text{ F} = 6 \text{ F}$$

and

$$2.5 \text{ C/V} = 2.5 \times 1 \text{ C/V} = 2.5 \times 1 \text{ F} = 2.5 \text{ F}$$

Because the farad is standard in electrical and electronics work, capacitance is specified in farads rather than coulombs per volt.

With Q in coulombs and V in volts, Eq. (9-1) automatically gives the answer in farads. So from now on, we need write only the numerical values of Q and V when setting up Eq. (9-1); the answer is always in farads. If Q is 10 C and V is 2 V, the capacitance is

$$C = \frac{10}{2} = 5 \text{ F}$$

Capacitance can be measured with an instrument such as a capacitance meter, a capacitance bridge, etc. These commercial instruments typically read in microfarads or picofarads.

EXAMPLE 9-1.

One plate of a capacitor has a stored charge of $+40 \mu C$. What is the charge stored on the other plate? The transferred charge? If the voltage between plates equals 2 V, what is the capacitance?

SOLUTION.

Charges on plates are always equal in magnitude and opposite in sign. Since the upper plate has a charge of $+40 \mu C$, the lower plate has a charge of $-40 \mu C$.

Transferred charge is the amount of negative charge that has moved from one plate to the other; it always equals the magnitude of the charge stored on either plate. In this example, the transferred charge is 40 μC.

Capacitance is the ratio of transferred charge to voltage between plates. With the given values,

$$C = \frac{Q}{V} = \frac{40(10^{-6})}{2} = 20(10^{-6}) \text{ F} = 20 \ \mu F$$

Incidentally, Figs. 9-2c and d show the schematic symbols for a capacitor. Both are widely used. For simplicity, we prefer Fig. 9-2c in this book.

Test 9-1

1. Transferred charge always equals
   (a) half the charge stored on a plate
   (b) the sum of the charges stored on both plates
   (c) the magnitude of the charge stored on either plate
   (d) twice the stored charge on the positive plate .... ( )
2. One plate of a capacitor has a stored charge of $-40 \mu C$. The other plate has a stored charge of
   (a) $-40 \mu C$ (b) $+40 \mu C$ (c) $+80 \mu C$ (d) $-80 \mu C$ .... ( )
3. One plate of a capacitor has a stored charge of $-40 \mu C$. The transferred charge equals
   (a) $-40 \mu C$ (b) $+40 \mu C$ (c) $40 \mu C$ (d) $\pm 40 \mu C$ .... ( )
4. We don't have to prove $C = Q/V$ because it's a
   (a) defining formula   (b) formula discovered by experiment
   (c) result of mathematical manipulation .... ( )
5. Which is closest in meaning to large capacitance?
   (a) high stored charge   (b) low voltage between plates
   (c) much transferred charge   (d) large transferred charge per volt .... ( )
6. Resistance is to ohm as capacitance is to
   (a) volt   (b) coulomb   (c) ampere   (d) farad .... ( )
7. Capacitance is the opposite of
   (a) resistance   (b) voltage   (c) resistor   (d) none of these .... ( )

9-2. LINEAR CAPACITANCE

The important questions answered in this section are

What is the capacitor law?
What is linear capacitance?
What are three ways to express the capacitor law?

Capacitor law

The capacitor law says the Q/V ratio of a capacitor is constant. This is equivalent to saying the capacitance is constant. For instance, if the transferred charge equals 6 C and the resulting voltage is 2 V, the capacitance is

$$C = \frac{Q}{V} = \frac{6}{2} = 3 \text{ F}$$

If the same capacitor is charged in a different circuit with $Q = 24$ C and $V = 8$ V, then

$$C = \frac{Q}{V} = \frac{24}{8} = 3 \text{ F}$$

Because the capacitance has the same value, the load obeys the capacitor law.

Linear capacitance

A load obeys the capacitor law if its capacitance is constant. We call such a load a *linear capacitor*. Most capacitors used in everyday work are linear capacitors, but there are exceptions like the *varactor* (a nonlinear capacitor discussed in electronics courses). Hereafter, every capacitor in this book is linear; its capacitance is constant.

Figure 9-3 shows the graph of Q versus V for a linear capacitor. As shown, charge is directly proportional to voltage. If the voltage doubles, the charge doubles; if the voltage increases 50 percent, the charge increases 50 percent.

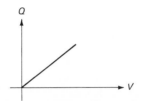

Figure 9-3. Linear capacitance.

Three useful formulas

The capacitor law may be expressed in any of three ways. With the understanding that C is constant for a given capacitor,

$$C = \frac{Q}{V} \qquad (9\text{-}2a)$$

or by rearranging,

$$V = \frac{Q}{C} \qquad (9\text{-}2b)$$

or by cross-multiplying,

$$Q = CV \qquad (9\text{-}2c)$$

These equations are useful for analyzing capacitive circuits; given two of the quantities, you can calculate the third.

EXAMPLE 9-2.
What does the capacitance equal in Fig. 9-4a?

SOLUTION.
The figure gives a Q of 50 μC and a V of 10 V. With Eq. (9-2a),

$$C = \frac{Q}{V} = \frac{50(10^{-6})}{10} = 5(10^{-6}) \text{ F} = 5 \ \mu\text{F}$$

EXAMPLE 9-3.
Figure 9-4b shows the capacitor used in the preceding example. What is the new value of voltage?

SOLUTION.
The stored charge now is 10 μC. The capacitance is still 5 μF. With Eq. (9-2b),

$$V = \frac{Q}{C} = \frac{10(10^{-6})}{5(10^{-6})} = 2 \text{ V}$$

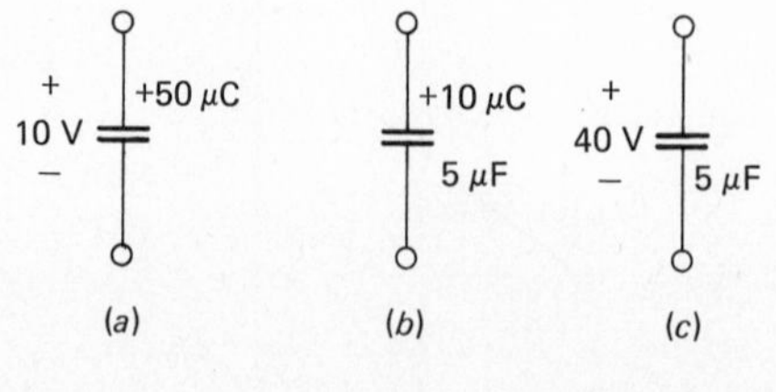

Figure 9-4. Applying capacitor law.

+ 10 V   +50 μC

(a)

+10 μC   5 μF

(b)

+ 40 V   5 μF

(c)

EXAMPLE 9-4.

Figure 9-4c shows the capacitor used in the two preceding examples. What is the charge stored on each plate?

SOLUTION.

The voltage has increased to 40 V. Capacitance still equals 5 μF. With Eq. (9-2c),

$$Q = CV = 5(10^{-6})40 = 200(10^{-6}) \text{ C} = 200 \ \mu\text{C}$$

So the upper plate has a stored charge of $+200 \ \mu$C, and the lower plate has $-200 \ \mu$C.

Test 9-2

1. Which of the following belongs least?
    (a) linear capacitor   (b) fixed C (c) Q/V constant
    (d) C proportional to V .... ( )
2. The following words can be rearranged to form a sentence: ARE LINEAR CAPAC-
    ITORS ALL. The sentence is
    (a) true   (b) false .... ( )
3. Capacitance is to resistance as Q/V is to
    (a) RI (b) V/I (c) V/R (d) CV .... ( )
4. Which of these is not valid?
    (a) $Q = C/V$ (b) $C = Q/V$ (c) $V = Q/C$ (d) $Q = CV$ .... ( )

9-3.   MORE ABOUT CAPACITANCE

Besides what you've learned about capacitance, you will need to know the answers to

> *How is capacitance related to plate area?*
> *To the distance between the plates?*
> *What is permittivity?*

Capacitance formula

In Fig. 9-5a, A stands for the area of a plate, and d is the distance between the plates. Given the dimensions of Fig. 9-5b,

$$A = 3 \text{ m} \times 2 \text{ m} = 6 \text{ m}^2$$

and $$d = 0.5 \text{ m}$$

The region between the plates may be a vacuum, air, or some other insulating material. From now on, we call this insulator the *dielectric*. The dielectric is important because it affects the value of capacitance.

Figure 9-5. Plate area and distance between plates.

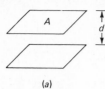
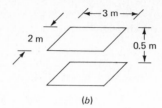

(a)   (b)

Starting with the defining formula

$$C = \frac{Q}{V}$$

plus other formulas, it is possible to derive this formula:

$$C = \epsilon \frac{A}{d} \tag{9-3}$$

where C = capacitance
  ϵ = a constant that depends on the dielectic
  A = plate area
  d = distance between plates

Capacitance proportional to area

Equation (9-3) makes sense. Capacitance is directly proportional to area, because more electrons can leave a larger area. In Fig. 9-6a the area is 4 m², but in Fig. 9-6b it is

Figure 9-6. Effect of area and distance on capacitance.

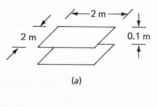
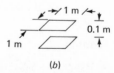

(a)   (b)

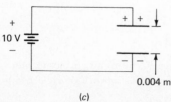
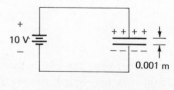

(c)   (d)

only 1 m². For a given voltage, four times as many electrons can be transferred in Fig. 9-6a as in Fig. 9-6b; therefore, the capacitance is four times greater.

Capacitance inversely proportional to distance

To pull an electron away from a core requires work. The greater the final distance between the electron and the core, the greater the work done. The transferred electrons of Fig. 9-6c are further away from the cores than the electrons of Fig. 9-6d. For this reason, it takes more work to transfer each electron in Fig. 9-6c. For the same voltage, therefore, *fewer* electrons are transferred in Fig. 9-6c than in 9-6d. This means capacitance decreases as the distance between plates increases.

Permittivity

In Eq. (9-3), ϵ is called the *permittivity;* its value depends on the dielectric between the plates. The permittivity of a vacuum has been measured. In the metric system,

$$\epsilon = 8.85(10^{-12})$$

This allows us to calculate the capacitance of two parallel plates in a vacuum. For instance, if the plates of Fig. 9-6a are in a vacuum, $\epsilon = 8.85(10^{-12})$, $A = 4$ m², $d = 0.1$ m, and

$$C = \epsilon \frac{A}{d} = 8.85(10^{-12}) \frac{4}{0.1} = 354(10^{-12}) \text{ F} = 354 \text{ pF}$$

Dielectrics such as paper, glass, plastics, etc. have larger values of permittivity than a vacuum. This allows manufacturers to produce a wide range of capacitance values. The next section discusses dielectric materials in detail, and Sec. 9-5 tells you about commercially available capacitors.

EXAMPLE 9-5.
Paper has a permittivity of approximately $35(10^{-12})$. If a paper capacitor has a plate area of 0.1 m² and a distance between plates of 0.001 m, what is the capacitance?

SOLUTION.
With Eq. (9-3),

$$C = \epsilon \frac{A}{d} = 35(10^{-12}) \frac{0.1}{0.001} = 3500(10^{-12}) \text{ F} = 3500 \text{ pF}$$

EXAMPLE 9-6.
Polystyrene (one kind of plastic) has a permittivity of approximately $23(10^{-12})$. If a polystyrene capacitor has the same plate area and distance between plates as the paper capacitor of the preceding example, what is the capacitance?

SOLUTION.

Again, use Eq. (9-3) to get

$$C = \epsilon \frac{A}{d} = 23(10^{-12}) \frac{0.1}{0.001} = 2300(10^{-12}) \text{ F} = 2300 \text{ pF}$$

Test 9-3

1. If plate area is tripled, the capacitance
    (*a*) doubles   (*b*) triples   (*c*) decreases by a factor of 3
    (*d*) decreases by a factor of 9 .... ( )
2. If the distance between plates is reduced by a factor of 3, the capacitance
    (*a*) doubles   (*b*) triples   (*c*) decreases by a factor of 3
    (*d*) decreases by a factor of 9 .... ( )
3. Resistance is to capacitance as resistivity is to
    (*a*) farad   (*b*) permittivity   (*c*) ohm   (*d*) area .... ( )
4. The expression $\epsilon A/d$ is analogous to
    (*a*) $\rho l/A$ (*b*) $I = V/R$ (*c*) $P = I^2 R$ (*d*) $G = I/R$ .... ( )
5. Capacitance is directly proportional to
    (*a*) transferred charge   (*b*) permittivity   (*c*) plate area
    (*d*) all of these .... ( )

9-4. THE DIELECTRIC

In the following discussion of the dielectric, be sure to get the answers to

> *What is polarization?*
> *How is dielectric constant defined?*
> *What does dielectric strength mean?*

Polarization

Figure 9-7*a* shows a single atom between a pair of charged plates. The outer-orbit electron is attracted by the upper plate and repelled by the lower plate. This changes the normally circular orbit into the elliptical orbit shown. Because the center of the ellipse is above the core, the electron spends more time above the core than below. The elliptical distortion and upward shift of the orbit is called *polarization*.

When a dielectric other than a vacuum is between the plates, the electron orbits of the dielectric are polarized. The amount of polarization depends on the permittivity of the dielectric. Because orbiting electrons spend more time above the core than below, the net effect is the same as having negative charges on top of positive charges (see Fig. 9-7*b*). Each polarized atom produces an upward electric field, symbolized by the small arrow between each core and electron.

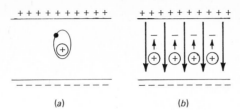

Figure 9-7. Polarized dielectric.

(a)　　　　　(b)

The total electric field between the plates is the superposition of the field produced by the plates and the field produced by the dielectric atoms. Since the dielectric field opposes the plate field, the net field between the plates is reduced. Because of this, it takes less work to transfer an electron from the upper to the lower plate. Stated another way, when we pull an electron away from the upper plate, this electron is subject to attraction of cores in the plate and the repulsion of polarized dielectric atoms. As a result, the same applied voltage can transfer more charge when a polarized dielectric is between the plates.

Dielectric constant

Suppose two capacitors have the same plate area and distance between plates. Then, the capacitances equal

$$C_1 = \epsilon_1 \frac{A}{d}$$

$$C_2 = \epsilon_2 \frac{A}{d}$$

and the ratio is

$$\frac{C_1}{C_2} = \frac{\epsilon_1}{\epsilon_2}$$

This says the ratio of capacitances equals the ratio of permittivities. If a dielectric has a permittivity five times greater than another dielectric, the first dielectric produces five times as much capacitance.

A vacuum is an ideal reference because it has the lowest permittivity of all dielectrics. To compare other dielectrics with a vacuum, we rewrite the foregoing equation as

$$\frac{C}{C_0} = \frac{\epsilon}{\epsilon_0} \tag{9-4}$$

where C and ϵ are for any dielectric, but C_0 and ϵ_0 are for a vacuum. If a dielectric has three times the permittivity of a vacuum, it increases capacitance by a factor of 3.

Because the ratio ϵ/ϵ_0 is used so much in practice, it has been named the *dielectric constant,* symbolized by ϵ_r. As a defining formula,

$$\epsilon_r = \frac{\epsilon}{\epsilon_0} \tag{9-5}$$

where ϵ_r = dielectric constant

ϵ = permittivity of any dielectric

ϵ_0 = permittivity of a vacuum, $8.85(10^{-12})$

Dielectric constant is sometimes called *relative permittivity,* because it indicates the effect of a dielectric relative to a vacuum.

Table 9-1 shows dielectric constants for materials used in modern capacitors. The values listed are approximate, because exact values depend on the grade of material, the production process, etc. Notice that air polarizes slightly. In all preliminary analysis, we approximate the dielectric constant of air as unity.

TABLE 9-1. DIELECTRIC CONSTANTS

DIELECTRIC	ϵ_r
Vacuum	1
Air	1.00054
Polystyrene	2.6
Polyester	3
Paper	4
Glass	4.2
Porcelain	6.2
Mica	7.5
Aluminum oxide	10
Tantalum oxide	11

Table 9-1 indicates how effective a dielectric is in increasing capacitance as compared with a vacuum. Polystyrene increases capacitance by a factor of 2.6, paper by a factor of 4, aluminum oxide by 10, and so on.

Dielectric strength

The greater the voltage between plates, the greater the polarization of dielectric atoms. If the voltage is too large, orbiting electrons can be *pulled away* from dielectric atoms. When this happens, the dielectric changes from an insulator to a conductor. Dislodging outer-orbit electrons with excessive voltage is called *voltage breakdown.*

The *dielectric strength* of a material is its ability to resist voltage breakdown. The defining formula is

$$\text{Dielectric strength} = \frac{V_{max}}{d} \tag{9-6}$$

where V_{max} = breakdown voltage

d = thickness of dielectric

When the distance between the plates is 0.01 m, air breaks down at 8000 V. Therefore, air has a dielectric strength of

$$\text{Dielectric strength} = \frac{8000 \text{ V}}{0.01 \text{ m}} = 800 \text{ kV/m}$$

Engineering handbooks list the dielectric strengths of different materials. Each dielectric can take only so much voltage for a given thickness before it breaks down. Because of this, a manufactured capacitor has a *voltage rating* which you cannot exceed without risking breakdown. In practice, therefore, the two most important things to know about a capacitor are its capacitance and its voltage rating.

EXAMPLE 9-7.

A capacitor has a plate area of 0.01 m² and a distance between plates of 0.001 m. What is the capacitance if the dielectric is air? Polyester? Aluminum oxide?

SOLUTION.

First, calculate the capacitance for an air dielectric. With Eq. (9-3),

$$C = \epsilon \frac{A}{d} = 8.85(10^{-12})\frac{0.01}{0.001} = 88.5 \text{ pF} \qquad \text{(air)}$$

Next, Table 9-1 gives an ϵ_r of 3 for polyester, and 10 for aluminum oxide. So,

$$C = 3 \times 88.5 \text{ pF} = 266 \text{ pF} \qquad \text{(polyester)}$$

$$C = 10 \times 88.5 \text{ pF} = 885 \text{ pF} \qquad \text{(aluminum oxide)}$$

Test 9-4

1. Which of these cannot have polarized atoms?
   (*a*) air   (*b*) mica   (*c*) paper   (*d*) a vacuum .... ( )
2. When a dielectric atom is polarized, its orbiting electrons will
   (*a*) be dislodged from the atom
   (*b*) follow a circular path
   (*c*) be attracted to the negative plate
   (*d*) spend more time nearer the positive plate .... ( )
3. Dielectric constant is not
   (*a*) a ratio   (*b*) measured in volts per meter
   (*c*) the same as relative permittivity   (*d*) equal to ϵ/ϵ_0 .... ( )
4. The greater the dielectric constant, the more
   (*a*) voltage the capacitor can withstand
   (*b*) energy needed to transfer electrons from one plate to the other
   (*c*) voltage needed per coulomb   (*d*) the transferred charge .... ( )

5.   Dielectric strength increases when you
   (*a*) double the thickness of the dielectric
   (*b*) use a dielectric half as thick   (*c*) use more plate area
   (*d*) none of these .... (   )

9-5.   MANUFACTURED CAPACITORS

With the dielectrics of Table 9-1, capacitors can be made with values from less than 1 pF to more than 10,000 μF.

Air capacitors

A *variable capacitor* can change capacitance over a 10:1 range (typical) by changing the plate area. Figure 9-8*a* shows minimum capacitance, Fig. 9-8*b* is intermediate capacitance, and Fig. 9-8*c* represents maximum capacitance. Often, several plates are paralleled to increase capacitance (Fig. 9-8*d*). Figure 9-8*e* is the schematic symbol for a variable capacitor.

A *trimmer capacitor* is a small variable capacitor that can change the distance between plates or the plate area. The typical way to use a trimmer capacitor is in parallel with another capacitor (see Fig. 9-8*f*). Varying the small trimmer capacitor, we can make fine adjustments on the equivalent capacitance; the total capacitance of Fig. 9-8*f* varies from 1001 to 1010 pF (proved later).

Air capacitors have values from less than 1 pF to more than 500 pF, with about a 10-to-1 range for a given variable capacitor (usually less range with trimmers).

Figure 9-8. Variable capacitors.

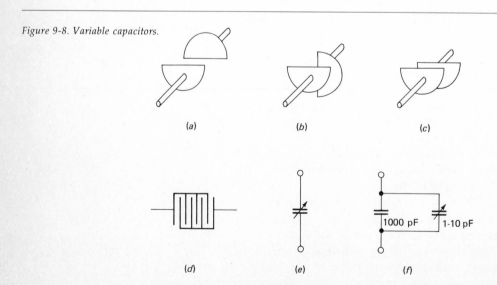

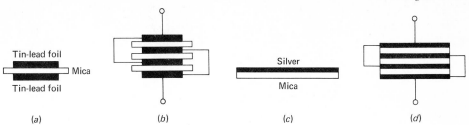

Figure 9-9. Mica capacitor.

(a)   (b)   (c)   (d)

Mica capacitors

Mica is a chemically inert mineral whose insulating properties change only slightly with time, temperature, moisture, etc. It is easily cut into sheets as thin as 0.0001 in. Because of this, mica was one of the first materials used as a dielectric.

Some mica capacitors look like Fig. 9-9a, a thin sheet of mica between two pieces of tin-lead *foil*. This sandwich acts like a parallel-plate capacitor. For added strength and capacitance, we can stack and parallel several sheets of mica and foil as shown in Fig. 9-9b. This is the way mica capacitors were made originally.

The modern way is better. A very thin layer of silver is bonded to the mica by furnace heating. The resulting *silvered-mica* sheet (Fig. 9-9c) has no air gaps, is more reliable, and is easier to work with. By stacking and paralleling these silvered-mica sheets (Fig. 9-9d), we can get a more reliable capacitor than the mica-foil capacitor previously described.

Silvered-mica capacitors have values from less than 1 pF to more than 10,000 pF.

Glass capacitors

Structurally, a glass capacitor is like the mica-foil capacitor of Fig. 9-9b, except sheets of glass (the dielectric) and aluminum foil (the plates) are used. Glass capacitors cost more than mica capacitors, but they have an advantage: they are moisture resistant. Used in high-humidity environments, glass capacitors show less change than any other type of capacitor.

Glass capacitors have values from less than 1 pF to more than 10,000 pF (the same as mica).

Paper capacitors

This type of capacitor uses sheets of paper dielectric and metal foil (Fig. 9-10a). Because the paper is flexible, the sheets can be rolled into a cylindrical structure like Fig. 9-10b. This paper-foil cylinder gives larger capacitance values than are possible with mica or glass capacitors.

Figure 9-10. Paper capacitor.

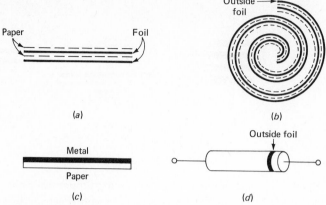

Some modern paper capacitors use *metallized paper,* made by depositing metal vapor on paper (Fig. 9-10c). The paper and its metal coating eliminate air gaps, improve reliability, etc. By rolling up sheets of metallized paper, we get a paper capacitor with greater reliability and higher capacitance than the paper-foil type.

Paper capacitors have a band or stripe on the end connected to the outside foil (Fig. 9-10d). When a schematic diagram shows one side of a paper capacitor grounded, most people ground the banded end. This is not necessary, but it has this advantage: it grounds the outside foil. The effect is to shield the inner foil from stray electric fields, unwanted signals, noise, etc.

As a guide, paper capacitors have values from less than 0.001 μF to more than 1 μF.

Plastic-film capacitors

Plastics are excellent dielectrics. The commonly used plastic dielectrics are *polyester* and *polystyrene.* These materials have high dielectric strength and can be produced in very thin sheets like film.[1]

Polyester-film capacitors use separate sheets of metal foil and polyester like Fig. 9-11a, or they use metallized polyester film like Fig. 9-11b. In either case, sheets of the flexible film can be rolled into a cylindrical structure to get large capacitance values (similar to Fig. 9-10b). Polyester film has a higher dielectric strength than paper. For the same voltage rating, thinner sheets can be used to get higher capacitance values than with paper.

The main disadvantage of polyester-film capacitors is their temperature sensitivity; large changes in capacitance occur when the temperature changes. Polystyrene is much less temperature sensitive than polyester. This is why *polystyrene-film capacitors* are preferred to polyester-film capacitors when large temperature variations are expected.

[1] Manufactured under tradenames such as Mylar, Kodak, etc.

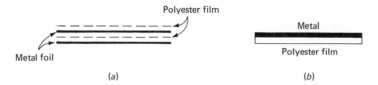

Figure 9-11. Plastic-film capacitor.

Polyester film

Metal foil

(a)

Metal

Polyester film

(b)

Polyester-film capacitors have values from less than 0.001 μF to more than 10 μF; polystyrene-film capacitors from less than 5 pF to more than 0.5 μF.

Ceramic capacitors

Clay is transformed by extreme heat. At temperatures near 1000°F, chemically combined water is driven off and the clay undergoes an irreversible chemical change. The resulting product is called a *ceramic*. Examples of ceramics are porcelain, china, brick, etc.

The main advantage of ceramic materials is their tremendous range in dielectric properties. For instance, porcelain has a dielectric constant of 6 (approximate); barium-strontium-titanate (another ceramic) has a dielectric constant of about 7500. By mixing ceramic materials, we can get a wide range in dielectric constant and other properties.

Using ceramic dielectrics, manufacturers can produce capacitors with positive, negative, or zero *temperature coefficients*. If you see a ceramic capacitor with P100 on it, this means it has a positive temperature coefficient of 100 *parts per million* (ppm) per degree. This is equivalent to

$$\frac{100}{1,000,000} = 10^{-4} = 0.01 \text{ percent}$$

per degree rise (Celsius). A ceramic capacitor with N200 means a negative temperature coefficient of 200 ppm per degree, equivalent to

$$\frac{-200}{1,000,000} = -2(10^{-4}) = -0.02 \text{ percent}$$

per degree rise. And with the right mix of materials, a ceramic capacitor can have a zero temperature coefficient, indicated by NP0; in this case, the capacitance doesn't change with temperature.

Ceramic capacitors have values from less than 10 pF to more than 1 μF.

Electrolytic capacitors

Electrolytic capacitors have more capacitance than any other type, because the dielectric is an extremely thin layer of *oxide* grown on one of the plates. Here's how an

Figure 9-12. Electrolytic capacitor.

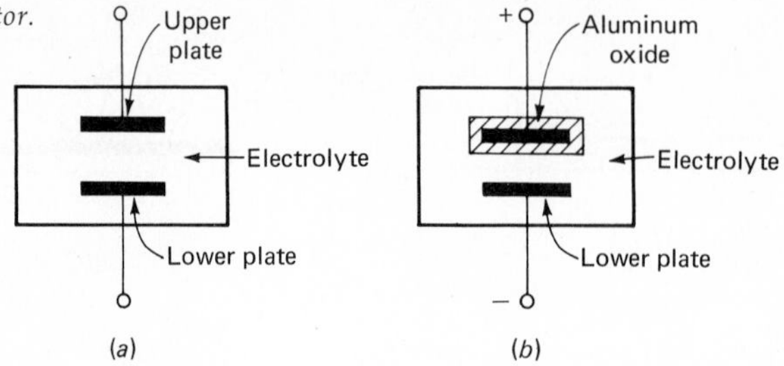

(a)   (b)

electrolytic capacitor is formed. Initially, two aluminum plates and an *electrolyte* (a conducting liquid or paste) are in a container (Fig. 9-12*a*). At first, this device does not act like a capacitor because the electrolyte shorts the two plates. But after voltage has been applied for a while, electrochemical action produces a layer of aluminum oxide on the positive plate (see Fig. 9-12*b*). The process of growing this layer is a slow one, and is referred to as *forming* the capacitor.

After the capacitor has been formed, the oxide layer acts like a dielectric between the upper aluminum plate and the electrolyte. The lower aluminum plate merely makes contact with the electrolyte, which functions as the negative plate.

The electrolytic capacitor of Fig. 9-12*b* works fine as long as the upper plate is positive and the lower one negative. But if you try to reverse the applied voltage, the oxide layer no longer acts like an insulator; instead you get a large current and a loss of normal capacitor action. For this reason, most electrolytic capacitors are *polarized;* they work properly only with one polarity of voltage. This is the reason a manufacturer puts a plus sign (or other marking) on the positive end of an electrolytic capacitor.

When an electrolytic capacitor is stored for long periods, the aluminum oxide layer may deteriorate. In this case, the capacitor has to be reformed before it can be used in a circuit. This requires slow charging of the capacitor to regrow the oxide layer. Having to reform the capacitor after long periods of storage is a nuisance, and has led to the development of *tantalum electrolytic capacitors.*

Tantalum capacitors can be stored indefinitely. They also have more capacitance than alumimum ones. For this reason, tantalum electrolytic capacitors are replacing aluminum electrolytic capacitors in many applications.

The thin oxide layer is the key to why electrolytic capacitors have more capacitance than other types. At the same time, however, the thin oxide layer has a disadvantage: its resistance is relatively low compared with other dielectrics. Instead of acting ideally like an open, it passes a small current (see Fig. 9-13*a*). We call this unwanted current the *leakage current.*

To account for leakage current, visualize a leakage resistance in parallel with the capacitance, as shown in Fig. 9-13*b*. The values given are representative of a 250-μF capacitor with 20 V across it; in this case, the leakage current is equivalent to having a 3-MΩ resistance across the capacitance. In general, the larger the capacitance, the

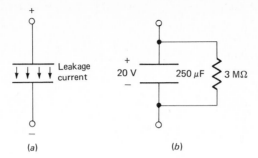

Figure 9-13. Leakage current.

(a)   (b)

smaller the leakage resistance; this leakage resistance can prevent some circuits from working properly.

Electrolytic capacitors have values from less than 1 μF to more than 10,000 μF.

Summary

We have covered most of the types of modern capacitors. As a summary, Table 9-2 shows the approximate range, main advantage, and main disadvantage of each type.

TABLE 9-2.   CAPACITOR TYPES

TYPE	RANGE (APPROX.)	ADVANTAGE	DISADVANTAGE
Air	1 to 500 pF	Variable	Low capacitance
Mica	1 to 10,000 pF	Stable values	Expensive
Glass	1 to 10,000 pF	Moisture resistant	Expensive
Paper	0.001 to 1 μF	Inexpensive	Temperature sensitive
Polyester	0.001 to 10 μF	Physically small	Temperature sensitive
Polystyrene	5 pF to 0.5 μF	Temperature stable	Less C than polyester
Ceramic	10 pF to 1 μF	Variety	Fragile
Electrolytic	1 to 10,000 μF	Large capacitance	Leakage resistance

As shown, mica capacitors are stable (little change with time, temperature, moisture, etc.), but are more expensive than other types. The ceramic capacitors offer the greatest range in capacitance values and temperature coefficients, but are somewhat fragile and may break during installation.

EXAMPLE 9-8.
A 100-pF ceramic capacitor has P50 on it. What value does this capacitor have at 125°C?

SOLUTION.
The value of 100 pF applies at room temperature, approximately 25°C. P50 means a positive temperature coefficient of 50 ppm per degree. This is equivalent to

$$50 \text{ ppm} = \frac{50}{1{,}000{,}000} = 50(10^{-6}) = 0.005 \text{ percent}$$

per degree rise.

The total rise in temperature is

$$T_1 - T_2 = 125°C - 25°C = 100°C$$

The total percent change is

$$50 \text{ ppm} \times (T_1 - T_2) = 0.005 \text{ percent} \times 100 = 0.5 \text{ percent}$$

Therefore, the total capacitance change is

$$0.5 \text{ percent} \times 100 \text{ pF} = 0.5 \text{ pF}$$

and the capacitance at 125°C is

$$C = 100 \text{ pF} + 0.5 \text{ pF} = 100.5 \text{ pF}$$

9-6. EQUIVALENT CAPACITANCE

Capacitances in series or parallel act like a single equivalent capacitance. This section answers

> *How do parallel capacitances combine?*
> *How do series capacitances combine?*

Parallel capacitances

Assume the capacitors of Fig. 9-14a are initially uncharged. After the switch closes, a total charge Q flows into the parallel connection of Fig. 9-14b. Q_1 goes to C_1, and Q_2 to C_2. Charges will flow as long as node B has less voltage than node A. For this reason, both capacitors charge until their voltages equal the source voltage.

Figure 9-14. Parallel capacitances.

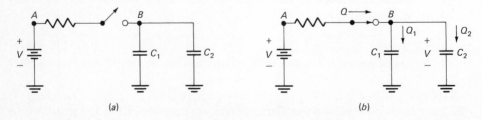

(a)   (b)

After the capacitors are fully charged, the capacitor law tells us

$$Q_1 = C_1V \qquad\qquad (9\text{-}7a)$$

and
$$Q_2 = C_2V \qquad\qquad (9\text{-}7b)$$

The total charge equals the sum of charges, so

$$Q = Q_1 + Q_2$$

With Eqs. (9-7a) and (9-7b), this becomes

$$Q = C_1V + C_2V$$

or
$$\frac{Q}{V} = C_1 + C_2$$

The ratio Q/V is the ratio of total charge to total voltage; therefore, Q/V is the total or equivalent capacitance. Because of this, the preceding equation becomes

$$C = C_1 + C_2 \qquad \text{(parallel)} \qquad\qquad (9\text{-}8)$$

This final result says the equivalent capacitance of two parallel capacitances equals the sum of capacitances. If C_1 is 500 pF and C_2 is 700 pF, the equivalent capacitance is

$$C = 500 \text{ pF} + 700 \text{ pF} = 1200 \text{ pF}$$

Or if $C_1 = 10 \ \mu\text{F}$ and $C_2 = 30 \ \mu\text{F}$,

$$C = 10 \ \mu\text{F} + 30 \ \mu\text{F} = 40 \ \mu\text{F}$$

Equation (9-8) makes sense. When you connect two capacitors in parallel, you are connecting the two upper plates and the two lower plates. The effect is the same as increasing plate area. Since capacitance is directly proportional to plate area, the equivalent capacitance increases.

By a similar proof, a circuit with n parallel capacitances has an equivalent capacitance of

$$C = C_1 + C_2 + \cdots + C_n \qquad\qquad (9\text{-}9)$$

This says you add parallel capacitances to get the equivalent capacitance.

Series capacitances

Figure 9-15 shows capacitors in series. Kirchhoff's current law says current has the same value at all points in a series circuit; therefore, each capacitor charge is the same and equals the total input charge. In symbols,

$$Q = Q_1 = Q_2 \qquad\qquad (9\text{-}10a)$$

Figure 9-15. Series capacitances.

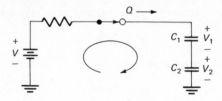

Applying Kirchhoff's voltage law to Fig. 9-15 gives

$$V = V_1 + V_2 \tag{9-10b}$$

Dividing this by total charge Q,

$$\frac{V}{Q} = \frac{V_1}{Q} + \frac{V_2}{Q}$$

Because of Eq. (9-10a),

$$\frac{V}{Q} = \frac{V_1}{Q_1} + \frac{V_2}{Q_2}$$

With the capacitor law, this becomes

$$\frac{1}{C} = \frac{1}{C_1} + \frac{1}{C_2}$$

Finally, algebra gives

$$C = \frac{C_1 C_2}{C_1 + C_2} \tag{9-11}$$

So, we wind up with the product-over-sum rule. Given two capacitances in series, we can lump them into a single equivalent capacitance equal to the product over the sum. If 10 pF is in parallel with 30 pF, the equivalent capacitance is

$$C = \frac{10 \text{ pF} \times 30 \text{ pF}}{10 \text{ pF} + 30 \text{ pF}} = 7.5 \text{ pF}$$

By a similar derivation, more than two capacitances combine by either of these two rules:

$$\frac{1}{C} = \frac{1}{C_1} + \frac{1}{C_2} + \cdots + \frac{1}{C_n} \tag{9-12a}$$

or

$$C = \frac{1}{1/C_1 + 1/C_2 + \cdots + 1/C_n} \tag{9-12b}$$

Either of these equations implies the equivalent capacitance of series capacitors is always smaller than the smallest capacitance.

EXAMPLE 9-9.

After the switch of Fig. 9-16a closes, what is the final charge on each capacitor? The total charge?

SOLUTION.

The capacitor law says the first capacitor has a charge of

$$Q_1 = C_1V = 3(10^{-6})10 = 30 \ \mu C$$

The second capacitor has a charge of

$$Q_2 = C_2V = 6(10^{-6})10 = 60 \ \mu C$$

The total charge in a parallel circuit always equals the sum of the charges. So,

$$Q = Q_1 + Q_2 = 30 \ \mu C + 60 \ \mu C = 90 \ \mu C$$

EXAMPLE 9-10.

After the switch closes in Fig. 9-16b, what is the total charge transferred from the source to the capacitors? How much voltage is across each capacitor?

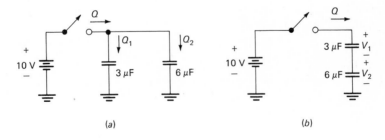

Figure 9-16. Examples of calculating equivalent capacitance.

(a)

(b)

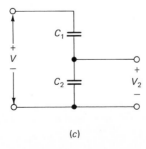

(c)

SOLUTION.

The equivalent capacitance of the series capacitors is

$$C = \frac{C_1 C_2}{C_1 + C_2} = \frac{3 \ \mu F \times 6 \ \mu F}{3 \ \mu F + 6 \ \mu F} = 2 \ \mu F$$

And the total charge is

$$Q = CV = 2 \ \mu F \times 10 \ V = 20 \ \mu C$$

As we know, $Q = Q_1 = Q_2$ in a series circuit. Therefore, applying the capacitor law to each capacitor gives

$$V_1 = \frac{Q_1}{C_1} = \frac{20(10^{-6})}{3(10^{-6})} = 6.67 \ V$$

and

$$V_2 = \frac{Q_2}{C_2} = \frac{20(10^{-6})}{6(10^{-6})} = 3.33 \ V$$

These results bring out an important idea. When you charge two series capacitors, the smaller capacitance ends up with more voltage. In fact, it is easy to show the capacitive voltage divider of Fig. 9-16c has this formula for its output voltage:

$$V_2 = \frac{C_1}{C_1 + C_2} \ V$$

EXAMPLE 9-11.

Suppose the 3-μF capacitor of Fig. 9-17 has an initial voltage of 10 V. With the switch open, this capacitor retains its charge indefinitely. But when the switch is closed, the capacitor will discharge into the other capacitor. What is the final voltage across the parallel connection after the switch is closed?

SOLUTION.

First, get the charge stored in the 3-μF capacitor before the switch is closed. With the capacitor law,

$$Q = CV = 3(10^{-6})10 = 30 \ \mu C$$

This means the upper plate has a charge of $+30 \ \mu C$, and the lower plate a charge of $-30 \ \mu C$.

After the switch is closed, the $+30 \ \mu C$ are distributed onto both upper plates; likewise, the $-30 \ \mu C$ spread over the two lower plates. Because of this, the total charge on

Figure 9-17. One capacitor discharging into another.

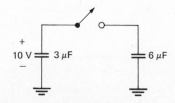

both capacitors is still 30 μC. This allows us to write

$$Q_1 + Q_2 = 30 \ \mu C$$

Applying the capacitor law to each capacitor, this becomes

$$C_1V + C_2V = 30 \ \mu C$$

(Voltage V is the same across both capacitors, because they are between the same pair of equipotential points.) Solving the preceding equation for V gives

$$V = \frac{30 \ \mu C}{C_1 + C_2} = \frac{30 \ \mu C}{3 \ \mu F + 6 \ \mu F} = 3.33 \ V$$

Test 9-5

1. The total charge flowing into two parallel capacitors equals the
   (a) sum of the charges into each capacitor   (b) difference of the charges
   (c) charge into either   (d) ratio of the first charge to the second charge ... ( )
2. Equivalent capacitance of parallel capacitors has no relation to the
   (a) kind of source used   (b) sum of capacitances
   (c) ratio of total charge to parallel voltage   (d) each capacitance .... ( )
3. Which of the following belongs least?
   (a) equivalent capacitance   (b) total capacitance   (c) Q/V ratio
   (d) charge .... ( )
4. Connecting capacitors in parallel has the same effect as
   (a) increasing plate area   (b) increasing distance between plates
   (c) decreasing plate area   (d) decreasing dielectric constant .... ( )
5. Connecting capacitors in series has the same effect as
   (a) increasing plate area   (b) increasing the distance between plates
   (c) decreasing the distance between plates
   (d) increasing the dielectric constant .... ( )

9-7. UNWANTED CAPACITANCE

Don't think you get capacitive effects only with a capacitor (a load optimized for its capacitance). Capacitance exists between any pair of conductors. Use your imagination, and you will realize all kinds of unwanted capacitance exist in a circuit. Because this unwanted capacitance is scattered throughout the circuit, it's called *stray capacitance*. This section discusses kinds of stray capacitance. Look for the answers to

What is stray wiring capacitance?
What is resistor capacitance?
What is node capacitance?
What is cable capacitance?

*Figure 9-18. Stray wiring
capacitance.*

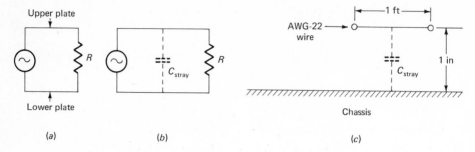

(a)   (b)   (c)

Stray wiring capacitance

Stray wiring capacitance is the capacitance between the connecting wires of a circuit. In Fig. 9-18a, the upper and lower wires act like the plates of a capacitor. The plate area is small, and the distance between wires is relatively large, but there still is some capacitance. At most frequencies, stray wiring capacitance has negligible effects. But at very high frequencies it does cause trouble and is taken into account.

Figure 9-18b shows how to indicate the presence of stray wiring capacitance. The dashed lines remind us it's not a separate component like a resistor, a capacitor, etc. but rather a stray or unwanted effect. Also, notice it appears in parallel with the resistance.

Often, the chassis is used as part of the circuit. The connecting wire between components may run approximately parallel to the chassis (see Fig. 9-18c). Because of this, the connecting wire acts like the upper plate of a stray capacitor, and the chassis like the lower plate. By an advanced derivation, the 1-ft piece of wire in Fig. 9-18c has a capacitance of 3.3 pF, equivalent to approximately 0.3 pF/in. Even though a connecting wire is not AWG 22, and may not be an inch away from the chassis, we can still use 0.3 pF/in as a rough estimate. For instance, a 3-in. piece of connecting wire has a stray capacitance of about 0.9 pF to the chassis. A 7-in piece of wire has about 2.1 pF, and so on.

For exact analysis, build the circuit and measure stray capacitance with a capacitance meter, a capacitance bridge, etc.

Resistor capacitance

A resistor has a lead on each end. These leads act like the plates of a small capacitor. Because of this, every resistor has some stray capacitance, as shown in Fig. 9-19a. The value of this stray capacitance depends on the length of the leads, the physical size of the resistor, etc.

The stray capacitance of most resistors is less than 1 pF. As a rough estimate, use 1

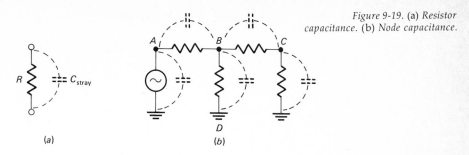

Figure 9-19. (a) *Resistor capacitance.* (b) *Node capacitance.*

pF for the stray capacitance of all resistors. For exact work, measure the stray capacitance of a resistor with a capacitance meter, a capacitance bridge, etc.

Node capacitance

Stray capacitances exist between the nodes of all circuits (see Fig. 9-19b). These stray capacitances are the sums of wiring and resistor capacitances. At lower frequencies, node capacitances have negligible effects. But at higher frequencies, they produce unwanted effects. We discuss these unwanted effects later (under *capacitive reactance*).

Coaxial-cable capacitance

Coaxial cable has an inner and outer conductor as shown in Figs. 9-20a and b. The inner conductor acts like one plate of a capacitor, and the outer conductor like the other. Therefore, a capacitance exists between the inner and outer conductor.

The longer the cable, the greater the capacitance (more plate area). Also, the smaller the distance between the inner and the outer conductors, the larger the capacitance. Every coaxial cable has a certain amount of capacitance per foot. These values, typically 10 pF to 50 pF per foot, are listed in engineering handbooks and manufacturers' catalogs.

One application in which you notice the effect of coaxial-cable capacitance is with oscilloscopes. Low-capacitance cables normally connect the load whose voltage is being measured to the input of the oscilloscope. Typically, a low-capacitance cable has approximately 10 pF per foot. Therefore, if you use a 3-ft cable, you are shunting 30 pF across the load whose voltage is being measured. At high frequencies, this added capacitance produces unwanted effects.

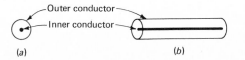

Figure 9-20. *Coaxial-cable capacitance.*

Test 9-6

1. As a rough estimate, a piece of wire 2½ ft long has a capacitance to the chassis of
    (a) 1 pF   (b) 8 pF   (c) 20 pF   (d) 50 pF .... ( )
2. A 6-in piece of wire connects two transistors. If you shorten this connecting wire to 2 in, the stray capacitance to ground decreases by a factor of
    (a) 3   (b) 2   (c) 1   (d) 0.5 .... ( )
3. Three resistors are in parallel. Their total stray capacitance is roughly
    (a) 0.5 pF   (b) 1 pF   (c) 3 pF   (d) 9 pF .... ( )
4. Two resistors are in series. The equivalent stray capacitance is roughly
    (a) 0.5 pF   (b) 1 pF   (c) 2 pF   (d) 4 pF .... ( )
5. RG-220/U is a coaxial cable with 29.5 pF/ft. How much capacitance does a 50-ft length of this cable have?
    (a) 295 pF   (b) 885 pF   (c) 1180 pF   (d) 1475 pF .... ( )

SUMMARY OF FORMULAS

DEFINING

$$C = \frac{Q}{V} \tag{9-1}$$

$$\epsilon_r = \frac{\epsilon}{\epsilon_0} \tag{9-5}$$

$$\text{Dielectric strength} = \frac{V_{max}}{d} \tag{9-6}$$

DERIVED

$$C = \epsilon \frac{A}{d} \tag{9-3}$$

$$\frac{C}{C_0} = \frac{\epsilon}{\epsilon_0} \tag{9-4}$$

$$C = C_1 + C_2 \quad \text{(parallel)} \tag{9-8}$$

$$C = \frac{C_1 C_2}{C_1 + C_2} \quad \text{(series)} \tag{9-11}$$

Problems

9-1. If the transferred charge is 0.025 C and the voltage between the plates is 5 V, what is the value of the capacitance?

9-2. A charge of 30 μC is transferred from one plate to another. If 10 V appears between the plates, what is the capacitance?

9-3. A charge of +40 pC is stored on one plate of a capacitor. What is the charge stored on the other plate? If the voltage between the plates is 8 V, what is the capacitance?

9-4. A 50-μF capacitor has 10 V between its plates. What is the value of charge stored on the positive plate?

9-5. A charge of +20 μC is stored on one plate of a capacitor. If the capacitance equals 4 μF, what is the voltage between the plates?

9-6. How much charge is stored in a 2000-pF capacitor for each of these voltages: 2 V, 10 V, 20 V, 100 V?

9-7. A parallel-plate capacitor has a plate area of 0.75 m² and a distance between plates of 0.3 m. What is the capacitance for a vacuum dielectric?

9-8. Use the dimensions as in the preceding problem. What is the capacitance for a paper dielectric? A polystyrene dielectric? (The permittivities are 35×10^{-12} and 23×10^{-12}, respectively.)

9-9. A capacitor has a capacitance of 250 pF in a vacuum. What is its capacitance if a polyester dielectric is used? If mica is used?

9-10. A capacitor with a polyester dielectric has a capacitance of 260 pF. What is the capacitance if the dielectric is changed to porcelain? To tantulum oxide?

9-11. A capacitor with a vacuum dielectric has a capacitance of 1000 pF. What is the value of the capacitance if each of the following dielectrics is used: polystyrene, polyester, paper, glass, mica, aluminum oxide?

9-12. A dielectric is 0.0005 m thick. If it breaks down when the voltage across it is 1000 V, what is the dielectric strength? If the thickness is reduced to 0.0001 m, what is the dielectric strength? For this smaller thickness, what is the breakdown voltage?

9-13. Some trimmer capacitors are made by varying the distance between plates. If the two plates of such a trimmer capacitor can have a distance from 2 mm to 10 mm, what is the ratio of maximum to minimum capacitance? If the minimum capacitance is 3 pF, what is the maximum capacitance?

9-14. A ceramic capacitor has a capacitance of 1000 pF at 25°C. What is its capacitance at 100°C if P200 is written on it? If N300 is on it?

9-15. What is the equivalent capacitance for the parallel connection of Fig. 9-21a?

9-16. What is the equivalent capacitance for the parallel connection of Fig. 9-21b?

9-17. What equivalent capacitance do the series capacitors of Fig. 9-21c have?

9-18. If a 10-V battery is connected across the circuit of Fig. 9-21a, what will the final stored charge be on each capacitor? The total charge?

9-19. How much voltage is there across each capacitor of Fig. 9-21b if a 20-V battery is connected between A and B? What is the stored charge on each capacitor?

9-20. If a 5-V battery is connected between A and B in Fig. 9-21c, what is the total charge delivered by the battery? The final charge on each capacitor? The voltage across each capacitor?

Figure 9-21.

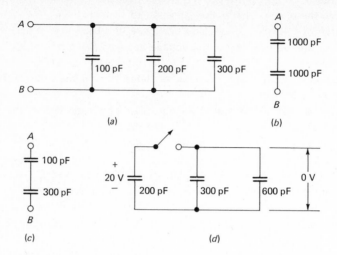

(a)

(b)

(c)

(d)

9-21.   The 200-pF capacitor of Fig. 9-21d has 20 V across it before the switch is closed. What is the voltage across the capacitors after the switch is closed? How much charge is stored on each capacitor after the switch is closed?

9-22.   It can be proved that the potential energy of a charged capacitor equals

$$E = \frac{1}{2} CV^2$$

What is the potential energy of a 100-μF capacitor with 25 V across it? If the voltage is doubled, what is the potential energy ?

9-23.   Roughly, how much stray wiring capacitance does a 6-in piece of wire have with respect to the chassis? An 18-in piece? A 2-in piece?

9-24.   What is the approximate stray capacitance across three parallel 100-kΩ resistors? If these resistors were in series, what would the approximate capacitance be from one end of the series connection to the other?

ANSWERS TO TESTS

9-1.   *c, b, c, a, d, d, d*
9-2.   *d, b, b, a*
9-3.   *b, b, b, a, d*
9-4.   *d, d, b, d, d*
9-5.   *a, a, d, a, b*
9-6.   *b, a, c, a, d*

10. Inductance

Inductance is the third property of all loads. It is related to the magnetic field around a load. This chapter discusses magnetic field, inductance, and other topics.

10-1. BASIC IDEAS

When magnets are close together, they attract or repel. In Fig. 10-1a, the *bar magnets* attract because a north (N) pole is opposite a south (S) pole. When N poles are opposite each other, the magnets repel, as shown in Fig. 10-1b. So the first thing to know about magnets is this: unlike poles always attract and like poles always repel.

As you read more about magnets, find the answers to

What is a magnetic field?
What is an inductor?
What is flux?

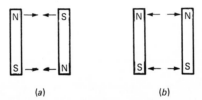

(a) (b)

Figure 10-1. Bar magnets.
(a) Attraction. (b) Repulsion.

The magnetic field

The needle of a compass acts like a small bar magnet. For this reason, its S pole is attracted by the earth's N pole; this is why the S pole of a compass needle points to the earth's N pole.

Bring a compass near a large bar magnet like Fig. 10-2, and the S pole of the needle no longer points to the earth's N pole. Instead the needle points in a direction that depends on its position with respect to the large bar magnet.

The space around the large magnet is called a *field*. Because of the force on a compass needle, we say a field of force surrounds the large magnet. To distinguish this field from electric and gravitational fields, we call it a magnetic field of force, or simply a *magnetic field*.

The direction of a magnetic field is defined as the direction in which the N pole of a compass needle points. If you explore the region around a large magnet, you will discover the magnetic field points in the directions shown in Fig. 10-2. The different lines in the field are called *lines of force*. Notice each line leaves the N pole of the large magnet and enters the S pole.

The inductor

During a classroom lecture in 1819, Oersted laid a compass near a conductor in which charges were flowing. When he later noticed the needle, he saw it pointed to the conductor rather than to the earth's N pole. This meant electricity and magnetism were somehow related. Countless experiments have since confirmed that a *magnetic field always exists around moving charges*.

The magnetic field is especially noticeable around an *inductor*, a coil of wire like Fig. 10-3a. The strength of the field depends on the current and number of turns of wire; increasing either increases the strength of the field.

Figure 10-2. Magnetic field around a bar magnet.

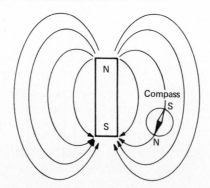

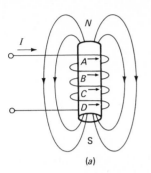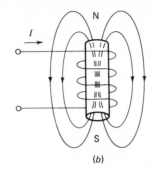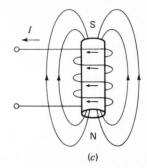

Figure 10-3. Field around an inductor. (a) N pole up. (b) Flux passes through inside. (c) N pole down.

(a) (b) (c)

Types of magnets

A bar magnet is a *permanent magnet*, because it retains its magnetic field indefinitely; that is, the strength of the field remains constant over long periods of time. An inductor is an *electromagnet*, because electric current produces the magnetic field; if there is no current, there is no field. (Recall that both types of magnets are used in a moving-coil meter.)

Flux

Flux is another name for lines of force. In Fig. 10-3a, the inductor is surrounded by flux, equivalent to saying it's surrounded by lines of force. The metric unit of measure for flux is the *weber* (abbreviated Wb). In formulas, Greek letter ϕ (*phi*) stands for flux. A flux of 5 webers is written as

$$\phi = 5 \text{ Wb}$$

Twice as much flux is written

$$\phi = 10 \text{ Wb}$$

Flux is inside an inductor as well as outside. Figure 10-3b shows this. Lines of force entering the S pole pass through the inside of the inductor and emerge from the N pole. Also notice each line of force is *closed* (no beginning or end).

Right-hand rule

In Fig. 10-3a, flux leaves the N pole and enters the S pole. If the current is reversed, the flux reverses as shown in Fig. 10-3c. To put it another way, when the current is reversed, the poles reverse.

The *right-hand rule* helps identify the N and S poles of an inductor. Here's how it works. In Fig. 10-3a, imagine curling your fingers around the inductor in the direction of conventional current. Then, your index finger points in the A direction, your next finger in the B direction, and so on. If you are visualizing this correctly, your thumb will point to the N pole.

Apply the right-hand rule to Fig. 10-3c. With the fingers curled in the same direction as conventional current in each winding, the thumb automatically points down toward the N pole. Coming discussions use the right-hand rule.

Test 10-1

1. Which belongs least in the following group?
    (*a*) electric field   (*b*) bar magnet   (*c*) electromagnet
    (*d*) magnetic field   .... ( )
2. The direction of a magnetic field is the same as the direction
    (*a*) of force on a negative charge   (*b*) of conventional current
    (*c*) in which the N pole of a compass needle points
    (*d*) of an electric field .... ( )
3. Electric field is to charge as magnetic field is to
    (*a*) mass   (*b*) current   (*c*) voltage   (*d*) resistance .... ( )
4. An inductor is
    (*a*) a permanent magnet   (*b*) optimized for its capacitance
    (*c*) an electromagnet   (*d*) nonmagnetic .... ( )
5. Which of the following does not belong?
    (*a*) charge at rest   (*b*) moving charge   (*c*) magnetic field
    (*d*) magnetic flux .... ( )
6. Flux is closest in meaning to
    (*a*) moving charge   (*b*) webers   (*c*) inductor   (*d*) lines of force ...... ( )
7. Resistance is to ohm as flux is to
    (*a*) current   (*b*) weber   (*c*) voltage   (*d*) magnetic field .... ( )

10-2.   DEFINITION OF INDUCTANCE

To pin down the magnetic effects produced by an inductor, we must define *inductance* and agree on a unit of measure. Therefore, the key questions in this section are

> *What is the defining formula for inductance?*
> *What is a henry?*

Defining formula

The magnetic effects of an inductor depend on the number of turns, the amount of flux, and the size of the current. All these factors are in the defining formula of inductance:

$$L = \frac{N\phi}{I} \qquad (10\text{-}1)$$

where L = inductance
 N = number of turns
 ϕ = flux
 I = current

The product $N\phi$ is often called the *flux linkage*. In words, Eq. (10-1) says inductance equals the flux linkage divided by the current.

Here are some examples. If an inductor with 100 turns and a current of 5 A produces a flux of 10 Wb (see Fig. 10-4a), the flux linkage equals

$$N\phi = 100 \times 10 \text{ Wb} = 1000 \text{ Wb}$$

and the inductance is

$$L = \frac{N\phi}{I} = \frac{1000 \text{ Wb}}{5 \text{ A}} = 200 \text{ Wb/A}$$

The inductor of Fig. 10-4b is different; it has 250 turns, a flux of 0.2 Wb, and a current of 4 A. Therefore, the flux linkage is

$$N\phi = 250 \times 0.2 \text{ Wb} = 50 \text{ Wb}$$

and the inductance is

$$L = \frac{N\phi}{I} = \frac{50 \text{ Wb}}{4 \text{ A}} = 12.5 \text{ Wb/A}$$

The henry

The *henry* (abbreviated H) is the metric unit of measure for inductance. It's defined as 1 weber per ampere. In symbols,

$$1 \text{ H} = 1 \text{ Wb/A}$$

Therefore, the preceding answers become 200 H and 12.5 H.

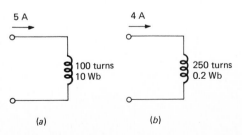

5 A

4 A

100 turns
10 Wb

250 turns
0.2 Wb

(a)

(b)

Figure 10-4. Inductance calculations.

Commercial instruments such as *inductance meters* and *inductance bridges* are used to measure inductance.

EXAMPLE 10-1.

A current of 5 mA flows through an inductor with 200 turns. If the flux equals $4(10^{-6})$ Wb, what is the inductance?

SOLUTION.

The defining formula gives

$$L = \frac{N\phi}{I} = \frac{200 \times 4(10^{-6})}{0.005} = 0.16 \text{ H} = 160 \text{ mH}$$

EXAMPLE 10-2.

A current of 50 mA flows through an inductor with 50 turns and an inductance of 200 mH. How much flux is there?

SOLUTION.

First, rearrange the defining formula to get

$$\phi = \frac{LI}{N}$$

Then,

$$\phi = \frac{0.2 \times 0.05}{50} = 0.0002 \text{ Wb}$$

Test 10-2

1. $L = N\phi/I$ cannot be proved because it is a
     (*a*) defining formula   (*b*) experimental formula   (*c*) derived formula ...   ( )
2.   Which is closest in meaning to large inductance?
     (*a*) large flux   (*b*) low current   (*c*) large flux linkage
     (*d*) large flux linkage per ampere   ....   ( )
3.   Voltage is to volt as inductance is to
     (*a*) henry   (*b*) ampere   (*c*) ohm   (*d*) weber   ....   ( )
4.   Which belongs least?
     (*a*) ohm   (*b*) farad   (*c*) henry   (*d*) inductance   ....   ( )
5.   If you double the number of turns but keep the flux and the current the same, the inductance will
     (*a*) stay the same   (*b*) halve   (*c*) double   (*d*) triple   ....   ( )

10-3. MORE ABOUT INDUCTANCE

This section answers

How is inductance related to length and area?
What does permeability mean?

Inductance formula

Figure 10-5*a* shows a *toroid,* a doughnut-shaped inductor. The *core* supports the turns of wire. By starting with the defining formula for inductance and other formulas, we can come up with this derived formula:

$$L = \mu \, \frac{N^2 A}{l} \qquad\qquad (10\text{-}2)$$

where μ = a constant that depends on the core material
N = number of turns
A = cross-sectional area of the core
l = length of the core

Figure 10-5*b* illustrates the area (shaded) and the length (dashed line).

This equation says inductance is directly proportional to the area and inversely proportional to the length. If you double the area of the core, you double the inductance. On the other hand, doubling the length of the core cuts the inductance in half.

Permeability

The core material can affect the flux; some materials result in more flux than others. The quantity μ, known as the *permeability,* indicates the effect of the core material. For instance, the permeability of a vacuum has been measured; in the metric system,

$$\mu = 1.26(10^{-6}) \qquad \text{(vacuum)}$$

Most materials have negligible effect on flux; they have a permeability of $1.26(10^{-6})$, the same as a vacuum. We refer to all materials with this value of permeability as *nonmagnetic.*

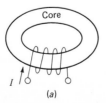

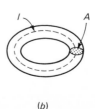

Figure 10-5. Toroid. (a) *Windings on core.* (b) *Length and cross-sectional area.*

(a)

(b)

A few materials are *diamagnetic;* their permeability is slightly less than $1.26(10^{-6})$. These materials slightly oppose the passage of flux. Because of this, a toroid with a diamagnetic core has slightly less inductance than one with a nonmagnetic core.

Some materials are *paramagnetic;* their permeability is slightly greater than a vacuum. A toroid with a paramagnetic core has slightly more inductance than it has with a nonmagnetic core.

A few materials are *ferromagnetic;* they have permeabilities much greater than a vacuum. Because of this, a toroid with a ferromagnetic core has much more inductance than one with a nonmagnetic core. Examples of ferromagnetic materials are iron, nickel, steel, and cobalt.

The solenoid

Figure 10-6a shows a *solenoid,* a cylindrically shaped inductor. Length l is the distance from one end to the other, and area A is the cross-sectional area. As an approximation, the inductance of a solenoid is given by

$$L \cong \mu \, \frac{N^2 A}{l} \qquad (10\text{-}3)$$

We use this equation as an approximation for all solenoids discussed in this book. Exact solenoid formulas are very complicated; you find them in engineering handbooks.

EXAMPLE 10-3.
Suppose a toroid has a nonmagnetic core. If $N = 100$, $A = 10^{-6}$ m², and $l = 0.01$ m, what is the inductance?

SOLUTION.
The permeability of a nonmagnetic material is $1.26(10^{-6})$, the same as a vacuum. With Eq. (10-2),

$$L = \mu \, \frac{N^2 A}{l} = 1.26(10^{-6}) \, \frac{100^2 (10^{-6})}{0.01} = 1.26 \ \mu\text{H}$$

EXAMPLE 10-4.
A solenoid has 150 turns, a length of 0.025 m, and an area of $25(10^{-6})$ m². What is its approximate inductance with a nonmagnetic core?

Figure 10-6. Solenoid.

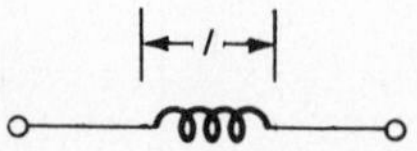

SOLUTION.

Use Eq. (10-3) as an approximation to get

$$L \cong 1.26(10^{-6}) \frac{150^2 \times 25(10^{-6})}{0.025} = 28.4 \ \mu H$$

Test 10-3

1. If the permeability of a toroidal core doubles, the inductance will
   (*a*) stay the same   (*b*) double   (*c*) quadruple   (*d*) drop by half ...... ( )
2. If the number of turns of a toroid doubles, the inductance will
   (*a*) stay the same   (*b*) double   (*c*) quadruple   (*d*) drop by half ...... ( )
3. If the area of a toroidal core doubles, the inductance will
   (*a*) stay the same   (*b*) double   (*c*) quadruple   (*d*) drop by half ...... ( )
4. If the length of a toroid doubles, the inductance will
   (*a*) stay the same   (*b*) double   (*c*) quadruple   (*d*) drop by half ...... ( )
5. Inductance is to capacitance as permeability is to
   (*a*) ferromagnetic   (*b*) farad   (*c*) permittivity   (*d*) resistivity .... ( )
6. Which of these belongs least?
   (*a*) $I = V/R$ (*b*) $R = \rho l/A$ (*c*) $C = \epsilon A/d$ (*d*) $L = \mu N^2 A/l$ .... ( )
7. Inductance is directly proportional to the
   (*a*) permeability   (*b*) square of the turns   (*c*) area of the core
   (*d*) all of these .... ( )

10-4.   CORE MATERIAL

You need to know more about the core material. In particular, after finishing this section you must know the answers to

> *Why does the core affect inductance?*
> *What is relative permeability?*
> *Why are ferrites important?*

Effect of core

Figure 10-7*a* shows an atom. The orbiting electron represents a small clockwise electron flow around the nucleus. The magnetic effects are the same as those produced by counterclockwise conventional current around the nucleus; this means the atom has a magnetic flux surrounding it. Furthermore, applying the right-hand rule to Fig. 10-7*a*, the upper end is the N pole and the lower end the S pole. Therefore, we can visualize an atom as a small bar magnet like Fig. 10-7*b*.

Ferromagnetic atoms have an unusual property: their magnetic poles can shift in the presence of another magnetic field. Figure 10-7*c* illustrates the idea. When a current

Figure 10-7. (a) Magnetic field of atom. (b) Equivalent bar magnet. (c) Alignment of core atoms.

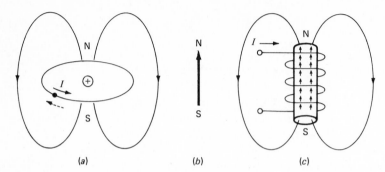

(a)   (b)   (c)

passes through the inductor, we get flux through the core. In addition, each core atom, acting like a small bar magnet, lines up with the flux as shown. Because each core atom has a small flux of its own, the aligned core atoms *increase* the total flux. Since inductance is proportional to flux, we get more inductance with a ferromagnetic core.

Nonmagnetic core atoms are different. Even though a strong magnetic field passes through them, their electron orbits cannot shift and add to the total flux. As a result, nonmagnetic cores have no effect on inductance.

Relative permeability

Relative permeability indicates how easily core atoms line up with a magnetic field. Nonmagnetic materials have low relative permeability, but ferromagnetic materials have high relative permeability. The defining formula for relative permeability is

$$\mu_r = \frac{\mu}{\mu_0} \tag{10-4}$$

where μ_r = relative permeability
μ = permeability of material
μ_0 = permeability of a vacuum

Relative permeability compares the permeability of any material to that of a vacuum (analogous to dielectric constant).

Table 10-1 shows relative permeabilities for some materials. The values are approximate, because exact values depend on the grade of the material, its heat treatment during manufacture, the amount of flux in the material, and other factors. As shown, air has a relative permeability very close to a vacuum; for all preliminary analysis, air is treated as nonmagnetic. Likewise, aluminum and most other conductors act like nonmagnetic materials.

TABLE 10-1. RELATIVE PERMEABILITY

MATERIAL	μ_r
Vacuum	1
Air	1.0000004
Aluminum	1.00000065
Nickel	50
Cobalt	60
Steel	300
Iron	5000
Ferrites	100 to 2500
Permalloy	25,000 to 100,000
Supermalloy	100,000 to 1,000,000

The best known magnetic materials are nickel, iron, steel, and cobalt. These materials have relative permeabilities from approximately 50 to 5000. By mixing materials, we can get alloys with even higher permeabilities.

Ferrites

Ferrites are examples of alloys with high permeabilities. A ferrite is a ceramic material made by combining Fe_2O_4 (iron and oxygen) with another element. For instance, $MnFe_2O_4$ is a combination of manganese and Fe_2O_4. Another widely used ferrite is $ZiFe_2O_4$, a combination of zinc and Fe_2O_4.

Ferrites are important in electronics. First, being ceramics, they have a wide variety of magnetic properties. Second, ferrites have very high resistivities (Sec. 2-5); this leads to extremely small *core losses* (unwanted power losses in the core). For these and other reasons, small ferrite cores are often used to store instructions and numbers inside digital computers.

EXAMPLE 10-5.
A toroid with a nonmagnetic core has an inductance of 100 μH. If the core is changed to a ferrite core with a relative permeability of 200, what is the new value of inductance.

SOLUTION.
Equation (10-2) says inductance is directly proportional to permeability. This means

$$L = \mu_r L_0 \qquad (10\text{-}5)$$

where L_0 is the inductance with a nonmagnetic core. Substituting the given values,

$$L = 200 \times 100 \ \mu H = 20 \ mH$$

Test 10-4

1. Which of these cores cannot have atoms that line up with a magnetic field?
   (*a*) air   (*b*) iron   (*c*) aluminum   (*d*) vacuum .... ( )
2. When a ferromagnetic atom is in the presence of a strong magnetic field, its own magnetic field
   (*a*) opposes the strong field   (*b*) adds to the total flux
   (*c*) has no effect on the flux   (*d*) subtracts from the flux .... ( )
3. Relative permeability is not
   (*a*) a ratio   (*b*) measured in henrys   (*c*) equal to μ/μ_0
   (*d*) related to core atoms lining up with a magnetic field .... ( )
4. The higher the relative permeability, the
   (*a*) greater the current in the inductor
   (*b*) harder it is for core atoms to align with the magnetic field
   (*c*) larger the number of turns   (*d*) greater the resulting flux .... ( )
5. Core material is to dielectric as relative permeability is to
   (*a*) dielectric constant   (*b*) relative permeance
   (*c*) relative resistivity   (*d*) dielectric strength .... ( )
6. Ferrites are
   (*a*) paramagnetic   (*b*) ferromagnetic   (*c*) diamagnetic
   (*d*) nonmagnetic .... ( )
7. Air capacitor is to ceramic capacitor as air-core inductor is to
   (*a*) air-core coil   (*b*) ferrite-core inductor
   (*c*) ferromagnetic inductor   (*d*) paramagnetic inductor .... ( )
8. Ferrites are important in electronics because of their
   (*a*) wide variety of magnetic properties   (*b*) use in computers
   (*c*) small core losses   (*d*) all of these .... ( )

10-5. NONLINEARITY AND RETENTIVITY

The alignment of ferromagnetic core atoms produces special effects. In what follows, look for the answers to

> *Why is the flux-current graph nonlinear?*
> *What is retentivity?*
> *What is hysteresis?*

Nonlinearity

The μ_r of a ferromagnetic core decreases when the flux increases. Why? In Fig. 10-8*a*, the flux is clockwise, evident from the right-hand rule. Core atoms will align themselves so as to add to the total flux. As we increase the current, more and more core atoms become aligned. Eventually, all core atoms are aligned. When this happens,

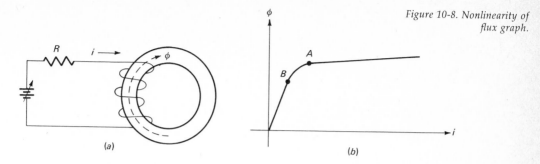

Figure 10-8. Nonlinearity of flux graph.

(a)

(b)

further increases in current produce only small increases in flux, because no more core atoms can be aligned. This is equivalent to saying the permeability of the core decreases at higher current levels.

Figure 10-8b summarizes the idea. Up to point B on the curve, increases in current produce large increases in flux, because additional core atoms can line up with the field. Further up the curve, however, the rate of increase in flux drops off. Beyond point A, all core atoms are aligned, and flux increases only slightly with continued increase in current.

The *saturation region* is the branch of the curve beyond point A where all core atoms have lined up. In the saturation region, the relative permeability is the same as in a nonmagnetic material, because all core atoms are already aligned.

Retentivity

The core atoms of some materials stay aligned after you remove the *magnetizing field* (the field that originally aligned the atoms). Because of this, a magnetized core can act like a permanent magnet. In fact, this is the reason permanent magnets exist in Nature; long ago, they were exposed to a strong magnetizing field; even though the original field has disappeared, the atoms in permanent magnets have remained in alignment.

Retentivity is the ability of core atoms to remain aligned after the magnetizing field is removed. *Hard iron* is an example of a material with high retentivity; permanent magnets often are made of hard iron. On the other hand, *soft iron* is an example of a material with low retentivity; remove the magnetizing field, and almost none of the core atoms stay aligned. Soft-iron magnets are sometimes called *temporary magnets*.

Figure 10-9a shows the graph of flux versus current for a soft-iron core. When current is increased, the core atoms line up and flux increases. When current is decreased, core atoms do not stay aligned and flux decreases. Similarly, if current is reversed, the flux reverses and we get the graph left of the origin. Especially important, since the core atoms do not remain in alignment after the magnetizing field is removed, the curve is *single-valued*; there's only one value of flux for each value of current.

Hysteresis

The graph for a high-retentivity core is different; it's *double-valued* (two flux values for some currents), as shown in Fig. 10-9b. The operating point on a graph like this depends on where you start. For instance, suppose there's 1 A through the inductor. Then, the operating point is at *A*. When current decreases to zero, the operating point follows the path from *A* to *B*. At point *B*, there's still flux in the core, because the core atoms have remained in alignment.

To demagnetize the core, we have to reverse the current. When current increases in the opposite direction, the operating point follows the path from *B* to *C*. The flux is zero at point *C*, because the negative current is sufficient to force the core atoms out of their original alignment.

Further increase in negative current forces the core atoms to align in the opposite direction. When the current equals −1 A, the operating point is at *D*; in this case, all core atoms are aligned in the opposite direction. If we reduce the current to zero, the operating point moves to *E*; the flux is negative, meaning the core is magnetized in the opposite direction.

Hysteresis is another name for double-valued. A curve like Fig. 10-9b is called a *hysteresis curve* because it is double-valued. You can immediately recognize a hysteresis curve because it encloses an area. In other words, loop *A-B-C-D-E-F-A* forms the outside boundary of an area centered on the origin. On the other hand, a single-valued curve like Fig. 10-9a is not a hysteresis curve, because it encloses no area.

Incidentally, the ferrite cores in digital computers have a square hysteresis loop like Fig. 10-9b. Each core is magnetized in one direction or the other (point *B* or *E*). A computer may have from 50,000 to more than 500,000 ferrite cores in its *memory*. By a

Figure 10-9. (a) Single-valued nonlinear graph. (b) Double-valued graph. (c) Linear graph.

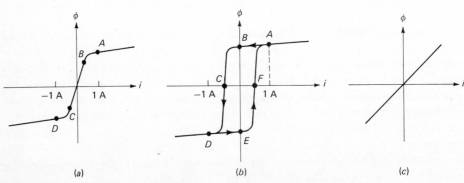

prearranged code, these magnetized cores store the binary numbers and instructions needed for computer operation.[2]

Nonmagnetic core

A final point. When a core is nonmagnetic, the relative permeability is unity and the retentivity is zero. This means core atoms do not add their fluxes to the magnetizing flux. Because of this, the permeability is constant, as in Fig. 10-9c. As shown, the curve is single-valued (no hysteresis) and linear.

Test 10-5

1. When a ferromagnetic core is used, the flux-current graph is nonlinear because
   (a) all core atoms eventually become aligned
   (b) it's always a double-valued curve
   (c) it's never a single-valued curve   (d) permeability remains constant ... ( )
2. A permanent magnet does not have
   (a) high retentivity   (b) high permeability   (c) hysteresis
   (d) soft iron .... ( )
3. High retentivity produces
   (a) a double-valued curve   (b) hysteresis
   (c) core atoms that stay aligned   (d) all of these .... ( )
4. Cause is to effect as retentivity is to
   (a) hysteresis   (b) permeability   (c) temporary magnet
   (d) ease of aligning core atoms .... ( )
5. Which of the following does not belong?
   (a) air core   (b) soft iron   (c) hard iron   (d) nonlinear .... ( )

10-6. COEFFICIENT OF COUPLING

When the magnetic field of an inductor passes through another inductor, unusual things may happen. This section discusses the interaction between two inductors. Your job is to learn the answers to

How is the coefficient of coupling defined?
What is loose coupling?
What is tight coupling?

[2] For a discussion of how these cores store numbers and instructions, see Malvino, A. P., and D. P. Leach: *Digital Principles and Applications,* McGraw-Hill Book Company, New York, 1969, Chap. 12.

Coefficient of coupling

Figure 10-10a shows a current I_1 in the left inductor. This creates a flux through the inductor designated as ϕ_1. Some of this flux passes through the second inductor; we designate this flux as ϕ_2.

The *coefficient of coupling* is defined as the ratio of the second flux to the first flux. In symbols, the defining formula is

$$k = \frac{\phi_2}{\phi_1} \qquad\qquad (10\text{-}6)$$

where k = coefficient of coupling
ϕ_1 = flux produced by first inductor
ϕ_2 = flux through second inductor

If ϕ_1 is 4 Wb and ϕ_2 is 0.1 Wb, the coefficient of coupling equals

$$k = \frac{0.1 \text{ Wb}}{4 \text{ Wb}} = 0.025$$

equivalent to 2.5 percent.

Tight coupling

When a ferromagnetic core is used, most of the flux is inside the core. Because of this, most of the flux produced by the first inductor of Fig. 10-10b passes through the second inductor. In other words, ϕ_2 almost equals ϕ_1. Typically, a ferromagnetic core has a coefficient of coupling in the interval

$$0.9 < k < 1 \qquad \text{(ferromagnetic core)}$$

Tight coupling means two inductors have a high value of k; most of the flux produced by one inductor passes through the other inductor. As a rule, two inductors on a ferromagnetic core are tightly coupled.

Figure 10-10. Coefficient of coupling.

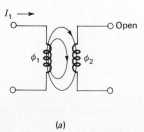

(a)

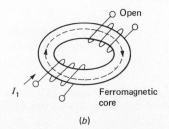

(b)

Loose coupling

If the core is made of plastic or some other nonmagnetic material, all it does is provide support for the windings. Because the core has no effect on the magnetic field, the flux is no longer confined to the core. In this case, the coefficient of coupling is low, typically in the interval

$$0 < k < 0.1 \qquad \text{(nonmagnetic core)}$$

This says less than 10 percent of the flux produced by the first inductor passes through the second when the core is nonmagnetic.

Loose coupling means the two inductors have a low value of k; only a fraction of the flux produced by one inductor passes through the other.

Test 10-6

1. What does $k = \phi_2/\phi_1$ have in common with $L = N\phi/I$?
   (*a*) they apply to any load   (*b*) you can calculate current with them
   (*c*) they're defining formulas   (*d*) they tell you what inductance is .... ( )
2. When every line of force produced by one inductor passes through the other inductor,
   (*a*) the coupling is loose   (*b*) k equals unity
   (*c*) the core is nonmagnetic   (*d*) a toroid is used for the core .... ( )
3. The following words can be rearranged into a sentence: UNITY k CANNOT THAN GREATER BE. The sentence is
   (*a*) true   (*b*) false .... ( )
4. High k is to low k as tight coupling is to
   (*a*) loose coupling   (*b*) intermediate coupling   (*c*) tight coupling
   (*d*) maximally flat coupling .... ( )
5. Which of the following belongs least?
   (*a*) loose coupling   (*b*) k approaching unity   (*c*) ϕ_2 almost equal to ϕ_1
   (*d*) ferromagnetic core .... ( )

10-7. INDUCTANCE OF TWO INDUCTORS

The inductance of a single inductor is defined as $N\phi/I$, the ratio of its self-produced flux linkage to its current. When two inductors are in series or in parallel, things get complicated because the flux of one inductor may pass through the other inductor. To calculate the equivalent inductance under these conditions, you have to learn the answers to

What is mutual inductance?
How do series inductances combine?
How do parallel inductances combine?

Figure 10-11. Mutual inductance.

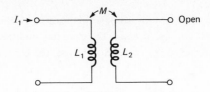

Mutual inductance

Given two inductors like Fig. 10-11, *mutual inductance* is defined as the flux linkage in the second inductor divided by the current in the first inductor. As a defining formula,

$$M = \frac{N_2 \phi_2}{I_1} \tag{10-7}$$

where M = mutual inductance

N_2 = turns on the second inductor
ϕ_2 = flux coupled into the second inductor
I_1 = current in the first inductor

If $N_2 = 100$, $\phi_2 = 0.2$ Wb, and $I_1 = 5$ A, then by definition the mutual inductance equals

$$M = \frac{N_2 \phi_2}{I_1} = \frac{100 \times 0.2}{5} = 4 \text{ H}$$

In Eq. (10-7), M is directly proportional to ϕ_2. In turn, ϕ_2 is related to the coefficient of coupling k. By an advanced derivation,

$$M = k \sqrt{L_1 L_2} \tag{10-8}$$

So if two inductors have a k of 0.95, and L_1 is 100 mH and L_2 is 400 mH, the mutual inductance equals

$$M = 0.95 \sqrt{0.1 \times 0.4} = 190 \text{ mH}$$

Series inductors

When two inductors are in series (Fig. 10-12*a*), some of the flux produced by each inductor passes through the other. Because of this, the *equivalent inductance* depends on each separate inductance and on the mutual inductance. The derivation of equivalent inductance is too complicated to go into here, but it can be shown that equivalent inductance L is given by

$$L = L_1 + L_2 \pm 2M \tag{10-9}$$

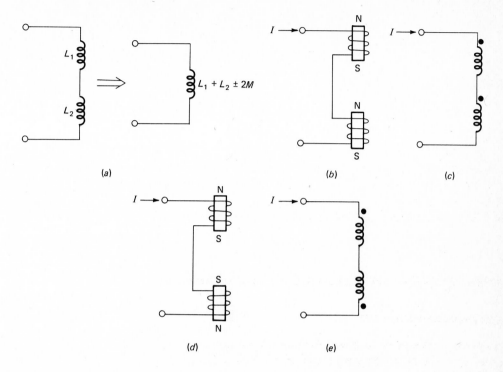

Figure 10-12. Inductors in series.

(a)   (b)   (c)

(d)   (e)

The final term may be plus or minus, depending on whether the fluxes in the inductors aid or oppose.

Figure 10-12b shows the case of aiding fluxes. Apply the right-hand rule to see that the flux through each inductor is upward. In other words, the upper end of each inductor is a N pole. On a schematic diagram, we indicate like poles by using dots, as shown in Fig. 10-12c. So if $L_1 = 100$ mH, $L_2 = 400$ mH, and $M = 190$ mH in Fig. 10-12c, Eq. (10-9) gives

$$L = L_1 + L_2 + 2M$$
$$= 100 \text{ mH} + 400 \text{ mH} + 2(190 \text{ mH}) = 880 \text{ mH}$$

Figure 10-12d illustrates the other case, opposing fluxes. The right-hand rule shows the fluxes point in opposite directions. Again, dots are used on a schematic diagram, as shown in Fig. 10-12e, to indicate similar poles. With the same values given in the preceding calculation, opposing inductors have an equivalent inductance of

$$L = L_1 + L_2 - 2M$$
$$= 100 \text{ mH} + 400 \text{ mH} - 2(190 \text{ mH}) = 120 \text{ mH}$$

Often, the series inductors are far apart and the coefficient of coupling is approximately zero. In this case, the equivalent inductance equals

$$L = L_1 + L_2 \qquad (10\text{-}9a)$$

This says you add the inductances of series inductors when there's negligible coupling between them. If the two inductors of the previous calculations are moved far enough apart, k drops to zero and the equivalent inductance becomes

$$L = L_1 + L_2$$
$$= 100 \text{ mH} + 400 \text{ mH} = 500 \text{ mH}$$

Parallel inductors

Inductors in parallel have an equivalent inductance of

$$L = \frac{L_1 L_2 - M^2}{L_1 + L_2 \pm 2M} \qquad (10\text{-}10a)$$

The $2M$ term in the denominator may be plus or minus, depending on whether the fluxes aid or oppose.

When the two inductors are far enough apart, k drops to zero and the equivalent inductance equals

$$L = \frac{L_1 L_2}{L_1 + L_2} \qquad (10\text{-}10b)$$

If $L_1 = 100 \text{ mH}$ and $L_2 = 400 \text{ mH}$, two parallel inductors with no mutual inductance have an equivalent inductance of

$$L = \frac{0.1 \times 0.4}{0.1 + 0.4} = 0.08 \text{ H} = 80 \text{ mH}$$

Test 10-7

1. $M = N_2 \phi_2 / I_1$ is
   (a) a definition   (b) based on experiment   (c) derived mathematically
   (d) a discovery .... ( )
2. Inductance is to one inductor as mutual inductance is to
   (a) one inductor   (b) two inductors   (c) flux   (d) turns .... ( )
3. When k equals zero, two inductors in series combine by
   (a) addition   (b) subtraction   (c) product-over-sum
   (d) multiplication .... ( )
4. Two parallel inductors with no mutual inductance combine by
   (a) addition   (b) subtraction   (c) product-over-sum
   (d) multiplication .... ( )

10-8. ELECTROMAGNETIC INDUCTION

Oersted's discovery showed moving charges produce a magnetic field. If the current is constant, the flux is constant; if the current is time-varying, the flux is time-varying.

Faraday discovered something even more exciting. He found a time-varying flux creates a voltage across an inductor. The faster the flux changes, the greater the voltage. Because electrical and magnetic quantities are involved, this phenomenon is called *electromagnetic induction*. To understand more about it, learn the answers to

> *What is Faraday's law?*
> *What does rate of flux change mean?*

Faraday's law

By experiment and observation, Faraday reached this conclusion: a changing flux through an inductor creates or *induces* a voltage across the inductor. The faster the flux changes, the greater the induced voltage. If the flux is constant in time, no voltage is induced.

The flux may be produced by the inductor itself, or by another inductor. For instance, Fig. 10-13a shows a time-varying current source driving an inductor. The changing current produces a time-varying flux in the inductor. In this case, a voltage v_1 is induced across the inductor.

Figure 10-13b shows a second way to get induced voltage. Here a time-varying current induces a voltage v_1 across the first inductor as before. In addition, some of the time-varying flux produced by the first inductor passes through the second inductor and creates a voltage v_2 across the inductor.

Faraday's law of electromagnetic induction is summarized by this experimental formula:

$$v = N\frac{d\phi}{dt} \tag{10-11}$$

Figure 10-13. Faraday's law. (a) Self-induced voltage. (b) Coupled voltage.

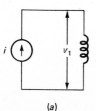

(a)

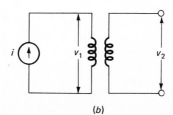

(b)

where $v =$ induced voltage
$N =$ number of turns
$d\phi/dt =$ rate of flux change

Rate of flux change

Here is a simplified explanation of what $d\phi/dt$ means. Figure 10-14a shows two points on a flux waveform. When the points are very close together, the special notation $d\phi$ is used to represent the flux change between the points, and dt is used for the time change between the points. The *rate of flux change* is defined as the flux change divided by the time change. As a defining formula,

$$\text{Rate of flux change} = \frac{d\phi}{dt} \tag{10-12}$$

Here's an example. The flux change between the first pair of points in Fig. 10-14b is

$$d\phi = 0.02 \text{ Wb}$$

and the time change is

$$dt = 1 \text{ ms}$$

Therefore, the rate of flux change is

$$\frac{d\phi}{dt} = \frac{0.02 \text{ Wb}}{0.001 \text{ s}} = 20 \text{ Wb/s}$$

As another example, the flux change between the second pair of points in Fig. 10-14b is

$$d\phi = 0.001 \text{ Wb}$$

and the time change is

$$dt = 1 \text{ ms}$$

So the rate of flux change is

$$\frac{d\phi}{dt} = \frac{0.001 \text{ Wb}}{0.001 \text{ s}} = 1 \text{ Wb/s}$$

The rate of flux change tells how fast the flux is changing. In the first of the preceding examples, $d\phi/dt$ equals 20 Wb/s, but in the second example it's only 1 Wb/s. The rate of flux change is much greater in the first case than in the second.

The rate of flux change is related to the *slope* or steepness of the waveform. The slope between the points of Fig. 10-14c is very steep, so $d\phi/dt$ has a large value. On the

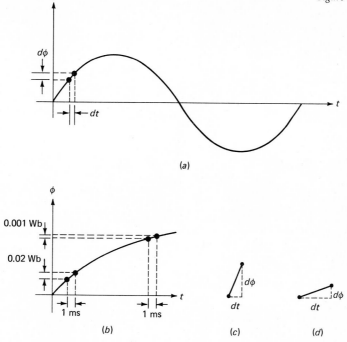

Figure 10-14. Rate of flux change.

other hand, the slope between the points of Fig. 10-14d is gentle, and $d\phi/dt$ has a small value. To put it another way, $d\phi/dt$ is large when the waveform rises quickly, and small when the waveform rises slowly.

For a sine wave like Fig. 10-14a, $d\phi/dt$ is maximum at the start of the wave (it is steepest here). Approaching the peak of the sine wave, the flux changes more slowly and $d\phi/dt$ is small. Right at the peak, the flux is momentarily unchanging and $d\phi/dt$ is zero. Beyond the peak, the flux decreases and $d\phi/dt$ has a negative value.

Induced voltage

Faraday's law says the voltage induced across an inductor equals

$$v = N \frac{d\phi}{dt}$$

This means induced voltage is directly proportional to N and to $d\phi/dt$. Doubling the number of turns doubles the induced voltage. Similarly, if $d\phi/dt$ doubles, the induced voltage doubles.

Test 10-8

1. The induced voltage across an inductor increases when
    (a) the number of turns decreases   (b) the flux is maximum
    (c) the flux is zero   (d) the rate of flux change increases .... (   )
2. If you double the number of turns, the same value of $d\phi/dt$ results in
    (a) the same voltage   (b) half as much voltage
    (c) twice as much voltage   (d) zero voltage .... (   )
3. The rate of change $d\phi/dt$ most closely means
    (a) how much flux there is   (b) how fast the flux is changing
    (c) the slope between two distant points on the waveform
    (d) horizontal change divided by vertical change .... (   )
4. If $N = 200$ and $d\phi/dt = 0.4$ Wb/s, the induced voltage is
    (a) 50 V   (b) 80 V   (c) 200 V   (d) 500 V .... (   )
5. Two inductors are wound on a toroidal ferromagnetic core. The first inductor has 200 turns and a $d\phi/dt$ of 0.4 Wb/s. If the second inductor has 125 turns and the coefficient of coupling is unity, how much voltage is induced in the second inductor?
    (a) 50 V   (b) 80 V   (c) 200 V   (d) 500 V .... (   )

10-9.   THE IDEAL TRANSFORMER

A *transformer* is two or more inductors whose coupling coefficient is greater than zero. The key questions in this section are

> *What is an ideal transformer?*
> *How are voltages and currents related?*
> *How are resistances related?*

What it is

An *ideal transformer* is a transformer with

1. Zero resistance in the windings
2. Zero core losses
3. Unity coefficient of coupling
4. Relative permeability of infinity

Manufacturers can't build an ideal transformer, but they can come close. They can choose wire sizes to get low winding resistance, produce cores with small core losses, optimize coupling to get k's near unity, and select core materials with high relative permeabilities.

   Because a manufacturer can approach the ideal transformer by optimizing wire size, core material, and other factors, most people treat the ferromagnetic-core transformer as an ideal transformer in preliminary analysis.

Ideal-transformer voltages

Figure 10-15a shows a transformer between a source and a load. On schematic diagrams the vertical lines between the windings stand for a ferromagnetic core. The dotted ends indicate the same polarity of voltage; if any dotted end is positive, all dotted ends are positive. The *primary winding* is the inductor on the source side, and the *secondary winding* is the inductor on the load side.

With an ideal transformer the only voltages across the windings are the induced voltages given by Faraday's law. In other words, a time-varying source results in a primary voltage of

$$v_1 = N_1 \frac{d\phi_1}{dt}$$

and a secondary voltage of

$$v_2 = N_2 \frac{d\phi_2}{dt}$$

These equations tell us the voltage across each winding equals the number of turns times the rate of flux change through the winding.

In an ideal transformer, the coefficient of coupling is unity, which means $\phi_1 = \phi_2$. Because of this, we can rewrite the equations as

$$v_1 = N_1 \frac{d\phi}{dt}$$

$$v_2 = N_2 \frac{d\phi}{dt}$$

Taking the ratio of these equations and rearranging gives

$$\frac{v_2}{v_1} = \frac{N_2}{N_1} \qquad \text{(ideal)} \tag{10-13}$$

This derived formula says the voltage ratio equals the turns ratio.

As an example, suppose a sine wave with a peak of 2 V drives the primary winding (Fig. 10-15b). Rather than give the actual turns in each winding, a schematic usually shows the ratio of turns. In Fig. 10-15b,

$$\frac{N_1}{N_2} = \frac{1}{10}$$

equivalent to

$$\frac{N_2}{N_1} = 10$$

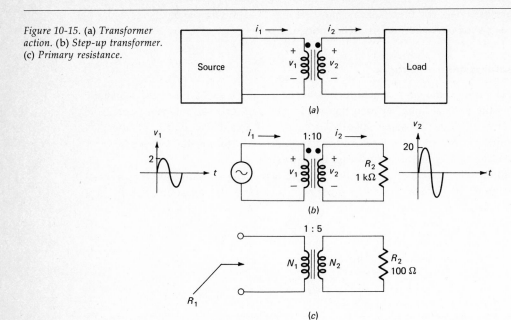

Figure 10-15. (a) Transformer action. (b) Step-up transformer. (c) Primary resistance.

With Eq. (10-13), the peak secondary voltage equals

$$v_2 = \frac{N_2}{N_1} v_1 = 10(2 \text{ V}) = 20 \text{ V}$$

Because Eq. (10-13) applies at each instant in time, the secondary voltage is proportional to the primary voltage. This is why a sine wave of primary voltage results in a sine wave of secondary voltage as shown in Fig. 10-15b.

In general, the similarity theorem applies to an ideal transformer; that is, the primary and secondary voltages have the same shape. Put a ramp into an ideal transformer and out comes a ramp; put in a triangular wave and out comes a triangular wave, etc.

When the turns ratio N_2/N_1 is greater than unity, the secondary voltage is greater than the primary voltage. A transformer with N_2/N_1 greater than unity is called a *step-up transformer*, because it steps up the voltage. If N_2/N_1 is less than unity, you call it a *step-down transformer*, because it reduces the voltage from input to output.

Ideal-transformer currents

There's a simple relation between currents in an ideal transformer. In Fig. 10-15a, the load power at any instant is

$$p_2 = v_2 i_2$$

The input power to the transformer is

$$p_1 = v_1 i_1$$

None of this input power is wasted in winding losses or core losses, because an ideal transformer has zero winding resistance and zero core losses. Therefore, in an ideal transformer all the input power is passed on to the output. In symbols,

$$p_1 = p_2$$

or
$$v_1 i_1 = v_2 i_2$$

which rearranges into

$$\frac{i_1}{i_2} = \frac{v_2}{v_1}$$

With Eq. (10-13), this becomes

$$\frac{i_1}{i_2} = \frac{N_2}{N_1} \qquad \text{(ideal)} \qquad (10\text{-}14)$$

This says the current ratio is the inverse of the turns ratio.

In Fig. 10-15b, Ohm's law gives the peak load current:

$$i_2 = \frac{v_2}{R_2} = \frac{20}{1000} = 20 \text{ mA}$$

With Eq. (10-14), the peak primary current is

$$i_1 = \frac{N_2}{N_1} i_2 = 10(20 \text{ mA}) = 200 \text{ mA}$$

So with a step-up transformer like Fig. 10-15b, the primary current is always larger than the secondary current.

Relation of resistances

Equation (10-13) may be inverted to get

$$\frac{v_1}{v_2} = \frac{N_1}{N_2}$$

Similarly, Eq. (10-14) may be inverted to get

$$\frac{i_2}{i_1} = \frac{N_1}{N_2}$$

The product of these two equations is

$$\frac{v_1 i_2}{v_2 i_1} = \left(\frac{N_1}{N_2}\right)^2$$

which rearranges into

$$\frac{v_1/i_1}{v_2/i_2} = \frac{R_1}{R_2} = \left(\frac{N_1}{N_2}\right)^2$$

or

$$R_1 = \left(\frac{N_1}{N_2}\right)^2 R_2 \tag{10-15}$$

Here is what it means. In Fig. 10-15c, a resistance R_2 is the load on the secondary winding. Since the transformer steps voltage up and current down, the resistance R_1 that loads a source connected to the primary winding will be different from R_2. Equation (10-15) gives the value of R_1. With the values shown in Fig. 10-15c,

$$R_1 = \left(\frac{N_1}{N_2}\right)^2 R_2 = \left(\frac{1}{5}\right)^2 100 = 4 \ \Omega$$

In other words, when a source is connected across the primary winding of Fig. 10-15c, the equivalent resistance that loads the source is 4 Ω, not 100 Ω.

Real transformers

In practice, real transformers are classified as *iron-core* (meaning ferromagnetic core) and *air-core* (which includes all nonmagnetic cores). For preliminary analysis, iron-core transformers are idealized; that is, people use Eqs. (10-13) through (10-15). In exact analysis, the relations are much more complicated; exact analysis is for specialists in transformer design.[1]

The air-core transformer cannot be idealized because its coefficient of coupling is much less than unity. Chapter 16 discusses air-core transformers further.

EXAMPLE 10-6.

The voltage from an ac outlet in a typical home is a sine wave with a peak of approximately 160 V and a frequency of 60 Hz. This voltage is known as *line voltage*. Figure 10-16a shows line voltage driving a 5-to-1 step-down transformer. What is the secondary voltage? The current in each winding? The primary resistance?

SOLUTION.

The iron-core transformer of Fig. 10-16a has a turns ratio of

$$\frac{N_2}{N_1} = \frac{1}{5}$$

Therefore, the peak secondary voltage is ideally

$$v_2 = \frac{N_2}{N_1} v_1 = \frac{1}{5} 160 = 32 \ V$$

---

[1] To see what exact analysis is like, you can start with *Reference Data for Radio Engineers,* Howard W. Sams & Company, Inc., New York, 1968, Chap. 12.

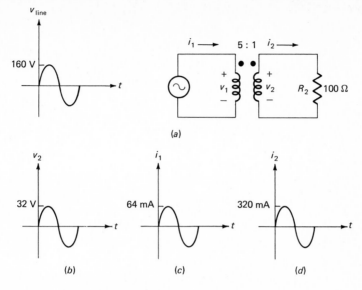

Figure 10-16. Example of transformer action.

Since the secondary voltage is one-fifth of the primary voltage at each instant, the secondary voltage is a sine wave with a peak of 32 V (see Fig. 10-16b). The dots of Fig. 10-16a tell us the secondary voltage is positive when the primary voltage is positive, and negative when the primary voltage is negative. The polarities shown are for the positive half cycle of input voltage.

At the positive peak of secondary voltage, the load current is

$$i_2 = \frac{v_2}{R_2} = \frac{32}{100} = 320 \text{ mA}$$

Equation (10-14) gives a primary peak current of

$$i_1 = \frac{N_2}{N_1} i_2 = \frac{320 \text{ mA}}{5} = 64 \text{ mA}$$

Figures 10-16c and d illustrate these current waveforms.

The primary resistance equals

$$R_1 = \left(\frac{N_1}{N_2}\right)^2 R_2 = (5)^2 \, 100 = 2500 \; \Omega$$

An alternative way to calculate this resistance is by taking the ratio of the primary voltage to the primary current. The peak primary voltage is 160 V and the peak primary current is 64 mA; therefore,

$$R_1 = \frac{v_1}{i_1} = \frac{160}{0.064} = 2500 \; \Omega$$

Test 10-9

1. Which belongs least?
   (a) air-core transformer   (b) ideal transformer
   (c) iron-core transformer   (d) ferromagnetic-core transformer .... ( )
2. The winding resistance and the core losses of an ideal transformer are
   (a) zero   (b) low   (c) high   (d) infinite .... ( )
3. In an ideal transformer, the coefficient of coupling is
   (a) zero   (b) low   (c) unity   (d) infinite .... ( )
4. Carbon-composition resistor is to perfect resistor as ferrite-core transformer is to
   (a) iron-core transformer   (b) air-core transformer
   (c) ideal transformer   (d) real transformer .... ( )
5. If the turns ratio N_2/N_1 of an ideal transformer is 3, the primary voltage is
   (a) larger than the secondary voltage   (b) three times the secondary voltage
   (c) nine times the secondary voltage
   (d) one-third of the secondary voltage .... ( )
6. In a step-up transformer, the secondary current is
   (a) smaller than the primary current   (b) larger than the primary current
   (c) equal to the primary current   (d) none of these .... ( )

10-10. LEAD INDUCTANCE

As a final point, don't think inductance exists only with coils of wire. Any conductor with current has a magnetic field around it. Therefore, even leads or connecting wires have a small amount of inductance. This unwanted inductance is called *lead inductance.*

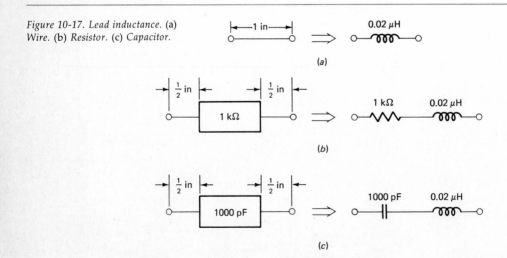

Figure 10-17. Lead inductance. (a) Wire. (b) Resistor. (c) Capacitor.

Lead inductance depends on wire diameter and length; AWG-18 wire, for instance, has an inductance of 0.018 μH/in; AWG-26 has close to 0.024 μH/in. As an approximation, we can estimate the lead inductance of connecting wires as 0.02 μH/in (Fig. 10-17a). A 3-in piece of straight wire, therefore, has a lead inductance of approximately 0.06 μH, a 12-in piece about 0.24 μH, etc.

The leads coming out of a resistor act like small inductors in series with the resistor. For instance, if we cut the leads to $\frac{1}{2}$ in on each end, a 1-kΩ resistor has an equivalent circuit like Fig. 10-17b. The 0.01-μH lead inductance on each end adds to a total of 0.02 μH, assuming negligible coupling ($k = 0$).

Similarly, the leads on a capacitor represent small inductors. A 1000-pF capacitor with $\frac{1}{2}$-in leads behaves like the equivalent circuit of Fig. 10-17c.

In all preliminary analysis, you can forget about lead inductance, because it's too small to have a significant effect on circuit action. Later, after studying reactance, you will understand why lead inductance causes trouble at very high frequencies.

SUMMARY OF FORMULAS

DEFINING

$$L = \frac{N\phi}{I} \tag{10-1}$$

$$\mu_r = \frac{\mu}{\mu_0} \tag{10-4}$$

$$k = \frac{\phi_2}{\phi_1} \tag{10-6}$$

$$M = \frac{N_2\phi_2}{I_1} \tag{10-7}$$

$$\text{Rate of flux change} = \frac{d\phi}{dt} \tag{10-12}$$

EXPERIMENTAL

$$v = N\frac{d\phi}{dt} \tag{10-11}$$

DERIVED

$$L = \mu\frac{N^2A}{l} \tag{10-2}$$

$$M = k \sqrt{L_1 L_2} \qquad\qquad\qquad (10\text{-}8)$$

$$L = L_1 + L_2 \qquad \text{(series)} \qquad\qquad (10\text{-}9a)$$

$$L = \frac{L_1 L_2}{L_1 + L_2} \qquad \text{(parallel)} \qquad\qquad (10\text{-}10b)$$

$$\frac{v_2}{v_1} = \frac{N_2}{N_1} \qquad \text{(ideal)} \qquad\qquad (10\text{-}13)$$

$$\frac{i_1}{i_2} = \frac{N_2}{N_1} \qquad\qquad\qquad (10\text{-}14)$$

$$R_1 = \left(\frac{N_1}{N_2}\right)^2 R_2 \qquad\qquad\qquad (10\text{-}15)$$

Problems

10-1.   An inductor has 500 turns. If a current of 2 A results in a flux of 0.01 Wb, what is the inductance?

10-2.   A current of 5 mA produces 0.01 Wb of flux through an inductor with eight turns. What does the inductance equal?

10-3.   An inductor with an inductance of 50 mH has a current of 10 mA through it. What is the value of the flux linkage?

10-4.   A toroid has a nonmagnetic core. If $N = 500$, $A = 0.001$ m², and $l = 0.04$ m, what does the inductance equal?

10-5.   A solenoid has 400 turns, a length of 0.01 m, and an area of $40(10^{-6})$ m. If the core is nonmagnetic, what is the approximate value of the inductance?

10-6.   A toroid has an inductance of 1 μH with a nonmagnetic core. If an iron core is used, what is the approximate value of the inductance? If a ferrite core with a permeability of 1000 is used, what is the inductance?

10-7.   An inductor with 50 turns has the flux-current graph shown in Fig. 10-18a. What does the inductance equal when the current is 1 A? When the current is 3 A?

10-8.   Figure 10-18b shows the flux-current graph for an inductor with 100 turns. What is the value of the inductance for a current of 1 A? For a current of 3 A?

10-9.   Half of the flux produced by one inductor passes through another inductor. What is the coefficient of coupling between the inductors? If the first inductor produces 0.048 Wb of flux, how much of this flux passes through the second inductor?

10-10. The coefficient of coupling between a pair of inductors is 0.2. If the flux passing through the second inductor is 0.48 Wb, how much flux does the first inductor produce?

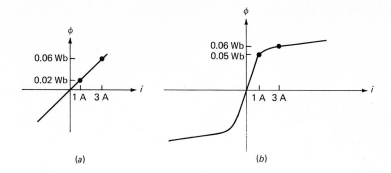

Figure 10-18.

(a)   (b)

10-11.   An inductor with a current of 2 A produces 0.01 Wb of flux. One-fifth of this flux passes through a second inductor with 50 turns. What is the mutual inductance between the inductors? The coefficient of coupling?

10-12.   A current of 5 A flows through an inductor with 250 turns and results in a total flux of 0.004 Wb; 0.001 Wb of the flux passes through a second inductor with 100 turns. What is the coefficient of coupling? The mutual inductance?

10-13.   Two tightly coupled inductors have $k = 0.95$, $L_1 = 40 \ \mu H$, and $L_2 = 100 \ \mu H$. What is the mutual inductance?

10-14.   Two loosely coupled coils have $k = 0.02$, $L_1 = 100 \ \mu H$, and $L_2 = 200 \ \mu H$. What does the mutual inductance equal?

10-15. $L_1 = 200 \ \mu H$, $L_2 = 1$ mH, and $M = 300 \ \mu H$. What does the coefficient of coupling equal?

10-16.   Two inductors are connected in series with values of $L_1 = 50 \ \mu H$, $L_2 = 50 \ \mu H$, and $M = 40 \ \mu H$. What is the equivalent inductance if they are connected as shown in Fig. 10-19a?

10-17.   An L_1 of 40 μH is in series with an L_2 of 100 μH. The coefficient of coupling equals 0.95. What is the minimum equivalent inductance? The maximum equivalent inductance?

10-18.   If the coupling between the coils of Fig. 10-19a is negligible, what is the equivalent inductance? For Fig. 10-19b?

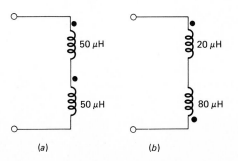

Figure 10-19.

50 μH

50 μH

20 μH

80 μH

(a)   (b)

Figure 10-20.

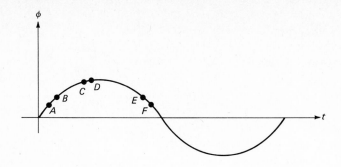

10-19.   The vertical change between two close points on a waveform is $d\phi = 0.002$ Wb. If the corresponding horizontal change is $dt = 0.005$ s, what is the rate of flux change?

10-20.   At a particular instant, $\phi = 2$ Wb and $t = 3$ s. At a slightly later instant, $\phi = 2.0005$ Wb and $t = 3.001$ s. Calculate the values of $d\phi$, dt, and $d\phi/dt$.

10-21.   Figure 10-20 shows a sine wave of flux versus time. The change in flux between A and B is 0.005 Wb. If $dt = 1$ μs, what does $d\phi/dt$ equal between A and B? If $dt = 1$ μs and $d\phi = 0.000001$ Wb between C and D, what is the rate of flux change? The flux at point E equals 2 Wb, and the flux at F is 1.9995 Wb. If the difference in time between E and F is 1 μs, what does the rate of change equal?

10-22.   Use the same information as in the preceding problem. If Fig. 10-20 is the graph of flux versus time for an inductor with 50 turns, what is the induced voltage in the vicinity of A? Near C? In the region of E?

Figure 10-21.

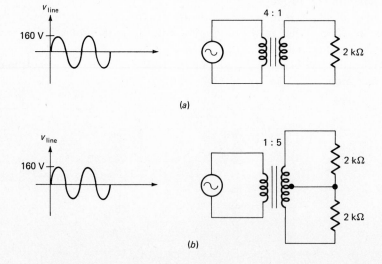

10-23. At the start of a particular waveform, $d\phi/dt$ equals 4 Wb/s. In this region, how much flux change is there for a dt of 1 μs?

10-24. A transformer has a turns ratio N_2/N_1 of 20. Ideally, what is the output voltage if the input voltage is 5 V? With a load resistance of 250 Ω, how much primary current would there be? What would the primary resistance be?

10-25. Line voltage drives the primary winding of Fig. 10-21a. Ideally, what is the secondary voltage? And what does the waveform of the load current look like? The waveform of the primary current? What does the primary resistance equal?

10-26. Figure 10-21b shows a *center-tapped* secondary winding, that is, a secondary winding with a lead connected between each half. The primary winding induces the same voltage in each half of the secondary. The turns ratio N_2/N_1 of 5 is the ratio of the total secondary turns to the primary turns. What is the voltage induced in each half of the secondary winding? What does the current waveform look like in each 2-kΩ resistor?

ANSWERS TO TESTS

10-1. *a, c, b, c, a, d, b*
10-2. *a, d, a, d, c*
10-3. *b, c, b, d, c, a, d*
10-4. *d, b, b, d, a, b, b, d*
10-5. *a, d, d, a, a*
10-6. *c, b, a, a, a*
10-7. *a, b, a, c*
10-8. *d, c, b, b, a*
10-9. *a, a, c, c, d, a*

11. Transients

Transients are time-varying voltages and currents produced by the closing or opening of a switch. In a purely resistive circuit the voltages and currents can change immediately. But when capacitance and inductance are present, the voltages and currents cannot change instantaneously because it takes time to charge a capacitor and to build up the field around an inductor.

11-1. THE *RC* SWITCHING PROTOTYPE

A *prototype* is the first or primary form of something. In circuit analysis it is a simple circuit that captures the key ideas behind a large group of related circuits.

This section is about the *resistive-capacitive (RC) switching prototype*, a series circuit containing a voltage source, a switch, a resistor, and a capacitor (Fig. 11-1a). Your job is to learn

What is an exponential wave?
How is a time constant defined?

The exponential wave

Figure 11-1a is the *RC* switching prototype. The components are ideal; that is, we neglect stray capacitance, lead inductance, etc. It's impossible to build this ideal circuit; nevertheless, the *RC* switching prototype is a good approximation for many circuits that can be built.

Suppose the capacitor of Fig. 11-1a is uncharged. Then it has 0 V across it as

Figure 11-1. Charging a capacitor.

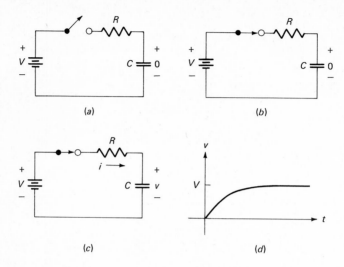

(a) (b)

(c) (d)

shown. At the instant the switch closes, the capacitor voltage is still zero (Fig. 11-1*b*) because electrons have not had a chance to move from one plate to the other. But as the electrons begin to transfer, voltage *v* appears across the capacitor (Fig. 11-1*c*).

The instantaneous current in Fig. 11-1*c* is given by

$$i = \frac{V - v}{R} \tag{11-1}$$

So as *v* increases, the instantaneous current decreases. This means the capacitor charges more slowly as the capacitor voltage increases. When the capacitor voltage approaches the source voltage, the current becomes very small. This is why the voltage across the capacitor changes very slowly as the capacitor approaches full charge (see Fig. 11-1*d*).

The wave of Fig. 11-1*d* is called an *exponential wave*. With such a wave, the rate of voltage change is rapid at first, but then slows down as the voltage approaches its final value. If you build the circuit of Fig. 11-1*a* and look at the capacitor voltage after the switch is closed, you will see an exponential wave like Fig. 11-1*d*.

Time constant

The *time constant* of an *RC* switching prototype is defined as the resistance times the capacitance. The defining formula is

$$\tau = RC \tag{11-2}$$

where τ = time constant
  R = resistance
  C = capacitance

τ always comes out in seconds, because the product of ohms and farads reduces to seconds as follows:

$$\text{Ohm} \times \text{farad} = \frac{\text{volt}}{\text{ampere}} \times \frac{\text{coulomb}}{\text{volt}} = \frac{\text{coulomb}}{\text{ampere}}$$

$$= \frac{\text{coulomb}}{\text{coulomb/second}} = \text{second}$$

Here are examples of calculating a time constant. In the circuit of Fig. 11-2a,

$$\tau = RC = 100(10^3) \times 100(10^{-6}) = 10 \text{ s}$$

In Fig. 11-2b, the time constant is

$$\tau = RC = 10^3 \times 10^{-6} = 1 \text{ ms}$$

The longer the time constant, the longer it takes to fully charge a capacitor. Here's why. Increasing either R or C increases τ. If R increases, the charging current into the capacitor decreases; therefore, the capacitor will take longer to reach its full charge. Similarly, if C increases, more charge is needed to attain the same voltage; in turn, this means it takes longer to reach full charge. So an increase in R or C results in a longer time constant, which means it takes longer to fully charge a capacitor.

Chapter 18 proves the following: an exponential wave is

63 percent finished after one time constant
86 percent finished after two time constants
95 percent finished after three time constants
98 percent finished after four time constants
99 percent finished after five time constants.

Figure 11-3 summarizes these values. One time constant after the switch is closed, that is, when $t = \tau$, the transient voltage equals $0.63V$ or 63 percent of the source voltage.

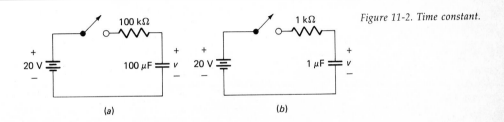

Figure 11-2. Time constant.

Figure 11-3. Exponential wave.

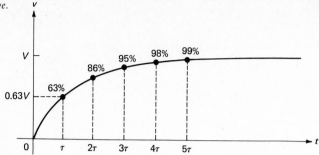

When $t = 2\tau$, the voltage equals $0.86V$ or 86 percent of the source voltage, and so on.

As a guide, most people approximate the exponential wave as finished after five time constants. If a circuit has a time constant of 1 ms, then the exponential wave is finished after approximately 5 ms. If a circuit has a time constant of 20 s, the exponential wave is finished after approximately 100 s.

The most important values to remember in Fig. 11-3 are these:

63 percent finished after one time constant
99 percent finished after five time constants.

These two values are useful in analyzing *RC* switching circuits.

EXAMPLE 11-1.
The capacitor of Fig. 11-4*a* is uncharged. Sketch the waveform of voltage v after the switch closes.

SOLUTION.
Unless otherwise indicated, $t = 0$ always represents the closing (or opening) of a switch. The capacitor voltage starts at zero and increases toward a final value of 10 V, as shown in Fig. 11-4*b*. The time constant equals

$$\tau = RC = 5(10^3) \times 10^{-6} = 5 \text{ ms}$$

Figure 11-4. Working out an exponential wave.

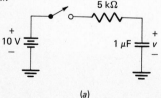

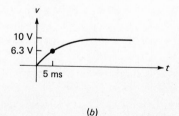

(a)

(b)

After one time constant, the capacitor voltage is

$$v = 0.63V = 0.63(10 \text{ V}) = 6.3 \text{ V}$$

as shown in Fig. 11-4b.

After five time constants, 25 ms in this case, the voltage has reached the 99 percent point.

EXAMPLE 11-2.
The capacitor of Fig. 11-5a is uncharged. Sketch the voltage waveform of v after the switch closes.

SOLUTION.
The time constant equals

$$\tau = RC = 10(10^3) \times 100(10^{-12}) = 1 \ \mu s$$

So the voltage waveform looks like the exponential shown in Fig. 11-5b. After one time constant, the voltage equals

$$v = 0.63V = 0.63(25 \text{ V}) = 15.8 \text{ V}$$

After five time constants (5 μs), the exponential wave is approximately finished.

EXAMPLE 11-3.
Before the switch closes in Fig. 11-6a, the capacitor has no charge. What does its voltage waveform look like after the switch closes?

SOLUTION.
First, Thevenize the circuit left of the switch. This results in Fig. 11-6b. Now it's clear the time constant equals

$$\tau = RC = 2(10^3) \times 2000(10^{-12}) = 4 \ \mu s$$

When $t = 4 \ \mu$s, the voltage equals 63 percent of the final value:

$$v = 0.63V = 0.63(12 \text{ V}) = 7.56 \text{ V}$$

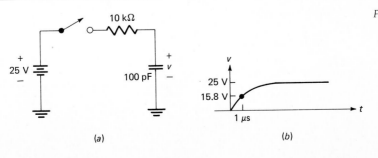

Figure 11-5. Example of exponential wave.

(a)

(b)

Figure 11-6. Another example of
an exponential wave.

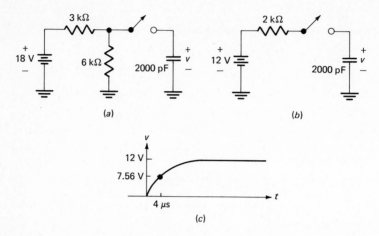

(a)

(b)

(c)

Figure 11-6c shows the exponential wave. It starts at zero, is 63 percent finished after one time constant, and 99 percent finished after five time constants.

Test 11-1

1. An exponential wave changes most rapidly
   (a) just after the switch closes
   (b) one time constant after the switch closes
   (c) five time constants after the switch has closed .... ( )
2. Which of these does not belong?
   (a) short time constant   (b) fast charging   (c) small value of RC
   (d) slow charging   .... ( )
3. In an RC switching prototype, the current is maximum when
   (a) $t = 0$ (b) $t = \tau$ (c) $t = 3\tau$ (d) $t = 5\tau$ .... ( )
4. An exponential wave changes least when
   (a) $t = 0$ (b) $t = \tau$ (c) $t = 3\tau$ (d) $t = 5\tau$ .... ( )
5. An RC switching prototype has how many loops?
   (a) 1   (b) 2   (c) 3   (d) 4 .... ( )
6. Which does not belong?
   (a) CR (b) τ (c) R/C (d) RC .... ( )
7. Which belongs least?
   (a) transient   (b) exponential wave
   (c) temporary time-varying voltage caused by switching
   (d) sine wave   .... ( )

11-2. CAPACITOR DISCHARGE

When a capacitor discharges, its voltage decreases. As you read about capacitor discharge, get the answers to

What is a decreasing exponential?
What is the discharging time constant?

Discharge

If the switch of Fig. 11-7a has been in position A for a long time, the capacitor is fully charged and its voltage equals V. When the switch is thrown to position B, the capacitor will discharge through R_2 (see Fig. 11-7b). This means conventional flow is out of the upper plate as shown. The current is largest at first, because the capacitor is just starting to discharge. Eventually the capacitor is completely discharged and the current is zero.

Figure 11-7c shows what the discharge waveform looks like. The rate of voltage change is greatest at first, because the current is maximum at the start of discharge. But as the capacitor discharges, the rate of flow decreases and the rate of voltage change decreases. In other words, the capacitor voltage changes more slowly. The wave of

Figure 11-7. (a) Charging capacitor. (b) Discharging capacitor. (c) Decreasing exponential. (d) Leading and trailing edges.

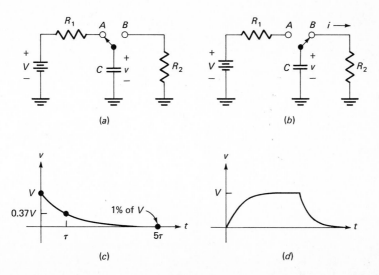

Fig. 11-7c is called a *decreasing exponential;* it changes rapidly at first but slows down as the capacitor discharges.

Chapter 18 proves the decreasing exponential of Fig. 11-7c is the inverted form of Fig. 11-3. In other words, after one time constant the capacitor discharge is 63 percent finished and the capacitor voltage equals $0.37V$. After five time constants the capacitor discharge is 99 percent finished and the capacitor voltage equals $0.01V$.

While the capacitor is discharging, it acts like a temporary source of energy. By an advanced derivation the energy stored in a capacitor equals

$$W = \frac{1}{2} CV^2 \tag{11-3}$$

If a 1-μF capacitor has 10 V across it, the stored energy is

$$W = \frac{1}{2} 10^{-6}(10^2) = 0.5(10^{-4}) \text{ J} = 50 \text{ }\mu\text{J}$$

In this case, the capacitor can deliver 50 μJ of energy to a load when its voltage decreases from 10 V to 0 V.

Time constant of one-loop circuits

The *time constant* of any one-loop RC circuit is defined as the product of resistance and capacitance:

$$\tau = RC$$

where R = equivalent resistance of loop
$\quad\;\; C$ = equivalent capacitance of loop

If there's more than one resistor in the loop, add the resistances to get the equivalent resistance. If there's more than one capacitor in the loop, combine the capacitances by the reciprocal rule for series capacitance.

In a circuit like Fig. 11-7a, two time constants exist because the capacitor is charging in position A but discharging in position B. With the switch in position A, the loop resistance is R_1 and the loop capacitance is C; therefore, the charging time constant is

$$\tau = R_1 C$$

When the switch is in position B, the loop resistance is R_2 and the loop capacitance is C. So the discharging time constant equals

$$\tau = R_2 C$$

The decreasing exponential of Fig. 11-7c is

63 percent finished after one time constant
99 percent finished after five time constants.

As shown in Fig. 11-7c, being 63 percent finished is the same as

$$v = 0.37V \qquad (t = \tau)$$

and being 99 percent finished means

$$v = 0.01V \qquad (t = 5\tau)$$

Charging and discharging

Suppose the capacitor of Fig. 11-7b is completely discharged. When the switch is thrown to position A (Fig. 11-7a), the capacitor will charge. This results in the *leading edge* of Fig. 11-7d. The time constant for this charging is R_1C. If the switch is left in position A for a long time, the voltage levels off at a final value of V.

If the switch is then thrown to position B (Fig. 11-7b), the capacitor discharges. This results in the *trailing edge* of Fig. 11-7d. The time constant for this discharging part of the waveform is R_2C. Eventually, the voltage approaches zero.

EXAMPLE 11-4.
The capacitor of Fig. 11-8a has an initial voltage of 10 V with the switch open. What does the voltage across the capacitor look like after the switch is closed?

SOLUTION.
The time constant after the switch is closed is

$$\tau = RC = 10(10^3) \times 10^{-6} = 10 \text{ ms}$$

The waveform will be a decreasing exponential like Fig. 11-8b. It starts at 10 V and *decays* (the same as decreases) to a final value of zero.

After one time constant, the transient voltage is

$$v = 0.37V = 0.37(10 \text{ V}) = 3.7 \text{ V}$$

After five time constants (50 ms), the transient voltage is

$$v = 0.01V = 0.01(10 \text{ V}) = 0.1 \text{ V}$$

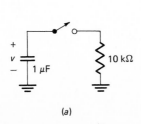

(a)

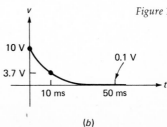

(b)

Figure 11-8. Example of decreasing exponential.

This voltage is so small compared to the starting voltage, that the waveform is considered finished after approximately five time constants (see Fig. 11-8b).

EXAMPLE 11-5.
The capacitor of Fig. 11-9a has 25 V across it before the switch is closed. What is the waveform of voltage across the capacitor after the switch closes? And the voltage waveform for v_R?

SOLUTION.
The capacitor works into an equivalent resistance of

$$R = 60,000 \parallel 30,000 = 20 \text{ k}\Omega$$

Figure 11-9b shows the equivalent circuit. The time constant during discharge is

$$\tau = RC = 20(10^3) \times 0.01(10^{-6}) = 0.2 \text{ ms}$$

Figure 11-9c shows the waveform of voltage across the capacitor; the voltage starts at 25 V and decays to a final value of zero. After one time constant,

$$v_c = 0.37V = 0.37(25 \text{ V}) = 9.25 \text{ V}$$

In Fig. 11-9a, it's clear that the voltage across the lower 15-kΩ resistor is half of the voltage across the capacitor (voltage-divider action). Therefore, each instantaneous value of v_R is half of v_C. This is why Fig. 11-9d shows the waveform of v_R starting at 12.5 V and decaying exponentially to zero.

Figure 11-9. Another decreasing exponential.

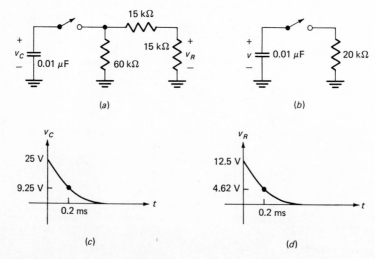

(a)

(b)

(c)

(d)

EXAMPLE 11-6.

The capacitor of Fig. 11-10a is initially uncharged. The switch is thrown to position A at $t = 0$. Then the switch is thrown to position C at $t = 100$ μs. Sketch the waveform of voltage across the capacitor.

SOLUTION.

The charging time constant equals

$$\tau = RC = 10(10^3) \times 10^{-9} = 10 \ \mu s$$

So the leading edge of the waveform reaches 6.3 V after the switch has been in position A for 10 μs (see Fig. 11-10b). After five time constants (50 μs) the charging is essentially over, and the capacitor voltage is approximately 10 V.

When the switch is thrown to position C at $t = 100$ μs, the capacitor voltage decays exponentially as shown in Fig. 11-10b. The discharging time constant equals

$$\tau = RC = 30(10^3) \times 10^{-9} = 30 \ \mu s$$

This is the reason the capacitor voltage equals 3.7 V when $t = 130$ μs.

This example and the earlier ones bring out an important idea about a capacitor. Whether it's being charged or discharged, it takes time for the voltage across a capacitor to change. In any circuit you build, there's always some resistance in series with the capacitor, even if it's only the resistance of the connecting wires. For this reason, the time constant never equals zero in a real circuit. Because of this, the *voltage across a capacitor cannot change instantaneously* in a real circuit. If there's 10 V across a capacitor just before a switch is thrown, there will be 10 V across the capacitor immediately after the switch is thrown.

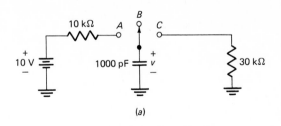

(a)

(b)

Figure 11-10. Charging and discharging in same circuit.

Test 11-2

1. A decreasing exponential changes most rapidly at
   (a) the instant the switch is thrown   (b) $t = \tau$ (c) $t = 5\tau$
   (d) long after the capacitor is discharged .... ( )
2. Which of these does not belong?
   (a) capacitor voltage equals 0.63V   (b) capacitor is discharging
   (c) decreasing exponential is 63 percent finished after one time constant
   (d) capacitor voltage is 0.37V .... ( )
3. After five time constants the voltage across a discharging capacitor equals
   (a) the initial voltage V   (b) 0.63V   (c) 0.37V   (d) 0.01V .... ( )
4. Two 1-μF capacitors are in parallel. If these capacitors discharge into two parallel
   1-kΩ resistors, the discharging time constant equals
   (a) 250 μs   (b) 500 μs   (c) 1000 μs   (d) 2000 μs .... ( )

11-3. INDUCTOR TRANSIENTS

As you read about inductor transients, be sure to find the answers to

> Why can't inductor current change instantaneously?
> How is time constant defined for RL circuits?

Initial current

As discussed earlier, the flux around an inductor is directly proportional to the current in the inductor; therefore, any change in the current produces a change in the flux.

The current in an inductor *cannot change instantaneously*. Here's why. Suppose the current in an inductor does change instantaneously, as shown in Fig. 11-11a. Then the flux also must change instantaneously, as shown in Fig. 11-11b. Since the flux has changed from ϕ_1 to ϕ_2 in zero time, the rate of flux change $d\phi/dt$ is infinite. According to Faraday's law, the induced voltage is

$$v = N \frac{d\phi}{dt} = N \times \infty = \infty$$

Figure 11-11. (a) *Current step.*
(b) *Corresponding flux step.*

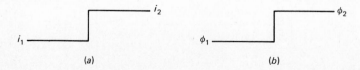

(a)   (b)

But infinite voltage is impossible in a real circuit (one you can build); therefore, the current in an inductor cannot change instantaneously as shown in Fig. 11-11a.

Remember this result. When a switch is thrown, the current in an inductor cannot change instantaneously. To put it another way, the current in an inductor has the same value just after a switch is thrown as it had just before the switch was thrown. For instance, if there's 5 mA in an inductor just before a switch is thrown, there will be 5 mA just after the switch is thrown.

Time constant

Figure 11-12a shows a *resistive-inductive* (RL) circuit. If the make-before-break switch has been in position B for a long time, the current through the inductor will be zero. If the switch is thrown to position A, the circuit appears like Fig. 11-12b. Since the current through the inductor was zero just before the switch was thrown, *i* still equals zero just after the switch is thrown. This explains why the transient current starts at zero as shown in Fig. 11-12c.

As the magnetic field around the inductor builds up, the current increases. Eventually, the magnetic field reaches its final value and the flux no longer changes; when this happens, the induced voltage decreases to zero and the inductor ideally appears like a short (neglecting its resistance). So the final current in Fig. 11-12b is

$$i = \frac{V}{R_1}$$

The *time constant* of a one-loop RL circuit is defined as the loop inductance divided

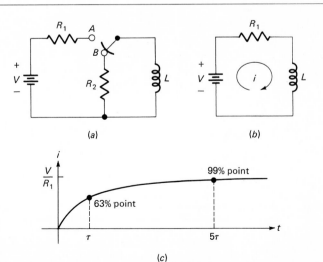

Figure 11-12. Charging an inductor.

by the loop resistance. As a defining formula,

$$\tau = \frac{L}{R} \tag{11-4}$$

Loop inductance is the equivalent inductance of the loop, the sum of all the series inductances. Loop resistance is the equivalent resistance of the loop, the sum of all the series resistances.

τ always comes out in seconds, because the ratio of inductance to resistance has the unit of seconds as follows:

$$\frac{\text{Henry}}{\text{Ohm}} = \frac{\text{weber/ampere}}{\text{volt/ampere}} = \frac{\text{weber}}{\text{volt}} = \frac{\text{volt second}}{\text{volt}} = \text{second}$$

Chapter 18 shows that the exponential wave of Fig. 11-12c is 63 percent finished after one time constant, and 99 percent finished after five time constants. As an approximation, the transient current in an inductor is essentially over after five time constants. In the charging circuit of Fig. 11-12b, the time constant is

$$\tau = \frac{L}{R_1}$$

So the current reaches a final value of V/R_1 after approximately $5L/R_1$.

Discharging an inductor

An inductor discharges when its magnetic field decreases. If the switch of Fig. 11-12a has been in position A for a long time, the inductor current equals V/R_1. If the switch is thrown to position B, the circuit appears like Fig. 11-13a. Current i still equals V/R_1 at the first instant the switch is thrown, because inductor current cannot change instantaneously.

But the current will decrease as the magnetic field begins to collapse. A decreasing magnetic field means $d\phi/dt$ is negative. In turn, the induced voltage is negative; its polarity is opposite that for charging. This is why a minus-plus voltage appears across the inductor as shown in Fig. 11-13b. During the transient, the inductor temporarily acts like an energy source because its collapsing magnetic field returns energy to the circuit. By an advanced derivation, the energy stored in the magnetic field of an inductor is

$$W = \frac{1}{2} LI^2 \tag{11-5}$$

The current during inductor discharge is a decreasing exponential wave like Fig. 11-13c. After one time constant, the wave is 63 percent finished, equivalent to a current of 37 percent of the initial current. After five time constants, the wave is 99 percent finished, which means a current of approximately 1 percent of the initial current.

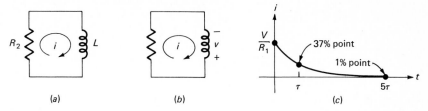

Figure 11-13. (a) *Steady current.*
(b) *Decreasing current induces
reverse voltage.* (c) *Decreasing
exponential.*

(a)   (b)   (c)

EXAMPLE 11-7.

Sketch the waveform of charging current through the inductor when the switch is
thrown to position A in Fig. 11-14a.

SOLUTION.
If the switch has been in position B for a long time, the inductor current equals zero
just before the switch is thrown to position A. Therefore, just after the switch is thrown
to A, the inductor current is still zero.

The charging time constant is

$$\tau = \frac{L}{R} = \frac{2}{1000} = 2 \text{ ms}$$

After the current reaches its final value, the field no longer changes and the induced
voltage is zero. So ideally the inductor looks like a short long after the switch is thrown
to position A. This means all the source voltage is across the resistor, and the current

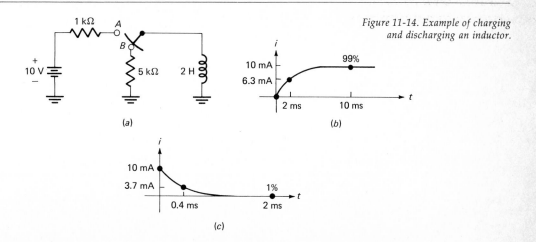

*Figure 11-14. Example of charging
and discharging an inductor.*

(a)   (b)

(c)

equals

$$i = \frac{V}{R} = \frac{10}{1000} = 10 \text{ mA}$$

Figure 11-14b shows the transient current through the inductor. It's an exponential wave starting at zero and approaching a final value of 10 mA. After one time constant (2 ms), the current has reached the 63 percent point.

EXAMPLE 11-8.
If the switch of Fig. 11-14a has been in position A for a long time, what does the discharging current through the inductor look like if the switch is thrown to position B?

SOLUTION.
Just before switching, the current is 10 mA. Just after switching the current is 10 mA, because inductor current cannot change instantaneously. This is why the first point on the current waveform is at 10 mA, as shown in Fig. 11-14c.

The discharging time constant equals

$$\tau = \frac{L}{R} = \frac{2}{5000} = 0.4 \text{ ms}$$

So after 0.4 ms, the current has decreased to the 37 percent point as shown in Fig. 11-14c. After five time constants (2 ms), the current is at the 1 percent point.

EXAMPLE 11-9.
Explain what happens when the switch of Fig. 11-15a is opened.

SOLUTION.
Before the switch is opened, the voltage across the inductor has the plus-minus polarity shown. If the magnetic field has reached its final value, the flux no longer changes and the induced voltage v is zero. Neglecting the resistance of the inductor, the current equals 2 A in Fig. 11-15a.

When the switch is opened, the circuit looks like Fig. 11-15b. At the first instant, the inductor current still equals 2 A because inductor current can never change instantaneously. Because the magnetic field starts to collapse, $d\phi/dt$ is negative and the induced voltage reverses polarity, as shown in Fig. 11-15b.

With the switch open, how can there be current? Either of two things can happen. First, the air between the switch contacts may break down if the induced voltage is enough to exceed the dielectric strength of air (Sec. 9-4). In this case, charges flow between the open contacts, as shown in Fig. 11-15c. Second, the inductor may discharge through its own stray capacitance (see Fig. 11-15d). Either or both of the foregoing possibilities will occur when you open the switch in Fig. 11-15a.

Figure 11-15. Interrupting inductor current. (a) Steady current. (b) Sudden switch opening. (c) Switch breakdown. (d) Discharge through stray capacitance.

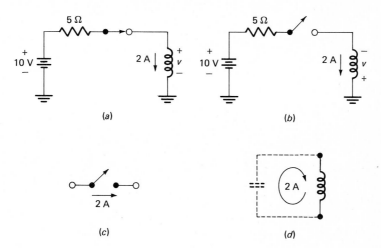

(a) (b)

(c) (d)

EXAMPLE 11-10.

If a 5-H inductor is used in Fig. 11-15a, how much energy is stored in the magnetic field when the current is 2 A?

SOLUTION.

With Eq. (11-5),

$$W = \frac{1}{2} LI^2 = \frac{1}{2} (5)2^2 = 10 \text{ J}$$

Test 11-3

1. Inductor current can't change instantaneously because
   (a) voltage can't be infinite (b) $d\phi/dt$ can't be infinite
   (c) the magnetic field can't change instantaneously
   (d) all of these .... ( )
2. Which does not belong?
   (a) RC (b) τ (c) L/R (d)R/L .... ( )
3. When an inductor discharges, which of the following occurs?
   (a) $d\phi/dt$ is positive (b) the flux decreases (c) ϕ is constant
   (d) the current increases .... ( )

4. If the magnetic field around an inductor is no longer changing, the inductor ideally appears like
    (*a*) a short   (*b*) an open   (*c*) a voltage source   (*d*) a capacitor .... ( )

SUMMARY OF FORMULAS

DEFINING

$$\tau = RC \tag{11-2}$$

$$\tau = \frac{L}{R} \tag{11-4}$$

DERIVED

$$W = \frac{1}{2} CV^2 \tag{11-3}$$

$$W = \frac{1}{2} LI^2 \tag{11-5}$$

Problems

11-1. Before the switch of Fig. 11-16*a* closes, the capacitor is uncharged. If the switch closes at $t = 0$, what is the current at the first instant?

11-2. Sketch the waveform of the voltage across the capacitor of Fig. 11-16*a*, assuming the capacitor is initially uncharged and the switch closes at $t = 0$.

11-3. Calculate the time constant for Fig. 11-16*a*. If the source voltage doubles, what happens to the time constant?

Figure 11-16.

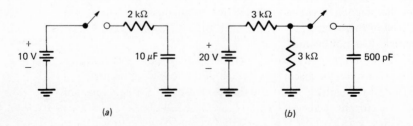

(a)   (b)

Figure 11-17.

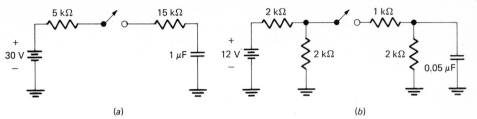

(a)   (b)

11-4. The capacitor is initially uncharged in Fig. 11-16*b*. What is the current into the capacitor just after the switch closes?

11-5. What is the time constant of Fig. 11-16*b* after the switch closes?

11-6. The capacitor of Fig. 11-16*b* is initially uncharged. Sketch the waveform of the voltage across the capacitor if the switch closes at $t = 0$.

11-7. What is the time constant in Fig. 11-17*a*? Sketch the voltage waveform across the capacitor, assuming the capacitor is initially uncharged and the switch closes at $t = 0$.

11-8. What is the time constant after the switch of Fig. 11-17*b* closes? The waveform across the capacitor if the capacitor is initially uncharged?

11-9. The capacitor of Fig. 11-18*a* has a voltage across it from a previous charging. How much energy is stored in it if 20 V are across it?

11-10. The capacitor of Fig. 11-18*a* is positively charged with 10 V across it. Sketch the waveform of the voltage after the switch closes.

11-11. What is the time constant in Fig. 11-18*b* after the switch closes? How much energy is stored in the capacitor if 20 V are across it?

11-12. Sketch the waveform of the voltage across the capacitor of Fig. 11-18*b* if the switch closes at $t = 0$ and the capacitor has an initial voltage of 12 V. What does the voltage across the 200-kΩ resistor look like?

11-13. If the switch of Fig. 11-19 is thrown to position *A*, approximately how long does it take the capacitor to reach full charge? Approximately how long will it take for the capacitor to discharge after the switch is thrown to position *C*?

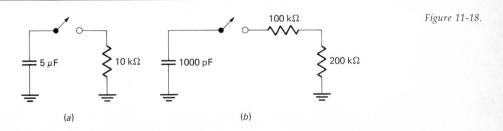

Figure 11-18.

(a)   (b)

Figure 11-19.

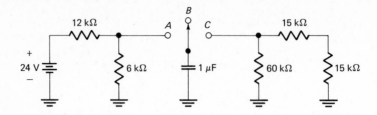

11-14. The capacitor of Fig. 11-19 is uncharged. The switch is thrown to position A at $t = 0$. Later, it's thrown to position C at $t = 50$ ms. Sketch the waveform of the voltage across the capacitor.

11-15. When the switch of Fig. 11-20a is opened, the capacitor charges. What is the final voltage across the capacitor? How long does it take the capacitor to reach the 63 percent point? Approximately how long does it take before the transient is finished?

11-16. The capacitor is uncharged in Fig. 11-20b. The switch closes at $t = 0$. What is the current into the capacitor just after the switch closes? The final capacitor voltage? The time constant?

11-17. What is the charging time constant in Fig. 11-21a (position A)? The discharging time constant (position B)?

11-18. There is no current in the inductor of Fig. 11-21a. Sketch the waveform of the current through the inductor after the switch is thrown to position A.

11-19. When the switch of Fig. 11-21a has been in position A for a long time, how much energy is stored in the magnetic field of the inductor?

11-20. The switch of Fig. 11-21a is thrown to position A at $t = 0$. Later, at $t = 1$ ms it's thrown to position B. Sketch the waveform of the current through the inductor.

11-21. What is the charging time constant in Fig. 11-21b (position A)? The discharging time constant (position B)?

Figure 11-20.

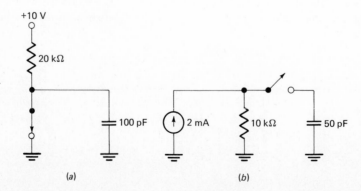

(a)   (b)

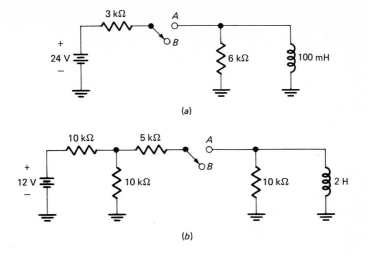

Figure 11-21.

(a)

(b)

11-22. If the switch of Fig. 11-21b is thrown to position A, what is the final current through the inductor? Approximately how long does it take for the current to reach this value?

11-23. How much energy is stored in the magnetic field of the inductor when the switch of Fig. 11-21b has been in position A for a long time?

11-24. The switch of Fig. 11-21b is thrown to position A and left there for a long time. Then it's thrown to position B. Sketch the inductor current during the discharge.

ANSWERS TO TEST

11-1. *a, d, a, d, a, c, d*

11-2. *a, a, d, c*

11-3. *d, d, b, a*

12. Reactance

This chapter is mainly about *reactance,* the opposition of an ideal capacitor or an ideal inductor to ac current.

12-1. VALUES OF A SINE WAVE

Chapter 8 introduced the sine wave (see Fig. 12-1*a*). As you read this section, get the answers to

> What is the peak-to-peak voltage?
> What is average voltage?
> What is rms voltage?

Instantaneous voltage

Chapter 8 discussed *instantaneous voltage,* the value of a time-varying voltage at a particular point in time. In Fig. 12-1*b*, for instance, the sine wave has these instantaneous values:

$v = 0$	when	$t = 0$
$v = 75$ V	when	$t = 2.5$ ms
$v = 0$	when	$t = 5$ ms
$v = -75$ V	when	$t = 7.5$ ms

and so on. You can measure instantaneous voltages with an oscilloscope.

Figure 12-1. Sine wave.

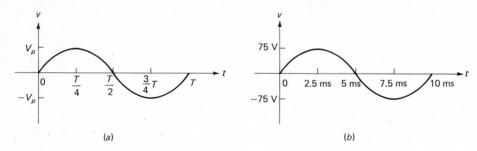

(a)   (b)

Peak voltage

The positive peak voltage is the maximum positive voltage of a sine wave, while the negative peak voltage is the maximum negative voltage. These two peak values have the same magnitude, but opposite sign. For example, the sine wave of Fig. 12-1b has a positive peak voltage of 75 V and a negative peak voltage of −75 V.

Peak-to-peak voltage

The *peak-to-peak voltage* of a sine wave is the total change in voltage between the positive and negative peak. You can calculate this change by taking the algebraic difference of the two peak values, that is

$$V_{pp} = V_p - (-V_p)$$

which equals

$$V_{pp} = 2V_p \qquad (12\text{-}1)$$

This says the peak-to-peak voltage of a sine wave is always double the peak voltage.

As an example, the sine wave of Fig. 12-1b has a peak-to-peak voltage of

$$V_{pp} = 2V_p = 2(75 \text{ V}) = 150 \text{ V}$$

Peak-to-peak voltage is easily measured with most oscilloscopes.

Average voltage

A sine wave like Fig. 12-1a has positive instantaneous values during the first half cycle, and negative instantaneous values during the second half cycle. Because a sine wave is symmetrical about the horizontal axis, the *average voltage* over a cycle is zero. In other words, if you add the instantaneous values for the entire cycle and divide by the number of samples, the average is zero:

$$V_{av} = 0 \qquad \text{(one cycle of a sine wave)} \qquad (12\text{-}2)$$

A practical application of Eq. (12-2) occurs with dc voltmeters. A dc voltmeter indicates the average voltage across its input terminals. Try measuring a sine wave like Fig. 12-1b with a dc voltmeter, and you will read zero.

There's an electronic circuit called a *half-wave rectifier* that removes or eliminates either half cycle. For instance, Fig. 12-2a shows a sine wave after the negative half cycle has been removed. Either experimentally or mathematically, it can be shown that

$$V_{av} = 0.318V_p \qquad \text{(half wave)} \qquad (12\text{-}3)$$

If the waveform of Fig. 12-2a is across a dc voltmeter, the reading will equal 31.8 percent of the peak voltage.

Another electronic circuit called a *full-wave rectifier* can invert either half cycle. For example, Fig. 12-2b shows a sine wave whose negative half cycles have been inverted. Either experimentally or mathematically, we can prove this formula:

$$V_{av} = 0.636V_p \qquad \text{(full wave)} \qquad (12\text{-}4)$$

A waveform like Fig. 12-2b across a dc voltmeter produces a reading of 63.6 percent of the peak voltage.

Rms voltage

A sine-wave voltage across a resistor produces heat, the same as a dc voltage. The *rms voltage* (also called the heating value or effective value) is defined as the dc voltage

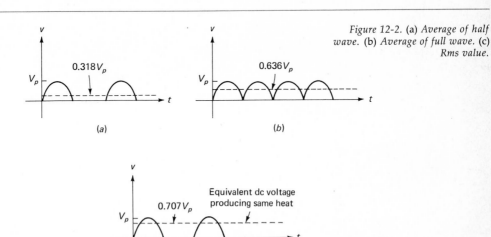

Figure 12-2. (a) *Average of half wave.* (b) *Average of full wave.* (c) *Rms value.*

that produces the same amount of heat as the sine wave (see Fig. 12-2c). Chapter 15 discusses rms voltage in detail, and derives this useful formula for a sine wave:

$$V_{rms} = 0.707V_p \qquad\qquad (12\text{-}5)$$

This says the rms voltage of a sine wave equals 70.7 percent of the peak voltage. If a sine wave has a peak of 100 V, the rms voltage is

$$V_{rms} = 0.707V_p = 0.707(100 \text{ V}) = 70.7 \text{ V}$$

If another sine wave has a peak of 250 V, the rms voltage is

$$V_{rms} = 0.707V_p = 0.707(250 \text{ V}) = 177 \text{ V}$$

Most ac voltmeters, whether the moving-coil or digital type, indicate the rms voltage of the sine wave across their input terminals. Because of this, it's convenient to rearrange Eq. (12-5) to get

$$V_p = \frac{V_{rms}}{0.707} = 1.414V_{rms} \qquad\qquad (12\text{-}6)$$

Use this formula when you want to convert an ac voltmeter reading to the peak voltage of the sine wave. For instance, if an ac voltmeter indicates 20 V rms, the peak voltage is

$$V_p = 1.414V_{rms} = 1.414(20 \text{ V}) = 28.3 \text{ V}$$

Currents

All the foregoing ideas apply to sine waves of current. That is, the peak-to-peak current equals

$$I_{pp} = 2I_p$$

The average current of a sine wave is

$$I_{av} = 0 \qquad \text{(one cycle of a sine wave)}$$

The average current for a half-wave signal is

$$I_{av} = 0.318I_p \qquad \text{(half wave)}$$

And for a full-wave signal,

$$I_{av} = 0.636I_p \qquad \text{(full wave)}$$

The rms current for a sine wave is

$$I_{rms} = 0.707I_p \qquad \text{(sine wave)}$$

EXAMPLE 12-1.
Figure 12-3 shows a sine wave with a peak of 50 V. What does each instrument indicate?

Figure 12-3. Measuring different
voltages.

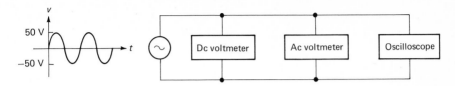

SOLUTION.
The dc voltmeter reads the average voltage which is

$$V_{av} = 0$$

The ac voltmeter reads the rms voltage which is

$$V_{rms} = 0.707V_p = 0.707(50 \text{ V}) = 35.4 \text{ V}$$

(Unless otherwise indicated, an ac voltmeter is always calibrated to read rms voltage.)
The oscilloscope displays the instantaneous voltages, including the positive and negative peaks. Because of this, the oscilloscope is ideal for measuring peak-to-peak voltages.

EXAMPLE 12-2.
An oscilloscope indicates a peak-to-peak voltage of 25 V for a sine wave. Work out the
peak voltage and the rms voltage.

SOLUTION.
Equation (12-1) rearranges into

$$V_p = \frac{V_{pp}}{2}$$

Therefore, the peak voltage is

$$V_p = \frac{25 \text{ V}}{2} = 12.5 \text{ V}$$

Next,

$$V_{rms} = 0.707V_p = 0.707(12.5 \text{ V}) = 8.84 \text{ V}$$

EXAMPLE 12-3.
A digital voltmeter reads 30 V rms for a sine-wave input. What is the peak voltage? The
peak-to-peak voltage?

SOLUTION.
With Eq. (12-6),

$$V_p = 1.414 V_{rms} = 1.414(30 \text{ V}) = 42.4 \text{ V}$$

And with Eq. (12-1),

$$V_{pp} = 2V_p = 2(42.4 \text{ V}) = 84.8 \text{ V}$$

Test 12-1

1. To measure instantaneous voltages, you need
    (*a*) a dc voltmeter   (*b*) an ac voltmeter   (*c*) an oscilloscope .... ( )
2. An ac voltmeter almost always indicates
    (*a*) average voltage   (*b*) dc voltage   (*c*) rms voltage
    (*d*) instantaneous voltage .... ( )
3. The peak-to-peak voltage of a sine wave is always greater than the peak voltage by what factor?
    (*a*) 0   (*b*) 0.707   (*c*) 1.414   (*d*) 2 .... ( )
4. A sine wave with a peak of 20 V is converted into a full-wave signal. If this signal is across a dc voltmeter, the reading is
    (*a*) 0   (*b*) 3.18 V   (*c*) 6.36 V   (*d*) 12.7 V .... ( )
5. An ac voltmeter indicates 20 V rms. On an oscilloscope, the same sine wave would have a peak-to-peak voltage of
    (*a*) 20 V   (*b*) 28.3 V   (*c*) 56.6 V   (*d*) 40 V .... ( )
6. The voltage out of an ac outlet is approximately 115 V rms. The peak voltage of the sine wave therefore is closest to
    (*a*) 80 V   (*b*) 160 V   (*c*) 180 V   (*d*) 230 V .... ( )

12-2. AC CURRENT IN AN INDUCTOR

Faraday's law says the voltage induced across an inductor equals the number of turns times the rate of flux change. Because of this, a sine wave of current produces a *cosine wave* of voltage. This section tells you why.

The ac-current source of Fig. 12-4*a* drives an ideal inductor. Since flux is directly proportional to current, its waveform will also be a sine wave (the solid graph of Fig. 12-4*b*). As you recall from the discussion of $d\phi/dt$ (Sec. 10-8), the rate of flux change is maximum at the start of the flux sine wave, zero at the peak, and so on. This is why the waveform of $d\phi/dt$ is a *cosine wave* (the dashed wave of Fig. 12-14*b*). This cosine wave has the same shape as a sine wave, but starts a quarter cycle earlier in time.

Faraday's law says

$$v = N \frac{d\phi}{dt}$$

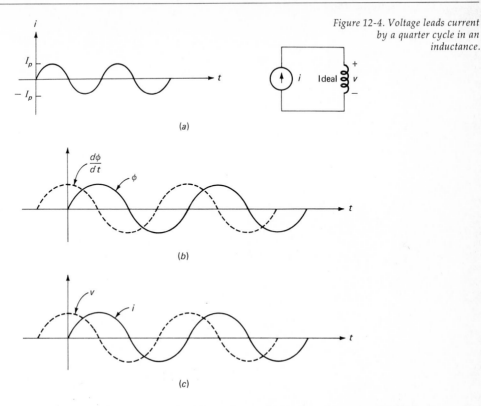

Figure 12-4. Voltage leads current by a quarter cycle in an inductance.

(a)

(b)

(c)

Since v is directly proportional to $d\phi/dt$, the waveform of v is a cosine wave (dashed wave of Fig. 12-4c). Therefore, the voltage across an ideal inductor starts a quarter cycle earlier in time than the current through the inductor. Sometimes we use the equivalent statement that the voltage *leads* current by a quarter cycle. Alternatively, it's all right to say the current *lags* the voltage by a quarter cycle.

The results are general; they apply to an ideal inductor in any kind of circuit. Whenever a sine wave of current is in an inductor, the induced voltage is always a cosine wave.

12-3. INDUCTIVE REACTANCE

The opposition of an ideal inductor to ac current is called *inductive reactance*, symbolized by X_L. This section answers

What is the defining formula for X_L?
How does X_L vary with frequency?
What is the key formula for X_L?

Defining formula

The defining formula for inductive reactance is

$$X_L = \frac{V_p}{I_p} \tag{12-7}$$

where V_p = peak voltage across the inductor
I_p = peak current through the inductor

For instance, if the induced voltage has a peak of 10 V and the current has a peak of 2 mA (see Fig. 12-5a), the inductive reactance is

$$X_L = \frac{V_p}{I_p} = \frac{10 \text{ V}}{2 \text{ mA}} = 5 \text{ k}\Omega$$

Or if the peak induced voltage is 30 V and the peak current is 10 mA (Fig. 12-5b),

$$X_L = \frac{V_p}{I_p} = \frac{30 \text{ V}}{10 \text{ mA}} = 3 \text{ k}\Omega$$

(Note that X_L has the unit of ohms, because it's the ratio of volts to amperes.)

If you are measuring rms values, it will be more convenient to use rms voltage and current. Equation (12-7) may be rearranged as

$$X_L = \frac{V_p}{I_p} = \frac{1.414 V_{\text{rms}}}{1.414 V_{\text{rms}}}$$

Figure 12-5. Examples of inductive reactance.

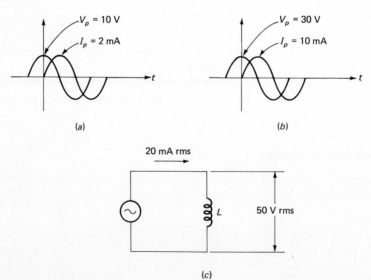

(a)

(b)

(c)

or
$$X_L = \frac{V_{rms}}{I_{rms}} \tag{12-8}$$

Given a circuit like Fig. 12-5c, with 50 V rms across the inductor and 20 mA rms through the inductor, you can calculate

$$X_L = \frac{V_{rms}}{I_{rms}} = \frac{50\ V}{20\ mA} = 2.5\ k\Omega$$

Variation with frequency

If a circuit like Fig. 12-6a is built, different rms voltages and currents occur when the source frequency is varied. By calculating the ratio of V_{rms} to I_{rms}, we can plot the values of X_L versus frequency. When this is done, a linear graph like Fig. 12-6b results. The most important thing the graph says is this: X_L is directly proportional to f; doubling f will double X_L; increasing f by a factor of 10 will increase X_L by the same factor.

This direct proportion makes sense. An increase in frequency means $d\phi/dt$ is larger at the start of the current sine wave. For a given peak value of current, therefore, more voltage is induced at higher frequencies. Since X_L equals V_{rms} divided by I_{rms}, it follows that X_L increases with frequency.

Variation with inductance

If the frequency is fixed but the inductance is increased, X_L will increase. This too makes perfect sense. For a given peak value of current, an increase in L means an increase in flux linkage; in turn, this induces more voltage. Therefore, X_L increases with L.

Key formula for X_L

Either experimentally or with advanced mathematics, we can prove this key formula:

$$X_L = 2\pi fL \tag{12-9}$$

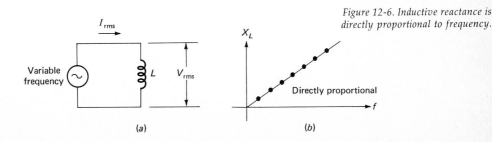

Figure 12-6. Inductive reactance is directly proportional to frequency.

(a)

(b)

This says inductive reactance equals 2π (approximately 6.28) times frequency times inductance. This is a very useful formula, because you often know the source frequency (it's given by the calibrated dial of the signal generator) and the inductance (it's on the schematic diagram, etc.).

Equation (12-9) is easy to use. Suppose $f = 10$ kHz and $L = 50$ mH. Then,

$$X_L = 2\pi fL = 6.28(10)10^3(50)10^{-3}$$
$$= 3.14 \text{ k}\Omega$$

Or if a signal generator has a frequency of 5 MHz and a schematic diagram shows an inductor value of 25 μH, then

$$X_L = 2\pi fL = 6.28(5)10^6(25)10^{-6}$$
$$= 785 \ \Omega$$

EXAMPLE 12-4.
What is the X_L of a 5-in piece of wire at 10 MHz?

SOLUTION.
Section 10-10 discussed *lead inductance,* the inductance of connecting wire. As you recall, lead inductance is approximately 0.02 μH/in. Therefore, a 5-in piece of wire has an inductance of approximately

$$L = 5 \times 0.02 \ \mu\text{H} = 0.1 \ \mu\text{H}$$

At 10 MHz,

$$X_L = 2\pi fL = 6.28(10)10^6(0.1)10^{-6}$$
$$= 6.28 \ \Omega$$

EXAMPLE 12-5.
Suppose the leads of a 1-kΩ resistor are cut to ½ in on each end. Calculate X_L at 300 MHz.

SOLUTION.
The total lead length is 1 in; therefore, we can estimate a lead inductance of 0.02 μH. Then,

$$X_L = 2\pi fL = 6.28(300)10^6(0.02)10^{-6}$$
$$= 37.7 \ \Omega$$

In this case, a 1-kΩ resistor acts like a resistance of 1 kΩ in series with an inductive reactance of 37.7 Ω.

Test 12-2

1.   Which of these does not belong?
     (*a*) X_L (*b*) ratio of peak voltage to peak current

(c) ratio of V_{rms} to I_{rms}

(d) average voltage divided by average current .... (   )

2. If the frequency is reduced by a factor of 2, X_L will

(a) double   (b) halve   (c) quadruple .... (   )

3. If the frequency is doubled, L will increase by

(a) 0   (b) 2   (c) 4 .... (   )

4. An oscilloscope indicates an induced voltage across an inductor of 28.3 V peak-to-peak. If the current through the inductor is 2 mA rms, the inductive reactance is approximately

(a) 5 kΩ   (b) 7 kΩ   (c) 10 kΩ   (d) 14 kΩ .... (   )

12-4. EQUIVALENT INDUCTIVE REACTANCE

How do you find the equivalent or total inductive reactance of two or more inductors in series or parallel?

Series inductors

Given two inductors in series like Fig. 12-7a, the equivalent inductance is

$$L = L_1 + L_2$$

After you have worked out the value of L, you can calculate X_L with

$$X_L = 2\pi f L$$

The same idea applies to any number of series inductors; add up the inductances to get the equivalent inductance; then calculate X_L for the equivalent inductance. This is almost always the easiest way to get the equivalent inductive reactance.

Occasionally, the inductive reactance of each series inductor may be given or already known. In this case, the equivalent or total inductive reactance of the two series inductors like Fig. 12-7a is

$$X_L = X_{L1} + X_{L2}$$

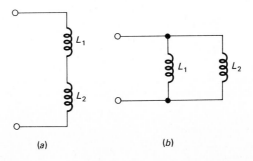

Figure 12-7. Equivalent inductive reactance.

(a)

(b)

Again, the same idea applies to any number of series inductors; you add the individual reactances to get the total or equivalent reactance.

Parallel inductors

If two inductors are in parallel like Fig. 12-7b, first combine them into an equivalent inductance using

$$L = \frac{L_1 L_2}{L_1 + L_2}$$

Then, apply the formula for inductive reactance:

$$X_L = 2\pi f L$$

If more than two inductors are in parallel, get the equivalent inductance with

$$L = \frac{1}{1/L_1 + 1/L_2 + 1/L_3 + \cdots}$$

With the value of L from this formula, you can then work out the X_L of the parallel combination.

If the individual reactances of each inductor in Fig. 12-7b are already known, you can use

$$X_L = \frac{X_{L1} X_{L2}}{X_{L1} + X_{L2}}$$

When more than two reactances are involved, use

$$X_L = \frac{1}{1/X_{L1} + 1/X_{L2} + 1/X_{L3} + \cdots}$$

Throughout the foregoing discussion, we have assumed negligible coupling between inductors. If there is significant coupling between inductors, mutual inductance has to be taken into account (Sec. 10-7). In practice, however, the coupling between separate inductors is usually negligible; so unless otherwise indicated, forget about mutual inductance.

12-5.　AC CURRENT IN A CAPACITOR

A sine wave of current through an ideal inductor produces a cosine wave of voltage across the inductor. Exactly the opposite happens with an ideal capacitor: a sine wave of voltage across a capacitor results in a cosine wave of current into the capacitor. This section tells you why.

Figure 12-8a shows a cosine wave of current driving a capacitor. If the capacitor is initially uncharged, the voltage across it is zero at $t = 0$ (the dashed waveform of Fig.

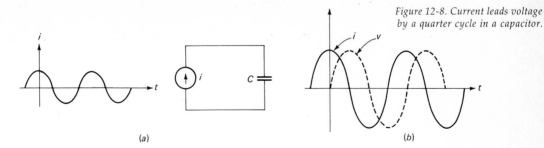

Figure 12-8. Current leads voltage by a quarter cycle in a capacitor.

(a)   (b)

12-8b). As charges flow into the capacitor, its voltage increases. The voltage continues to increase, as long as the current is positive. When the current is zero (quarter cycle after $t = 0$), the voltage is maximum. Beyond this point, the current is negative, meaning the capacitor is discharging; so it follows that the capacitor voltage then decreases. Continuing in this way, we can conclude the waveform of voltage *lags* the waveform of current by a quarter cycle.

Either experimentally or mathematically, we can extend the idea to an ideal capacitor in any kind of circuit. In other words, the result is always true: the ac current into an ideal capacitor always *leads* the ac voltage by a quarter cycle.

As you recall from Chap. 8, the ac voltage and current of a resistor are *in phase*, meaning there's no shift between the two waves; a sine wave of current produces a sine wave of voltage, and vice versa. To summarize the shifts for basic components:

Resistance: ac voltage is in phase with current.
Inductance: ac voltage leads current by a quarter cycle.
Capacitance: ac voltage lags current by a quarter cycle.

12-6.   CAPACITIVE REACTANCE

The opposition of an ideal capacitor to ac current is called *capacitive reactance*, symbolized X_C. This section answers

> *What is the defining formula for X_C?*
> *How does X_C vary with frequency?*
> *What is the key formula for X_C?*

Defining formula

The defining formula for capacitive reactance is

$$X_C = \frac{V_p}{I_p}$$

(12-10a)

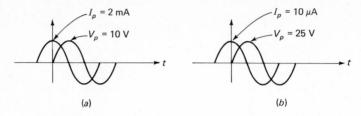

Figure 12-9. Examples of capacitive reactance.

(a)

(b)

where V_p = peak voltage across the capacitor
I_p = peak current into the capacitor

For example, if the peak voltage is 10 V and the peak current is 2 mA (see Fig. 12-9a),

$$X_C = \frac{V_p}{I_p} = \frac{10 \text{ V}}{2 \text{ mA}} = 5 \text{ k}\Omega$$

Or if the peak voltage equals 25 V and the peak current is 10 μA (Fig. 12-9b),

$$X_C = \frac{V_p}{I_p} = \frac{25 \text{ V}}{10 \text{ }\mu\text{A}} = 2.5 \text{ M}\Omega$$

(Note that the unit of capacitive reactance is ohms, because X_C is the ratio of voltage to current.)

Since rms values are proportional to peak values, we can also use

$$X_C = \frac{V_{\text{rms}}}{I_{\text{rms}}} \qquad (12\text{-}10b)$$

If 10 V rms are across the capacitor and 2 mA rms are into it,

$$X_C = \frac{V_{\text{rms}}}{I_{\text{rms}}} = \frac{10 \text{ V}}{2 \text{ mA}} = 5 \text{ k}\Omega$$

Variation with frequency

When the source frequency of Fig. 12-10a is varied, different rms voltages and currents occur. By calculating the ratio of $V_{\text{rms}}/I_{\text{rms}}$, we can plot X_C versus frequency. When this is done, a graph like Fig. 12-10b results. The most important feature of the curve is this: X_C is inversely proportional to f; when f is doubled, X_C drops by half; if f is increased by a factor of 10, X_C decreases by a factor of 10; and so on.

This inverse proportion makes sense. An increase in frequency means the capacitor has less time to charge during each cycle. For the same peak current, an increase in frequency means a smaller voltage. With X_C equal to V_p divided by I_p, it follows that X_C decreases when the frequency increases.

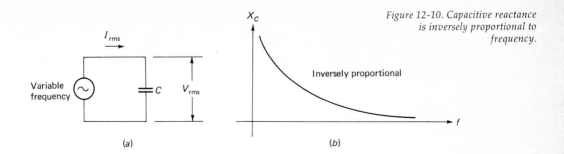

Figure 12-10. Capacitive reactance is inversely proportional to frequency.

(a)   (b)

Variation with capacitance

If the frequency is fixed but the capacitance is increased, X_C will decrease. Here's why. For a given peak voltage, more charge is stored in a larger capacitor; this means the peak current is larger. Since X_C equals V_p/I_p, it follows that X_C decreases when the capacitance increases.

Key Formula for X_C

By experiment or by advanced mathematics, we can prove this formula:

$$X_C = \frac{1}{2\pi fC} \qquad (12\text{-}11)$$

This says capacitive reactance is the reciprocal of 2π (approximately 6.28) times frequency times capacitance. This is a very important formula, because you often will know the frequency of an ac source and the capacitance.

Here are examples. If $f = 1$ kHz and $C = 5$ μF,

$$X_C = \frac{1}{2\pi fC} = \frac{1}{6.28(10^3)5(10^{-6})}$$
$$= 31.8 \ \Omega$$

Or if a signal generator has a frequency of 20 MHz and a schematic diagram shows a capacitance value of 50 pF,

$$X_C = \frac{1}{2\pi fC} = \frac{1}{6.28(20)10^6(50)10^{-12}}$$
$$= 159 \ \Omega$$

EXAMPLE 12-6.

What is the X_C between a 4-in piece of wire and the chassis at 5 MHz?

SOLUTION.

Section 9-7 discusses stray wiring capacitance. As you recall, the stray wiring capacitance can be roughly estimated as 0.3 pF/in with respect to the chassis. Therefore, a 4-in piece of wire has a stray capacitance of

$$C = 4 \times 0.3 \text{ pF} = 1.2 \text{ pF}$$

At 5 MHz, the reactance between this wire and the chassis is

$$X_C = \frac{1}{2\pi f C} = \frac{1}{6.28(5)10^6(1.2)10^{-12}}$$
$$= 26.5 \text{ k}\Omega$$

EXAMPLE 12-7.

A 1-MΩ resistor has a stray capacitance of 1 pF. Calculate X_C at 1 MHz.

SOLUTION.

$$X_C = \frac{1}{2\pi f C} = \frac{1}{6.28(10^6)10^{-12}}$$
$$= 159 \text{ k}\Omega$$

In this case, a 1-MΩ resistor acts like a resistance of 1 MΩ shunted by a capacitive reactance of 159 kΩ.

Test 12-3

1. Which of these does not belong?
    (a) average voltage divided by average current   (b) X_C
    (c) V_{rms} divided by I_{rms} (d) ratio of V_p to I_p .... ( )
2. If the frequency is tripled, X_C
    (a) triples   (b) decreases by a factor of 3   (c) remains the same
    (d) drops by a factor of 9 .... ( )
3. If C is doubled, X_C will
    (a) double   (b) halve   (c) triple   (d) quadruple .... ( )
4. An oscilloscope indicates a peak voltage of 50 mV across a capacitor. If an ac ammeter indicates 10 μA rms into the capacitor, X_C is closest to
    (a) 2 kΩ (b) 3.5 kΩ (c) 5 kΩ (d) 20 kΩ .... ( )
5. The following words can be rearranged into a sentence: WITH X_C DECREASES FREQUENCY HIGHER. The sentence is
    (a) true   (b) false .... ( )
6. X_L is to f as X_C is to
    (a) L   (b) C   (c) f (d) 1/f .... ( )
7. Suppose the graph of X_L versus f (Fig. 12-6b) is plotted on the same set of axes as the graph of X_C versus f (Fig. 12-10b). How many points of intersection are there?
    (a) 0   (b) 1   (c) 2   (d) 3 .... ( )

12-7. EQUIVALENT CAPACITIVE REACTANCE

Suppose capacitors are connected in series or in parallel. How do you calculate the equivalent capacitive reactance?

Series capacitors

Given two capacitors in series like Fig. 12-11a, the equivalent capacitance is

$$C = \frac{C_1 C_2}{C_1 + C_2}$$

With the value of C obtained from this product-over-sum, you can then calculate the equivalent or total capacitive reactance with

$$X_C = \frac{1}{2\pi f C}$$

The same idea applies to more than two capacitors in series. First, combine the capacitances with the reciprocal rule:

$$C = \frac{1}{1/C_1 + 1/C_2 + 1/C_3 + \cdots}$$

Second, calculate the equivalent capacitive reactance with

$$X_C = \frac{1}{2\pi f C}$$

If the individual reactances of the capacitors are already known, it's easier to use

$$X_C = X_{C1} + X_{C2}$$

for two series capacitors, and

$$X_C = X_{C1} + X_{C2} + X_{C3} + \cdots$$

for more than two series capacitors.

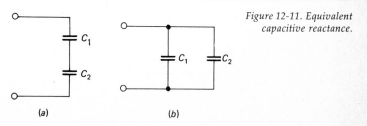

Figure 12-11. Equivalent capacitive reactance.

(a)   (b)

Parallel capacitors

If two capacitors are in parallel like Fig. 12-11b, first combine them into an equivalent capacitance using

$$C = C_1 + C_2$$

Then apply the standard formula

$$X_C = \frac{1}{2\pi f C}$$

When several capacitors are in parallel, add to get the equivalent capacitance; then use the formula for X_C.

If individual reactances are already known, use

$$X_C = \frac{X_{C1} X_{C2}}{X_{C1} + X_{C2}}$$

for two parallel capacitors, and

$$X_C = \frac{1}{1/X_{C1} + 1/X_{C2} + 1/X_{C3} + \cdots}$$

for more than two parallel capacitors.

12-8.   COUPLING CAPACITOR AND CHOKE

This section is about the typical use of capacitors and inductors in electronics. In particular, these questions are answered:

> *What is a coupling capacitor?*
> *What is a choke?*

Coupling capacitor

You often encounter the circuit of Fig. 12-12a; either you will see it exactly as shown, or you will be able to reduce more complicated circuits to this series RC circuit. Because X_C decreases when f increases, the capacitor appears like a short at high frequencies. On account of this, the only opposition to current at high frequencies is the resistance R. Therefore, the rms current is

$$I_{rms} = \frac{V_{rms}}{R} \quad \text{(high frequencies)}$$

On the other hand, when the frequency approaches zero, X_C approaches infinity and the capacitor appears like an open. In this case, the rms current is approximately

$$I_{rms} = 0 \quad \text{(low frequencies)}$$

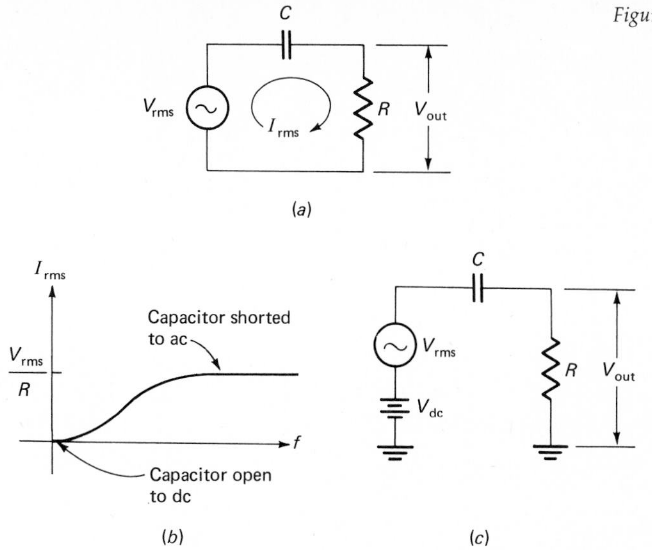

Figure 12-12. Coupling capacitor.

By building a circuit like Fig. 12-12a, we can measure I_{rms} at different frequencies. By plotting I_{rms} versus f, a typical graph like Fig. 12-12b results. Note that the capacitor appears open at low frequencies, and this forces the current to approach zero. But at high frequencies the capacitor appears like a short, and the current has a value of V_{rms}/R.

When the capacitor appears like a short, the output voltage in Fig. 12-12a equals

$$V_{out} = V_{rms} \qquad \text{(high frequencies)}$$

Because of this, we say the capacitor *couples* the source signal to the output. When a capacitor is used in this way, it's called a *coupling capacitor*. In fact, Chap. 14 proves good coupling action occurs when

$$X_C < \frac{R}{10} \tag{12-12}$$

For example, if R is 10 kΩ, X_C has to be less than 1 kΩ at the frequency you're trying to couple.

Often, the input signal will have a dc and ac component as shown in Fig. 12-12c. The capacitor appears like an open to dc because the frequency is zero; therefore, the capacitor blocks the dc voltage from the output. On the other hand, if the capacitance is large enough to make X_C less than one-tenth of R at the frequency of the ac source, then the capacitor appears approximately like an ac short. In this case, almost all of the ac source voltage appears across the resistor. Remember this basic idea: a coupling capacitor blocks dc but passes ac when condition (12-12) is satisfied.

Quadratic sum

A *quadratic sum* is defined as the square root of the sum of squares. For instance, given two numbers like 3 and 4, the quadratic sum is

$$\sqrt{3^2 + 4^2} = 5$$

Or given R and X_C, the quadratic sum is

$$\sqrt{R^2 + X_C^2}$$

Quadratic sums are common in ac analysis. As proved in Chap. 14, resistance and reactance in series always add quadratically, never algebraically. Basically, the reason is that current and voltage are a quarter cycle apart in inductors and capacitors, but are *in phase* (no separation) in resistors. More is said about this later.

Rms current in series *RC* circuit

Chapter 14 will prove this important formula for any one-loop *RC* circuit like Fig. 12-13*a*:

$$I_{rms} = \frac{V_{rms}}{\sqrt{R^2 + X_C^2}} \tag{12-13}$$

This says the rms current in a one-loop *RC* circuit equals the rms source voltage divided by the quadratic sum of resistance and reactance. If there's more than one resistor in the loop, or more than one capacitor, use the equivalent resistance and equivalent reactance.

Here's an example. Figure 12-13*b* has these values:

$$V_{rms} = 10 \text{ V}$$

$$R = 1 \text{ k}\Omega$$

$$X_C = 1 \text{ k}\Omega$$

Figure 12-13. Calculating rms current in series RC circuit.

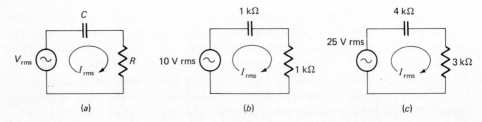

(a) (b) (c)

Substituting into Eq. (12-13) gives

$$I_{rms} = \frac{V_{rms}}{\sqrt{R^2 + X_C^2}} = \frac{10}{\sqrt{1000^2 + 1000^2}} = \frac{10}{1414}$$
$$= 7.07 \text{ mA}$$

Notice that if you try to add resistance and reactance algebraically instead of quadratically, you will get the wrong answer. As already mentioned, you must use the quadratic sum of R and X_C because of the quarter-cycle shift in voltage and current waveforms (discussed further in Chap. 14).

As another example, the rms current in Fig. 12-13c is

$$I_{rms} = \frac{V_{rms}}{\sqrt{R^2 + X_C^2}} = \frac{25}{\sqrt{3000^2 + 4000^2}} = \frac{25}{5000}$$
$$= 5 \text{ mA}$$

Choke

Another common circuit encountered in electronics is the series RL circuit of Fig. 12-14a. Because X_L increases with an increase in frequency, the inductor appears like an open at high frequencies. As a result, the current is approximately zero for large values of f (see Fig. 12-14b). On the other hand, at low frequencies X_L approaches zero and the inductor ideally appears like a short. For this reason, the current equals V_{rms}/R for low values of f (Fig. 12-14b).

At high frequencies the inductor appears open, and it effectively blocks the source

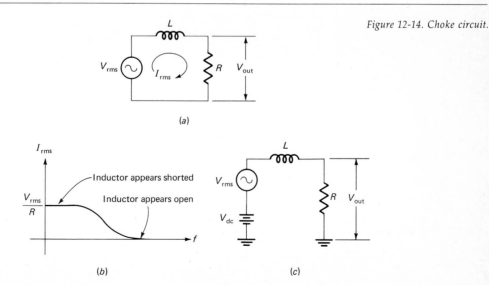

Figure 12-14. Choke circuit.

signal from the output resistance. When used in this way, the inductor is called a *choke*. As will be proved in Chap. 14, good choking action occurs when

$$X_L > 10R \qquad (12\text{-}14)$$

For instance, if R is 1 kΩ, X_L has to be more than 10 kΩ at the frequency you're trying to block from the output.

Usually, the source will have a dc and ac component as shown in Fig. 12-14c. The inductor passes the dc, but blocks the ac. In this case, almost all the dc voltage appears across the output resistor.

Rms current in series *RL* circuit

Again, the quarter-cycle shift in voltage and current in an inductor means you cannot algebraically add R and X_L in a series circuit. You have to use the quadratic sum. On this basis, it can be shown that the rms current in a one-loop RL circuit is

$$I_{\text{rms}} = \frac{V_{\text{rms}}}{\sqrt{R^2 + X_L{}^2}} \qquad (12\text{-}15)$$

When there's more than one resistor or inductor in the loop, use equivalent resistance and equivalent reactance.

As an example, suppose the circuit of Fig. 12-14a has these values: $V_{\text{rms}} = 6$ V, $R = 2$ kΩ, and $X_L = 4$ kΩ. Then,

$$I_{\text{rms}} = \frac{6}{\sqrt{2000^2 + 4000^2}} = 1.34 \text{ mA}$$

EXAMPLE 12-8.
Figure 12-15 shows an RC circuit. The source frequency may be anywhere from 20 Hz to 20 kHz. Select a size for the capacitor to get good coupling action.

SOLUTION.
When there's a range in source frequency, the lowest frequency is always the hardest one to couple. In other words, if the lowest frequency is well coupled, all higher frequencies are automatically well coupled to the output. In Fig. 12-15, R is 10 kΩ;

Figure 12-15. Example of selecting coupling capacitor.

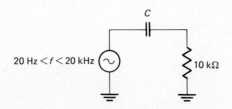

therefore, X_C must be less than 1 kΩ at 20 Hz. So we proceed like this:

$$X_C = \frac{1}{2\pi f C}$$

$$1000 = \frac{1}{6.28(20)C}$$

Solving for C gives

$$C = \frac{1}{6.28(20)1000} = 7.96 \ \mu\text{F}$$

In practice, you would select the nearest larger standard size, which is 8 μF. With this capacitor in Fig. 12-15, all source frequencies greater than 20 Hz would be well coupled to the output.

Test 12-4

1. Which of these belongs least?
   (a) open to ac   (b) shorted to ac   (c) open to dc
   (d) coupling capacitor .... ( )
2. Coupling capacitor is to choke as ac short is to
   (a) dc open   (b) ac open   (c) resistance   (d) inductance .... ( )
3. The quadratic sum of 3 and 4 is
   (a) 1   (b) 5   (c) 7   (d) 8 .... ( )
4. When used at appropriate frequencies, a choke
   (a) blocks dc   (b) blocks ac   (c) passes ac
   (d) removes the dc component .... ( )
5. In a series RC circuit the ac voltage across the capacitor
   (a) lags the ac current   (b) leads the ac current   (c) neither .... ( )

12-9. SERIES RESONANCE

When $R, L,$ and C are in series as shown in Fig. 12-16a, something unusual happens. In what follows, find the answers to

What is the resonant frequency?
What is the formula for resonant frequency?

Series RLC circuit

Look at Fig. 12-16a. At low frequencies the capacitor appears open, and blocks the source voltage from the resistor. At high frequencies the inductor looks open, and likewise blocks the source voltage from the output resistor. But between these two ex-

Figure 12-16. Series resonance.

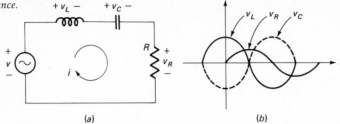

(a)   (b)

tremes, there is a single frequency called the *resonant frequency*, where the L and the C pass the source voltage on to the output resistor. In formulas we designate the resonant frequency as f_r.

In fact, at the resonant frequency

$$X_L = X_C$$

A circuit satisfying this condition is said to be *series resonant*. Because the same current is in all parts of a series circuit, the peak inductor voltage has the same value as the peak capacitor voltage. Furthermore, a sine wave of current will result in a sine wave of voltage across the resistor, a cosine wave of voltage across the inductor, and an inverted cosine wave of voltage across the capacitor as shown in Fig. 12-16*b*. At the resonant frequency, the sum of v_L and v_C is zero because the waves are equal in magnitude but a half cycle apart. Because of this cancelation, the series *RLC* circuit of Fig. 12-16*a* acts the same as the circuit of Fig. 12-17*a* at the resonant frequency. This means all the source voltage is passed on to the output resistor.

Building the circuit of Fig. 12-16*a* and plotting the rms current at different source frequencies results in a curve like Fig. 12-17*b*. At low and high frequencies the current approaches zero, because either the capacitor or the inductor appears open. But in between these extremes, we will experimentally find a frequency f_r where the rms current reaches a maximum value. This rms current will equal

$$I_{rms} = \frac{V_{rms}}{R} \qquad \text{(series resonance)}$$

Figure 12-17. (a) At resonance. (b) Resonance curve.

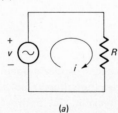

(a)

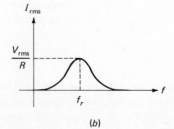

(b)

because all the source voltage will appear across the output resistor at the resonant frequency. A check of X_L and X_C at this frequency will experimentally prove

$$X_L = X_C \qquad \text{(series resonance)}$$

A circuit like Fig. 12-16a is useful when you are trying to pass a single frequency or a narrow band of frequencies.

Formula for resonant frequency

Here's how to get a useful formula for the resonant frequency. The defining formula for series resonance is

$$X_L = X_C \qquad\qquad (12\text{-}16)$$

This may be rewritten as

$$2\pi f_r L = \frac{1}{2\pi f_r C}$$

Multiplying both sides by f_r and rearranging,

$$f_r^2 = \frac{1}{(2\pi)^2 LC}$$

Taking the square root of both sides gives

$$f_r = \frac{1}{2\pi \sqrt{LC}} \qquad\qquad (12\text{-}17)$$

This is an important formula. By looking at a schematic diagram, you can read the values of L and C. With Eq. (12-17), you can then calculate the resonant frequency, the frequency where all the source voltage appears across the output resistor. For instance, in Fig. 12-18, $L = 20\ \mu\text{H}$ and $C = 500\ \text{pF}$. The resonant frequency of this circuit is

$$f_r = \frac{1}{2\pi \sqrt{LC}} = \frac{1}{6.28 \sqrt{20(10^{-6})500(10^{-12})}}$$
$$= 1.59\ \text{MHz}$$

At this frequency, the current is maximum and all the source voltage appears across the

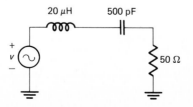

20 μH   500 pF

50 Ω

Figure 12-18. Calculating the resonant frequency.

50-Ω resistor; above and below this frequency, the current and the output voltage will decrease.

Test 12-5

1. Series resonance occurs at the frequency where
   (a) $L = C$ (b) $X_L = X_C$ (c) C is open   (d) L is open   .... ( )
2. Above and below the resonant frequency, the current
   (a) decreases   (b) increases   (c) stays the same .... ( )
3. At the resonant frequency the voltage waves across the inductor and capacitor are a
   (a) quarter cycle apart   (b) half cycle apart   (c) full cycle apart .... ( )
4. The voltage wave across the resistor always leads the current by
   (a) a quarter cycle   (b) a half cycle   (c) no separation .... ( )

12-10.   PARALLEL RESONANCE

Figure 12-19a shows a *parallel resonant circuit*, a parallel connection of R, L, and C. Most of the source current i goes through the inductor at low frequencies, because X_L approaches zero; on the other hand, most of the source current is into the capacitor at high frequencies because X_C approaches zero. In between these extremes there's a frequency where all the source current goes through the resistor.

Current i in Fig. 12-19a produces a common voltage v across the circuit. If v is a sine wave of voltage, the current i_R is a sine wave of current, because voltage and cur-

---

Figure 12-19. Parallel resonance.

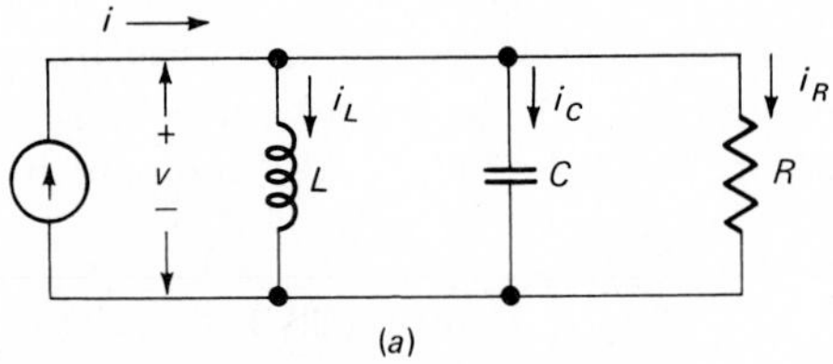

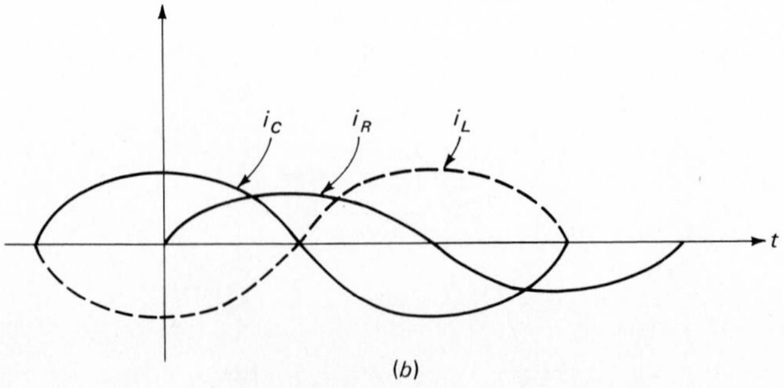

(a)

(b)

rent are in phase in a resistor (see Fig. 12-19b). Capacitor current i_C leads v by a quarter cycle, while inductor current i_L lags v by a quarter cycle as shown. Resonance occurs when

$$X_L = X_C$$

For this condition, the peak of i_C equals the peak of i_L. Since these two waves are a half cycle apart, their sum is zero at each point in time. Because of this, all the source current passes through the resistor.

Since resonance occurs when $X_L = X_C$, the derived formula for f_r is the same as that given in the preceding section:

$$f_r = \frac{1}{2\pi \sqrt{LC}}$$

This formula assumes an ideal inductor, one whose resistance and stray capacitance are negligible. Chapter 16 discusses the formula for a nonideal inductor.

We can build a parallel resonant circuit like Fig. 12-20a, using a transistor for a current source. By plotting the rms voltage across the circuit for different source frequencies, the typical *resonance curve* of Fig. 12-20b results. This agrees with the explanation given earlier; the voltage approaches zero at low frequencies (inductor-shorted) and at high frequencies (capacitor-shorted).

Parallel resonant circuits are important because they are used to tune in different stations in a radio or TV receiver. In other words, because of the *filtering* action implied in Fig. 12-20b, a parallel resonant circuit can select one frequency from the many that are received by an antenna.

EXAMPLE 12-9.
What is the resonant frequency of the circuit shown in Fig. 12-21?

SOLUTION.
Since $L = 50$ μH and $C = 1000$ pF,

$$f_r = \frac{1}{2\pi \sqrt{LC}} = \frac{1}{6.28 \sqrt{50(10^{-6})10^{-9}}}$$
$$\cong 710 \text{ kHz}$$

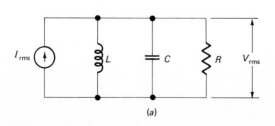

(a)

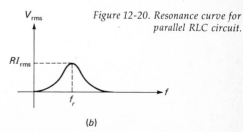

Figure 12-20. Resonance curve for parallel RLC circuit.

(b)

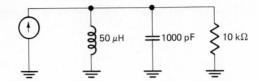

Figure 12-21. Calculating resonant frequency of parallel RLC circuit.

The AM broadcast band covers frequencies from 550 to 1600 kHz; so the circuit of Fig. 12-21 is tuned to a station whose broadcast frequency is 710 kHz. By using a variable capacitor, we can build circuits to tune in different stations.

Test 12-6

1. When driven by a current source, the rms voltage across a parallel resonant circuit reaches a maximum value at
   (a) low frequencies   (b) the resonant frequency   (c) high frequencies   ( )
2. Neglecting the resistance of the inductor, parallel resonance occurs when
   (a) the source frequency approaches zero   (b) f is high
   (c) $X_L = X_C$ .... ( )
3. The waveforms of the inductor current and the capacitor current in a parallel resonant circuit are
   (a) in phase   (b) a quarter cycle apart   (c) a half cycle apart
   (d) a full cycle apart .... ( )
4. A current source drives a parallel resonant circuit. Which of these does not belong?
   (a) low voltage   (b) resonance
   (c) all source current through the resistor   (d) maximum voltage .... ( )

SUMMARY OF FORMULAS

DEFINING

$$V_{pp} = 2V_p \tag{12-1}$$

$$X_L = \frac{V_p}{I_p} \tag{12-7}$$

$$X_C = \frac{V_p}{I_p} \tag{12-10a}$$

$$X_C < \frac{R}{10} \quad \text{(good coupling)} \tag{12-12}$$

$$X_L > 10R \qquad \text{(good choking)} \qquad\qquad (12\text{-}14)$$

$$X_L = X_C \qquad \text{(resonance)} \qquad\qquad (12\text{-}16)$$

DERIVED

$$V_{\text{av}} = 0.318V_p \qquad \text{(half wave)} \qquad\qquad (12\text{-}3)$$

$$V_{\text{av}} = 0.636V_p \qquad \text{(full wave)} \qquad\qquad (12\text{-}4)$$

$$V_{\text{rms}} = 0.707V_p \qquad \text{(sine wave)} \qquad\qquad (12\text{-}5)$$

$$X_L = \frac{V_{\text{rms}}}{I_{\text{rms}}} \qquad\qquad (12\text{-}8)$$

$$X_L = 2\pi fL \qquad\qquad (12\text{-}9)$$

$$X_C = \frac{V_{\text{rms}}}{I_{\text{rms}}} \qquad\qquad (12\text{-}10b)$$

$$X_C = \frac{1}{2\pi fC} \qquad\qquad (12\text{-}11)$$

$$I_{\text{rms}} = \frac{V_{\text{rms}}}{\sqrt{R^2 + X_C{}^2}} \qquad\qquad (12\text{-}13)$$

$$I_{\text{rms}} = \frac{V_{\text{rms}}}{\sqrt{R^2 + X_L{}^2}} \qquad\qquad (12\text{-}15)$$

$$f_r = \frac{1}{2\pi \sqrt{LC}}$$

Problems

12-1.   A sine wave has a peak value of 50 mV. What is its peak-to-peak value? Its rms value?

12-2.   An oscilloscope shows a sine wave with a peak-to-peak value of 250 mV. What is the peak value? The rms value?

12-3.   An ac voltmeter reads 30 mV rms when driven by a sine wave. What is the peak value? The peak-to-peak value?

12-4.   What is the rms value of a sine wave if

    a. $V_p = 100 \ \mu V$

    b. $V_{pp} = 450$ mV

    c.   A peak value of 25 mV

    d.   A peak-to-peak value of 15 V

Figure 12-22.

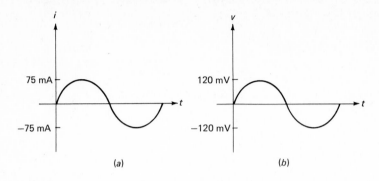

(a)   (b)

12-5. A sine wave of current has a peak value of 20 mA. What is the rms value? The peak-to-peak value?

12-6. What is the peak-to-peak value of the sine wave of current shown in Fig. 12-22a? The rms value?

12-7. Calculate the rms value of the sine wave of voltage in Fig. 12-22b. What is the peak-to-peak value?

12-8. What is the inductive reactance of each of these:
   a. $f = 1$ kHz and $L = 200$ mH
   b. $f = 5$ MHz and $L = 100$ μH
   c. $f = 30$ MHz and $L = 2$ μH
   d. $f = 75$ MHz and $L = 0.2$ μH

12-9. If 115 V rms are across a 30-H inductor at 60 Hz, what is the rms current?

12-10. Channel 4 of a TV receiver operates at a frequency of 69 MHz. How much inductive reactance does a 3-in piece of wire have at this frequency? (Use 0.02 μH/in.)

12-11. A 1-Ω resistor has a lead inductance of 0.02 μH. What does X_L equal at 10 MHz? At what frequency does X_L equal 0.1 Ω?

12-12. A 1-MΩ resistor has a stray capacitance of 1 pF. At what frequency does X_C equal 1 MΩ?

12-13. If the source frequency of Fig. 12-23a equals 100 kHz, what is the rms current through the inductor? If the frequency is changed to 500 kHz, what is the new value of the rms current?

Figure 12-23.

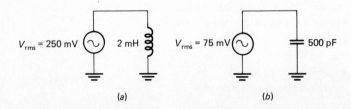

$V_{rms} = 250$ mV   2 mH $V_{rms} = 75$ mV   500 pF

(a)   (b)

12-14. Calculate the rms current into the capacitor of Fig. 12-23b for each of these source frequencies:
  a. 100 kHz
  b. 1 MHz
  c. 10 MHz

12-15. At what frequency does a 50-μF capacitor have a reactance of 20 Ω?

12-16. What value should C have to make X_C equal 1 kΩ at 1 MHz?

12-17. An inductor has 250 mV rms across it and 50 mA rms through it. What is the value of X_L?

12-18. A capacitor has 25 mV peak-to-peak across it and 2 μA rms into it. What is X_C?

12-19. Three 500-pF capacitors are in series. What is the equivalent capacitive reactance at 1 MHz?

12-20. If three capacitors have reactances of 25 Ω, 50 Ω, and 100 Ω, what is the equivalent reactance when these capacitors are in series? In parallel?

12-21. Two inductors with values of 20 mH and 50 mH are in series. What is the equivalent reactance at 500 kHz?

12-22. A capacitor is in series with a 10-kΩ resistor. To get good coupling at 1 kHz, what size should the capacitor have?

12-23. A 1-μF capacitor is in series with a 2-kΩ resistor. You will get good coupling as defined by condition (12-12) down to what frequency?

12-24. The source frequency of Fig. 12-24a can be from 10 Hz to 50 kHz. What size should C have to get good coupling?

12-25. Good coupling occurs in Fig. 12-24b provided the frequency is greater than what value?

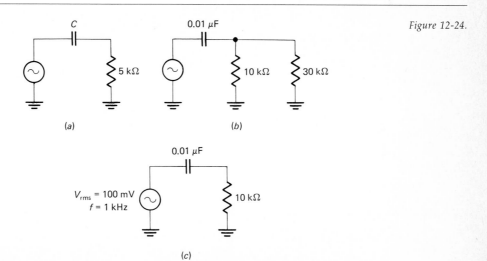

Figure 12-24.

(a)

(b)

(c)

Figure 12-25.

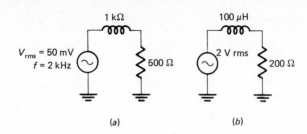

(a)   (b)

12-26.   Work out the quadratic sums of
   a.   1 and 1
   b.   3 and 4
   c.   5 and 7
   d.   4 and 8

12-27.   A 5-kΩ resistance is in series with a capacitive reactance of 5 kΩ. If the source voltage is 20 mV rms, what is the rms current?

12-28.   Calculate the rms current through the 10-kΩ resistor of Fig. 12-24c.

12-29.   An inductor is in series with a 1-kΩ resistor. To get good choking action, what size should X_L be? If the source frequency is 120 Hz, what value should L have?

12-30.   A resistance of 500 Ω is in series with an inductive reactance of 700 Ω. If the source voltage equals 100 mV rms, what is the rms current through the resistor?

12-31. $X_L = 10$ kΩ and $R = 1$ kΩ. If these are in series with a 250-mV rms source, what is the rms current?

12-32.   What is the rms current in Fig. 12-25a? The value of L?

12-33.   If the frequency of Fig. 12-25a is increased to 20 kHz, what is the new value of X_L? The new rms current?

12-34.   The source of Fig. 12-25b has a frequency of 1 MHz. What is the rms current?

12-35.   A 100-μH inductor is in series with a 400-pF capacitor and a 75-Ω resistor. What is the resonant frequency?

12-36.   A 20-μH inductor is in series with a capacitor. If the resonant frequency equals 2 MHz what is the value of C?

12-37.   A 1000-pF capacitor and an inductor have a resonant frequency of 5 MHz. What is the value of L?

Figure 12-26.

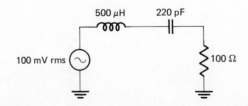

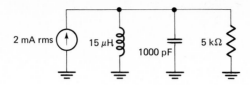

Figure 12-27.

12-38.   In Fig. 12-26, what is the resonant frequency? The rms current at resonance? The rms voltage across the 100-Ω resistor?

12-39.   What is the resonant frequency for a parallel connection of 200 mH and 2500 pF?

12-40.   A 5-mA rms current source drives a parallel resonant circuit with a resistance of 1 kΩ. What is the rms voltage across the circuit at resonance? The peak-to-peak voltage?

12-41.   What is the resonant frequency in Fig. 12-27? The rms voltage across the circuit?

ANSWERS TO TESTS

12-1.   *c, c, d, d, c, b*
12-2.   *d, b, a, a*
12-3.   *a, b, b, b, a, d, b*
12-4.   *a, b, b, b, a*
12-5.   *b, a, b, c*
12-6.   *b, c, c, a*

13. Codes, Numbers, and Sinusoids

Before you can understand complete ac analysis, you will have to learn about *codes, numbers,* and *sinusoids.* Once this is accomplished, we can extend almost all the theory about dc circuits to ac circuits.

13-1. CODES

As you read this section, get the answers to

> *What is a code?*
> *What is a one-to-one code?*

Definition of code

A *set* is a collection of similar but distinct objects such as letters, numbers, symbols, etc. These objects are referred to as *elements* of the set. Examples of sets are the English alphabet, Roman numerals, positive numbers, etc.

A *code* is any set whose elements stand for the elements of another set. The Morse code is a familiar example. As you know, dots and dashes stand for letters or numbers. Figure 13-1a illustrates the idea of the Morse code; dot-dash stands for A, dash-dot-dot-dot for B, and so on.

One-to-one codes

A code is a *one-to-one* code if each code element stands for one and only one element in another set (Fig. 13-1b). The Morse code is a one-to-one code, because each group of dots and dashes represents a unique letter or number.

Figure 13-1. (a) *Morse code*. (b) *One-to-one code*.

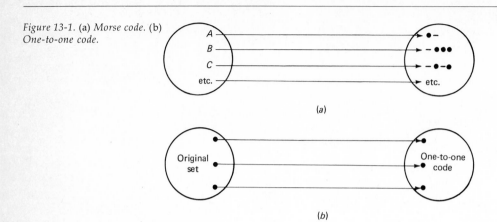

(a)

(b)

The *color code* is another example of a one-to-one code (see Fig. 13-2a). Each element in the color code is an arrangement of colors that stands for a particular resistor value.

Integers (whole numbers) are another one-to-one code (Fig. 13-2b). Here each integer (1, 2, 3, etc.) stands for a unique amount (•, • •, • • •, and so on). With the integer code, we can count; this leads to arithmetic and higher forms of mathematics.

Figure 13-2. (a) *Color code*. (b) *Integer code*. (c) *Not a one-to-one code*.

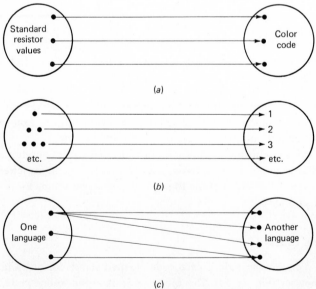

(a)

(b)

(c)

Not all codes are one-to-one. Languages, for example, are sets of words. Each word in one language may be translated into one or more words in another language, and vice versa (Fig. 13-2c). For this reason, even though languages are codes, they do not have the one-to-one property.

Test 13-1

1. Which of the following does not belong?
    (*a*) ambiguous   (*b*) one-to-one   (*c*) unique code element
    (*d*) unambiguous .... (   )
2. Element is to set as human being is to
    (*a*) child   (*b*) animal   (*c*) mankind   (*d*) trees .... (   )
3. The following words can be rearranged into a sentence: ARE ONE-TO-ONE CODE A SURNAMES. The sentence is
    (*a*) true   (*b*) false .... (   )
4. Which belongs least?
    (*a*) color code   (*b*) integer code   (*c*) Morse code   (*d*) language ...... (   )

13-2.   NUMBERS

Your job in this section is to find the answers to

> *What are real numbers?*
> *What are imaginary numbers?*
> *What are complex numbers?*

Real numbers

Real numbers are a code for the points along a line. Figure 13-3a illustrates the idea. The number 0 stands for the origin, a reference point from which all other points are

Figure 13-3. Real-number code.

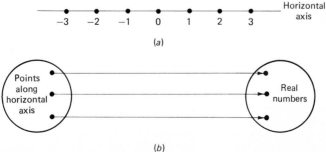

(*a*)

(*b*)

measured. Points right of the origin are labeled with positive numbers; those left of the origin with negative numbers.

Real numbers are a one-to-one code, as shown in Fig. 13-3*b*. Each real number stands for a unique point along the horizontal axis. Because of this, we automatically know that 3 represents the point that's three units right of the origin, -2 stands for the point two units left of the origin, and so on.

Imaginary numbers

Somewhere along the line you've heard of imaginary numbers like $\sqrt{-1}$, $\sqrt{-4}$, $\sqrt{-9}$, etc. And you already know, there's no way to get the square root of a negative number. In other words, there's no real number that multiplied by itself results in a negative number. Because it's impossible to extract the square root of a negative number, the early mathematicians called numbers like $\sqrt{-1}$, $\sqrt{-4}$, $\sqrt{-9}$, etc. *imaginary numbers.*

Imaginary numbers had no practical value until Gauss (1799) decided to use them as a code for points along a vertical axis. In Gauss' code, $\sqrt{-1}$ represents the unit vertical distance as shown in Fig. 13-4*a*. All other points have proportional labels; this is why $2\sqrt{-1}$ stands for the point that's two units above the origin, $3\sqrt{-1}$ for the point three units above the origin, and so on. Similarly, $-\sqrt{-1}$ represents the point one unit below the origin, $-2\sqrt{-1}$ is for the point two units below the origin, $-3\sqrt{-1}$ for the point three units below the origin, etc.

Gauss' imaginary-number code is a one-to-one code (Fig. 13-4*b*), because each

Figure 13-4. Imaginary-number code.

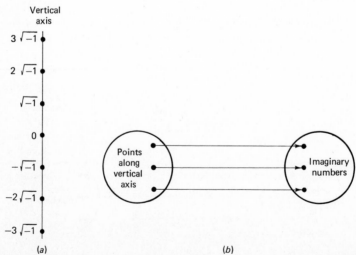

(a)   (b)

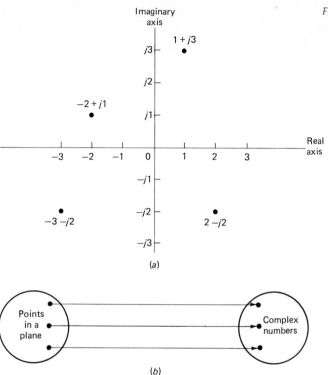

Figure 13-5. *Complex-number code.*

imaginary number stands for a unique point along the vertical axis. Each imaginary number represents one and only one point along this axis.

In Fig. 13-4a the basic unit for the entire code is $\sqrt{-1}$; all other code elements are multiples of this basic unit. But $\sqrt{-1}$ is somewhat awkward to use in diagrams and equations. For this reason, most people use j instead of $\sqrt{-1}$. As a defining equation,

$$j = \sqrt{-1}$$

So, $2\sqrt{-1}$ can be written as $2j$; normally, the j is put in front as an immediate reminder that an imaginary number is involved; therefore, $2\sqrt{-1}$ is conventionally written as $j2$.

Complex numbers

A *complex number* is the algebraic sum of a real number and an imaginary number. Examples of complex numbers are

$$1 + j3$$

$$-2 + j1$$

$$-3 - j2$$

$$2 - j2$$

Numbers like these were worthless until Gauss used them as a code for points in a plane.

Here's the idea behind Gauss' complex-number code. Points on the horizontal axis are coded with real numbers, as shown in Fig. 13-5a; for this reason, the horizontal axis is called the *real axis*. Similarly, points along the vertical axis are labeled with imaginary numbers, and this axis is known as the *imaginary axis*. A complex number like $1 + j3$ stands for the point that's one unit to the right of the origin and three units up (see Fig. 13-5a).

Likewise, complex number $-2 + j1$ represents the point two units left of the origin and one unit up; $-3 - j2$ is for the point three units left of the origin and two units down; $2 - j2$ stands for the point two units right of the origin and two units down, and so forth. In general, $x + jy$ represents the point x units right of the origin and y units up. (If x is negative, move to the left of the origin; if y is negative, move down.)

Complex numbers are a one-to-one code (Fig. 13-5b), because each complex number represents one and only one point in the plane.

Test 13-2

1. The following words can be rearranged into a sentence: ONE-TO-ONE CODE A REAL NUMBERS ARE. The sentence is
   (*a*) true   (*b*) false   .... ( )
2. Real number is to imaginary number as point on a horizontal axis is to point on
   (*a*) a horizontal axis   (*b*) a vertical axis   (*c*) both axes
   (*d*) a plane .... ( )
3. Imaginary numbers stand for points
   (*a*) on a horizontal axis   (*b*) along a vertical axis   (*c*) in a plane .... ( )
4. Imaginary numbers don't exist
   (*a*) at all   (*b*) on the imaginary axis   (*c*) on the real axis   .... ( )
5. Real number is to complex number as a point on the real axis is to a
   (*a*) complex number   (*b*) point on the real axis
   (*c*) point on the imaginary axis   (*d*) point in the plane   .... ( )

13-3.   RECTANGULAR, POLAR, AND VECTOR CODES

Gauss' complex-number code revolutionized mathematics and led to other useful codes. In reading this section, find the answers to

What is a rectangular number?
What is a polar number?
What is a vector?

Rectangular numbers

In the preceding section, each complex number was written in the form

$$x + jy$$

A complex number like this is called a *rectangular number,* because the coded point lies x units right of the origin and y units up (assuming positive x and y values).

The first half of a rectangular number is called the *real part,* and the second half is known as the *imaginary part.* So x is the real part, and jy is the imaginary part. Although jy is an imaginary number, y itself is a real number. For instance, given a number like $2 + j3$, notice that 2 and 3 are real numbers; $j3$ is an imaginary number.

Polar numbers

Figure 13-6a shows a point $x + jy$. It's located x units right of the origin and y units up. By drawing a hypotenuse from the origin to the point, we get a right triangle. With the Pythagorean theorem, the length of this hypotenuse is

$$M = \sqrt{x^2 + y^2} \tag{13-1}$$

(Notice that y is used in this equation, not jy.) So, given a rectangular number like $3 + j4$, the length of the hypotenuse is

$$M = \sqrt{3^2 + 4^2} = 5$$

as shown in Fig. 13-6b.

You've already studied trigonometry elsewhere, and know that the tangent of an angle is the side opposite the angle divided by the adjacent side. In Fig. 13-6a,

$$\tan \phi = \frac{y}{x} \tag{13-2}$$

(Notice y is used in this equation, not jy.) In terms of the inverse function, Eq. (13-2) may be written

$$\phi = \arctan \frac{y}{x} \tag{13-3}$$

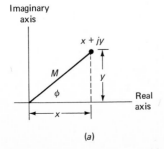

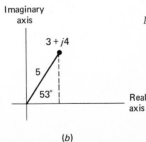

(a) (b)

Figure 13-6. Applying the Pythagorean theorem and the arctangent formula.

Figure 13-7. Polar numbers.

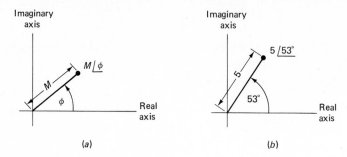

(a)   (b)

So the point coded $3 + j4$ in Fig. 13-6b results in an angle of

$$\phi = \arctan \frac{4}{3} = \arctan 1.33 = 53°$$

(In this book all angles are rounded off to the nearest degree.)

Think. Is it not true that we can code any point in the plane by specifying its M and ϕ values rather than its x and y values? In other words, each point in the plane has a unique pair of M and ϕ values. Because of this, we can code each point by specifying the distance M from the origin and the angle ϕ between the hypotenuse and the horizontal axis (see Fig. 13-7a).

The standard way to specify the M and ϕ values of a point is to label the point with

$$M \angle \phi$$

where M is the distance from the origin to the point, and ϕ is the angle between the hypotenuse and the horizontal axis. The foregoing label is read "M at an angle of ϕ." For instance, Fig. 13-7b shows a point labeled $5 \angle 53°$; you read this as "5 at an angle of 53°." This label tells you the point is located 5 units from the origin at an angle of 53°.

Numbers in the form of $M \angle \phi$ are called *polar numbers*. Numbers like these are classified as complex numbers, because they are code elements for points in the plane. To put it another way, complex numbers may be written in either of two ways: as rectangular numbers or as polar numbers. Rectangular number $3 + j4$ (Fig. 13-6b) and polar number $5 \angle 53°$ (Fig. 13-7b) are both complex numbers. Furthermore, since they each represent the same point in the plane, they are equal and we may write

$$3 + j4 = 5 \angle 53°$$

Polar numbers as a code

Polar numbers may be thought of as a one-to-one code for rectangular numbers (see Fig. 13-8a). Each polar number stands for a unique rectangular number. To convert from a rectangular to a polar number, Eq. (13-1) gives the M value and Eq. (13-3) gives the ϕ value. Because of this, the rectangular-to-polar conversion formula is

$$M \angle \phi = \sqrt{x^2 + y^2} \angle \text{ arctan } y/x \qquad (13\text{-}4)$$

Here's an example of using Eq. (13-4). Given $3 + j2$ (Fig. 3-8b), the corresponding polar number is

$$M \angle \phi = \sqrt{3^2 + 2^2} \angle \text{arctan } 2/3 = 3.61 \angle 34°$$

As another example, $2 - j4$ (Fig. 3-8c) converts to polar number

$$M \angle \phi = \sqrt{2^2 + (-4)^2} \angle \text{arctan}(-4/2) = 4.47 \angle -63°$$

In Fig. 13-9,

$$\sin \phi = \frac{y}{M}$$

and

$$\cos \phi = \frac{x}{M}$$

These equations rearrange into

$$y = M \sin \phi$$

$$x = M \cos \phi$$

Figure 13-8. Polar numbers are a one-to-one code for rectangular numbers.

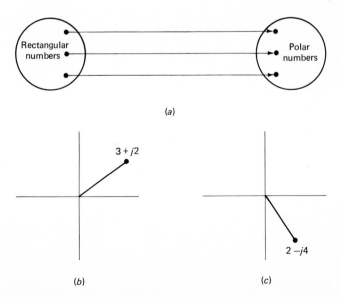

(a)

(b)   (c)

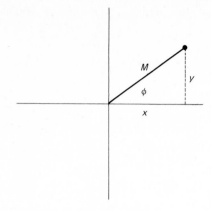

Figure 13-9. Converting from polar to rectangular numbers.

Therefore, the polar-to-rectangular conversion formula is

$$x + jy = M \cos \phi + jM \sin \phi \qquad (13\text{-}5)$$

For instance, given $4\underline{/\,30°}$, the corresponding rectangular number is

$$x + jy = 4 \cos 30° + j4 \sin 30° = 3.46 + j2$$

Or given $7\,\underline{/-60°}$, the corresponding rectangular number is

$$x + jy = 7 \cos (-60°) + j7 \sin (-60°)$$
$$= 7 \cos 60° - j7 \sin 60° = 3.5 - j6.06$$

Become adept at using Eqs. (13-4) and (13-5); they are essential for ac analysis. Equation (13-4) is the one to use when you want to convert a rectangular number to the equivalent polar number. Conversely, Eq. (13-5) is for converting a polar number to the equivalent rectangular number.

Vectors

A *vector* is an arrow from the origin to a point in the plane (see Fig. 13-10a). The magnitude M of a vector is the same as its length, and the angle ϕ of a vector is equal to the angle between the vector and the horizontal axis. By definition, an angle is positive when measured counterclockwise from the positive real axis. For instance, Fig. 13-10b shows a vector with a magnitude of 5 and an angle of 60°; this angle is positive because it's measured in a counterclockwise direction.

Angles are negative if measured clockwise from the positive real axis. Figure 13-10c shows a vector with a magnitude of 5 and an angle of −60°; the angle is negative, because it's measured clockwise from the positive real axis.

When specifying the angle of a vector, you are free to measure the angle in either

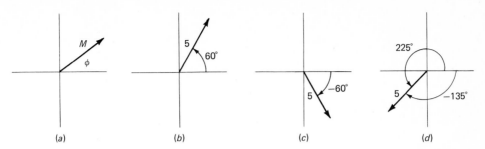

Figure 13-10. Vectors.

(a)   (b)   (c)   (d)

direction. In Fig. 13-10d, the vector has a magnitude of 5. The angle can be specified either as 225° or as −135°; these angles are equivalent, because either pins down the position of the vector.

Vectors are a one-to-one code for complex numbers (Fig. 13-11). Since each vector has a unique value of M and φ, the point at the tip of the vector is M∠φ. This means each distinct vector is associated with one and only one complex number. Because of this, vectors are a *visual* code for complex numbers; they help us visualize complex numbers.

So, now we have three ways to represent complex numbers: by rectangular numbers, by polar numbers, and by vectors. In ac analysis rectangular and polar numbers are used to calculate currents and voltages; vectors are used to draw diagrams as a visual aid in the analysis.

EXAMPLE 13-1.
The vectors of Fig. 13-12a point at j3 and −j2. Convert these rectangular numbers to polar numbers.

SOLUTION.
You can convert to polar numbers by using Eq. (13-4), but when the answer is geometrically obvious you don't need to use Eq. (13-4). Look at the upper vector in Fig. 13-12a. The magnitude of the vector is 3, because this is the distance from the origin to the point; the angle is 90°. Therefore, by inspection you can write

$$j3 = 3 \ \underline{/90°}$$

Figure 13-11. Vectors are a one-to-one code for complex numbers.

Complex numbers   Vectors

Figure 13-12. Examples.

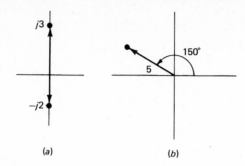

(a)   (b)

Likewise, the lower vector in Fig. 13-12a has a magnitude of 2 and an angle of $-90°$; so

$$-j2 = 2 \; \underline{/-90°}$$

EXAMPLE 13-2.

What complex number does the vector of Fig. 13-12b represent?

SOLUTION.

The vector has a magnitude of 5 and an angle of 150°; therefore, the point at the tip of the vector is associated with complex number $5\underline{/150°}$. If necessary, this polar number may be converted to rectangular form using Eq. (13-5):

$$x + jy = M \cos \phi + jM \sin \phi$$
$$= 5 \cos 150° + j5 \sin 150°$$
$$= -4.33 + j2.5$$

Test 13-3

1. Rectangular numbers have
   (a) magnitude and angle   (b) real and imaginary parts
   (c) magnitude only   (d) real part only .... ( )
2. Which belongs least?
   (a) real number   (b) vector   (c) rectangular number
   (d) polar number .... ( )
3. A vector that points at a positive imaginary number must have an angle of
   (a) 0°   (b) 90°   (c) 180°   (d) 270° .... ( )
4. If a vector points at -4, it has an angle of
   (a) 0°   (b) 90°   (c) 180°   (d) $-90°$ .... ( )
5. A vector points at $-j3$. This vector has an angle of
   (a) 0°   (b) 90°   (c) 180°   (d) $-90°$ .... ( )

13-4. THE ARITHMETIC OF COMPLEX NUMBERS

Before you can analyze ac circuits, you need to know the answers to

How do you add and subtract complex numbers?
How do you multiply and divide complex numbers?

Addition and subtraction

You can't add or subtract complex numbers if they are in polar form; it's necessary to express complex numbers in rectangular form before you can add or subtract. Given two complex numbers

$$x_1 + jy_1$$

$$x_2 + jy_2$$

the defining formula for the sum of these complex numbers is

$$(x_1 + jy_1) + (x_2 + jy_2) = x_1 + x_2 + j(y_1 + y_2) \qquad (13\text{-}6)$$

All this says is add the real parts and then add the imaginary parts; the result is the sum of the original numbers.

For instance, given

$$2 + j3$$

$$4 + j7$$

the sum is

$$(2 + j3) + (4 + j7) = 6 + j10$$

When we say sum, we mean *algebraic* sum. So if the original numbers are

$$-3 + j4$$

$$5 - j2$$

the sum is

$$(-3 + j4) + (5 - j2) = 2 + j2$$

Another example, if the original numbers are

$$5 - j6$$

$$-8 + j7$$

the sum is

$$(5 - j6) + (-8 + j7) = -3 + j1$$

Subtraction is similar. If the original numbers are

$$x_1 + jy_1$$

$$x_2 + jy_2$$

the algebraic difference is

$$(x_1 + jy_1) - (x_2 + jy_2) = x_1 - x_2 + j(y_1 - y_2) \tag{13-7}$$

This defining formula says the difference of two complex numbers is the difference of the real parts and the difference of the imaginary parts. Given

$$5 - j6$$

$$-8 + j3$$

the difference is

$$(5 - j6) - (-8 + j3) = 13 - j9$$

Multiplication and division

It's usually easiest to multiply and divide complex numbers when they are in polar form. Given two polar numbers like

$$M_1 \underline{/\phi_1}$$

$$M_2 \underline{/\phi_2}$$

the defining formula for the product is

$$M_1 \underline{/\phi_1} \times M_2 \underline{/\phi_2} = M_1 M_2 \underline{/\phi_1 + \phi_2} \tag{13-8}$$

This says you multiply the magnitudes and add the angles algebraically.

For instance, given

$$5 \underline{/30°}$$

$$10 \underline{/60°}$$

the product of these two complex numbers is

$$5 \underline{/30°} \times 10 \underline{/60°} = 50 \underline{/90°}$$

We've multiplied the magnitudes and added the angles. As another example, if the two numbers are

$$4 \underline{/60°}$$

$$2 \underline{/-90°}$$

the product is

$$4 \,\underline{/60°} \times 2 \,\underline{/-90°} = 8 \,\underline{/-30°}$$

Notice the angles are added algebraically.

Division is defined in a similar way. If the two complex numbers are

$$M_1 \,\underline{/\phi_1}$$

$$M_2 \,\underline{/\phi_2}$$

the quotient is defined as

$$\frac{M_1 \,\underline{/\phi_1}}{M_2 \,\underline{/\phi_2}} = \frac{M_1}{M_2} \,\underline{/\phi_1 - \phi_2} \qquad (13\text{-}9)$$

This defining formula says you divide the magnitudes and algebraically subtract the angles.

So if the two numbers are

$$12 \,\underline{/50°}$$

$$3 \,\underline{/-15°}$$

the quotient is

$$\frac{12 \,\underline{/50°}}{3 \,\underline{/-15°}} = 4 \,\underline{/65°}$$

Properties of j

Notice that

$$j^2 = j \times j = \sqrt{-1} \times \sqrt{-1} = -1$$

So from now on, j^2 may be simplified by replacing it with -1. Also notice that

$$j^3 = j^2 \times j = -1 \times j = -j$$

Therefore, j^3 may be replaced by $-j$. Finally, note that

$$j^4 = j^2 \times j^2 = -1 \times -1 = 1$$

Whenever you see j^4, you can replace it with 1.

Another property of j is this:

$$\frac{1}{j} = -j$$

This is easy to prove as follows.

$$\frac{1}{j} = \frac{1 \,\underline{/0°}}{1 \,\underline{/90°}} = 1 \,\underline{/-90°} = -j$$

In summary, remember these simplification formulas:

$$j^2 = -1$$

$$j^3 = -j$$

$$j^4 = 1$$

$$\frac{1}{j} = -j$$

Rectangular multiplication

Even though multiplication is usually easier when the complex numbers are in polar form, occasionally you may need to multiply two rectangular numbers directly. Given

$$3 + j4$$

$$2 + j3$$

you get the product, the same as in ordinary algebra:

$$
\begin{aligned}
(3 + j4)(2 + j3) &= 6 + j9 + j8 + j^2 12 \\
&= 6 + j17 - 12 \\
&= -6 + j17
\end{aligned}
$$

As another example, if the two numbers are $2 - j5$ and $4 + j9$, the product is

$$
\begin{aligned}
(2 - j5)(4 + j9) &= 8 + j18 - j20 - j^2 45 \\
&= 8 - j2 + 45 \\
&= 53 - j2
\end{aligned}
$$

Rectangular division

If a complex number is in the rectangular form

$$x + jy$$

the *complex conjugate* is defined as

$$x - jy$$

So given $3 + j4$, the complex conjugate is $3 - j4$. Or given $2 - j5$, the complex conjugate is $2 + j5$. To get the complex conjugate of any rectangular number, therefore, all you do is change the sign of the imaginary part.

Even though division of complex numbers is usually easier when the numbers are in polar form, occasionally you may need to divide complex numbers in rectangular form. Here's how it's done. Suppose you have

$$\frac{3 + j4}{2 + j3}$$

Division is accomplished by multiplying the numerator and the denominator by the complex conjugate of the denominator as follows:

$$\frac{3 + j4}{2 + j3}\frac{2 - j3}{2 - j3}$$

In this first step both the numerator and the denominator are being multiplied by the complex conjugate of the denominator which is $2 - j3$. The next step is to carry out the rectangular multiplication as follows:

$$\frac{3 + j4}{2 + j3}\frac{2 - j3}{2 - j3} = \frac{6 - j9 + j8 - j^2 12}{4 - j6 + j6 - j^2 9} = \frac{18 - j1}{13}$$
$$= 1.38 - j0.0769$$

This may not look like division, but nevertheless it is. We divided $3 + j4$ by $2 + j3$ and got an answer of $1.38 - j0.0769$. The key idea in dividing two rectangular numbers is to multiply the numerator and the denominator by the complex conjugate of the denominator; after carrying out the multiplication and simplifying as much as possible, you have the quotient of the original numbers. Sometimes this process of division is called *rationalization*.

Test 13-4

1. Addition of complex numbers is usually easier when the numbers are in
   (*a*) rectangular form   (*b*) polar form .... ( )
2. Multiplication of complex numbers is usually easier when the numbers are in
   (*a*) rectangular form   (*b*) polar form .... ( )
3. When you multiply polar numbers, the angles are always algebraically
   (*a*) added   (*b*) subtracted   (*c*) multiplied   (*d*) divided .... ( )
4. Which of these does not belong?
   (*a*) division of polar numbers   (*b*) algebraic difference of angles
   (*c*) quotient of angles   (*d*) quotient of magnitudes .... ( )
5. Which of the following is the same as j^7?
   (*a*) j (*b*) -1 (*c*) $-j$ (*d*) 1 .... ( )

13-5. SINUSOIDS

Figure 13-13*a* shows a *sine wave,* the graph of any equation in the form

$$y = M \sin \theta$$

You've already learned in other courses how to look up the sine, cosine, and tangent of any angle. So it should be clear why the waveform has these values:

$$y = 0 \qquad \text{when} \qquad \theta = 0°$$

$$y = M \qquad \text{when} \qquad \theta = 90°$$

$$y = 0 \qquad \text{when} \qquad \theta = 180°$$

$$y = -M \qquad \text{when} \qquad \theta = 270°$$

and so on. For any other value of θ, you can calculate the value of y by looking up the sine function in a trigonometry table, or by using a slide rule.

Figure 13-13b shows a *cosine wave*, the graph of any equation in the form

$$y = M \cos \theta \tag{13-10}$$

With the well-known trigonometric identity

$$\cos \theta = \sin (\theta + 90°)$$

Eq. (13-10) may be written as

$$y = M \sin (\theta + 90°) \tag{13-11}$$

Whether you graph Eq. (13-10) or (13-11), the result is the same; you get the cosine wave shown in Fig. 13-13b.

Notice that a cosine wave *leads* a sine wave by 90°, that is, it starts 90° sooner than a sine wave.

Figure 13-13. (a) *Sine wave.* (b) *Cosine wave.*

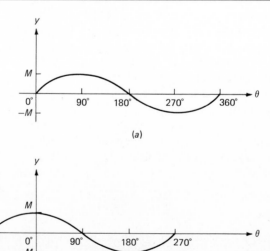

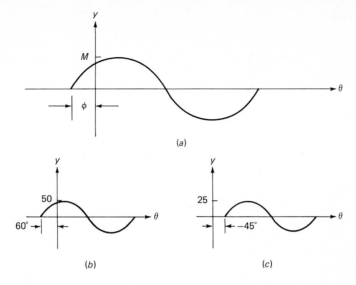

Figure 13-14. Sinusoid.

Figure 13-14*a* shows a waveform that leads a sine wave by ϕ degrees. The angle ϕ is called the *phase angle*. When ϕ equals 0°, the waveform is a sine wave; when ϕ equals 90°, the waveform is a cosine wave. For any value of ϕ, we refer to the waveform by the general term *sinusoid*.

A sinusoid, therefore, is any waveform with the shape of a sine wave but with an arbitrary phase angle. In fact, the equation for any sinusoid has the form

$$y = M \sin (\theta + \phi) \tag{13-12}$$

So given an equation like

$$y = 50 \sin (\theta + 60°)$$

we know immediately that the graph of this equation has a peak value of 50 and a phase angle of 60°, as shown in Fig. 13-14*b*.

If ϕ is negative, the sinusoid lags instead of leads. For instance, given

$$y = 25 \sin (\theta - 45°)$$

the peak value is 25 and the phase angle is −45°. Therefore, the sinusoid looks like the waveform of Fig. 13-14*c*; it has a peak value of 25 but lags a sine wave by 45°.

To summarize the important ideas we need for ac analysis:

1. A sinusoid is a waveform described by the general equation

$$y = M \sin (\theta + \phi)$$

2. M is the peak value and ϕ is the phase angle.

Figure 13-15. Examples of peak value and phase angle.

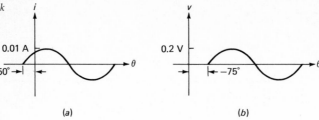

3.   A positive value of ϕ means the sinusoid leads a sine wave; a negative value of ϕ means the sinusoid lags a sine wave.

EXAMPLE 13-3.

Write the equations for the sinusoids shown in Figs. 13-15a and b.

SOLUTION.

a.   The peak value is 0.01 and the phase angle is 50°. Therefore,

$$i = 0.01 \sin (\theta + 50°)$$

b.   In this case, the voltage has a peak value of 0.2 V and a phase angle of $-75°$. So,

$$v = 0.2 \sin (\theta - 75°)$$

Incidentally, *sinusoidal* is the adjective form of sinusoid. Waveforms like Fig. 13-15a and b are described as sinusoidal current and sinusoidal voltage.

Test 13-5

1.   A sine wave has a phase angle of
     (a) 0°   (b) 30°   (c) 90°   (d) $-90°$ .... ( )
2.   A cosine wave has a phase angle of
     (a) 0°   (b) $-90°$ (c) 90°   (d) 180° .... ( )
3.   When a phase angle is negative, the sinusoidal wave
     (a) leads a sine wave   (b) lags a sine wave
     (c) is in phase with a sine wave .... ( )

SUMMARY OF FORMULAS

DEFINED

$$(x_1 + jy_1) + (x_2 + jy_2) = x_1 + x_2 + j(y_1 + y_2) \tag{13-6}$$

$$(x_1 + jy_1) - (x_2 + jy_2) = x_1 - x_2 + j(y_1 - y_2) \tag{13-7}$$

$$M_1 \ \underline{/\phi_1} \times M_2 \ \underline{/\phi_2} = M_1M_2 \ \underline{/\phi_1 + \phi_2} \tag{13-8}$$

$$\frac{M_1 \ \underline{/\phi_1}}{M_2 \ \underline{/\phi_2}} = \frac{M_1}{M_2} \ \underline{/\phi_1 - \phi_2} \tag{13-9}$$

DERIVED

$$M = \sqrt{x^2 + y^2} \tag{13-1}$$

$$\phi = \arctan \frac{y}{x} \tag{13-3}$$

$$M \ \underline{/\phi} = \sqrt{x^2 + y^2} \ \underline{/\arctan y/x} \tag{13-4}$$

$$x + jy = M \cos \phi + jM \sin \phi \tag{13-5}$$

Problems

13-1. Work out the magnitude of the polar number corresponding to each of these rectangular numbers:
  a. $2 + j5$
  b. $-3 + j6$
  c. $-4 - j4$
  d. $5 - j2$

13-2. Calculate the angle of the polar number corresponding to each rectangular number in the preceding problem.

13-3. Convert each of the following rectangular numbers to a polar number:
  a. $1 + j1$
  b. $-2 + j3$
  c. $-3 - j1$
  d. $4 - j2$

13-4. Convert each of the following rectangular numbers to a polar number:
  a. $150 + j300$
  b. $-200 + j350$
  c. $-375 - j250$
  d. $465 - j750$

13-5. Convert each of the following polar numbers to a rectangular number:
  a. $4 \ \underline{/60°}$
  b. $7 \ \underline{/130°}$
  c. $5 \ \underline{/250°}$
  d. $8 \ \underline{/330°}$

13-6. Convert each of the following polar numbers to a rectangular number:
  a. $2000 \ \underline{/45°}$
  b. $1500 \ \underline{/125°}$

    c. 3000 $\underline{/-60°}$

    d. 4500 $\underline{/-130°}$

13-7. Draw the vectors for each rectangular number of Problem 13-1.

13-8. Draw the vectors for each rectangular number of Problem 13-3.

13-9. Draw the vectors for each polar number of Problem 13-5.

13-10. Draw the vectors for each polar number of Problem 13-6.

13-11. Add the following complex numbers:

    a. $(2 + j5) + (-3 + j6)$

    b. $(-4 - j4) + (5 - j2)$

    c. $(-3 + j6) + (-4 - j4)$

    d. $(5 - j2) + (3 + j8)$

13-12. Subtract the following complex numbers:

    a. $(3 - j2) - (4 + j7)$

    b. $(4 - j5) - (-6 + j8)$

    c. $(-5 + j3) - (3 - j6)$

    d. $(-7 + j2) - (-3 - j9)$

13-13. Multiply the following polar numbers:

    a. 5 $\underline{/30°}$ × 3 $\underline{/45°}$

    b. 7 $\underline{/120°}$ × 4 $\underline{/90°}$

    c. 3 $\underline{/65°}$ × 8 $\underline{/-36°}$

    d. 5 $\underline{/-35°}$ × 6 $\underline{/-30°}$

13-14. Divide the following polar numbers:

    a. $\dfrac{5\ \underline{/30°}}{2\ \underline{/70°}}$

    b. $\dfrac{7\ \underline{/120°}}{3\ \underline{/-30°}}$

    c. $\dfrac{10\ \underline{/-50°}}{2\ \underline{/20°}}$

    d. $\dfrac{12\ \underline{/-60°}}{3\ \underline{/-30°}}$

13-15. Simplify each of these:

    a. $j^2 18$

    b. $j^3 7$

    c. $j^5 9$

    d. $j^6 10$

13-16. Multiply each of the following:

    a. $(2 + j3) \times (3 + j5)$

    b. $(-4 + j5) \times (-5 - j2)$

    c. $(-3 - j2) \times (-6 - j7)$

    d. $(7 - j2) \times (5 + j3)$

13-17. Divide each of the following using rationalization:

    a. $\dfrac{4 + j2}{2 + j3}$

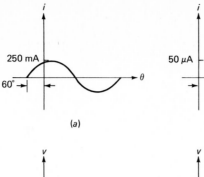

(a)

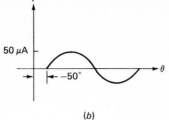

(b)

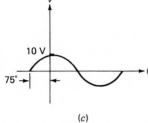

(c)

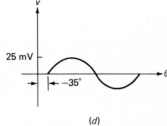

(d)

Figure 13-16.

b. $\dfrac{3 + j1}{2 - j4}$

c. $\dfrac{5 - j3}{6 + j4}$

d. $\dfrac{6 - j5}{3 - j2}$

13-18. What is the peak value and phase angle for each of these sinusoids:
  a. $v = 50 \sin (\theta + 65°)$
  b. $i = 25 \sin (\theta - 50°)$
  c. $v = 12 \sin (\theta + 120°)$
  d. $i = 0.5 \sin (\theta - 45°)$

13-19. What is the peak value and phase angle for each sinusoidal waveform shown in Fig. 13-16? Write the equation for each.

ANSWERS TO TESTS

13-1. *a, c, b, d*
13-2. *a, b, b, c, d*
13-3. *b, a, b, c, d*
13-4. *a, b, a, c, c*
13-5. *a, c, b*

14. Phasor Analysis

Most of the ideas about dc circuits apply to ac circuits, provided you use complex numbers when calculating currents and voltages.

14-1. PHASOR CURRENT

The general equation for any sinusoid has the form

$$y = M \sin (\theta + \phi)$$

For a sinusoidal current, the equation becomes

$$i = I_p \sin (\theta + \phi) \tag{14-1}$$

where i = instantaneous current
I_p = peak current
ϕ = phase angle

(Sometimes I_m is used instead of I_p.) Figure 14-1a shows the graph of this sinusoidal current.

Phasor current is defined as the polar number whose magnitude equals the rms current, and whose angle equals the phase angle. As a defining formula,

$$\mathbf{I} = I \; \underline{/\phi} \tag{14-2}$$

where $\mathbf{I}$ = phasor current
I = rms current
ϕ = phase angle

Figure 14-1. Phasor current.

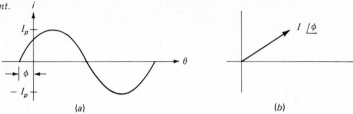

(a)   (b)

Notice that phasor current is printed as a boldface **I**; this is done to indicate it's a complex number. On the other hand, rms current is printed as an italic I. Watch for these differences in later formulas. Since phasor current is a complex number, we can visualize it as a vector (see Fig. 14-1*b*).

Here's an example. Suppose a sinusoidal current has an equation of

$$i = 10 \sin (\theta + 30°)$$

The rms current is

$$I = 0.707I_p = 0.707(10 \text{ A}) = 7.07 \text{ A}$$

Therefore, the phasor current is

$$\mathbf{I} = I \underline{/\phi} = 7.07 \text{ A} \underline{/30°}$$

Read this as 7.07 A at an angle of 30°. Figure 14-2*a* shows the waveform, and Fig. 14-2*b* shows the phasor current.

Figure 14-2. Examples of phasor current.

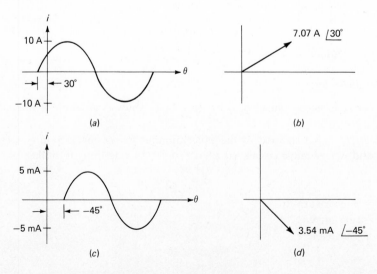

(a)   (b)

(c)   (d)

As another example, given a sinusoidal current like Fig. 14-2c, you can calculate an rms current of

$$I = 0.707I_p = 0.707(5 \text{ mA}) = 3.54 \text{ mA}$$

Then, the phasor current is

$$\mathbf{I} = I \; \underline{/\phi} = 3.54 \text{ mA} \; \underline{/-45°}$$

Figure 14-2d shows how to visualize this phasor current as a vector.

14-2. PHASOR VOLTAGE

The general formula

$$y = M \sin(\theta + \phi)$$

is written as follows for a sinusoidal voltage:

$$v = V_p \sin(\theta + \phi) \tag{14-3}$$

where v = instantaneous voltage
V_p = peak voltage
ϕ = phase angle

(Sometimes V_m or E_m is used instead of V_p.)

Phasor voltage is defined as the polar number whose magnitude equals the rms voltage, and whose angle equals the phase angle. In symbols,

$$\mathbf{V} = V \; \underline{/\phi} \tag{14-4}$$

where $\mathbf{V}$ = phasor voltage
V = rms voltage
ϕ = phase angle

Again notice the difference in the printing of phasor voltage and rms voltage; phasor voltage is printed as a boldface $\mathbf{V}$, while rms voltage is printed as an italic V.

If you see a waveform like Fig. 14-3a on an oscilloscope, you can calculate an rms voltage of

$$V = 0.707V_p = 0.707(20 \text{ mV}) = 14.1 \text{ mV}$$

Since the phase angle is 60°, the phasor voltage is

$$\mathbf{V} = V \; \underline{/\phi} = 14.1 \text{ mV} \; \underline{/60°}$$

Read this as 14.1 mV at an angle of 60°. Figure 14-3b shows this phasor voltage as a vector.

EXAMPLE 14-1.
A cosine wave has a peak-to-peak value of 20 V. What is the phasor voltage?

Figure 14-3. Phasor voltage.

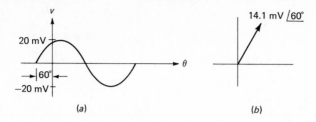

(a)   (b)

SOLUTION.

The peak value is half the peak-to-peak value. So,

$$V = 0.707V_p = 0.707(10\ V) = 7.07\ V$$

As explained in the preceding chapter, a cosine wave always has a phase angle of 90°. Therefore, the phasor voltage is

$$\mathbf{V} = V\underline{/\phi} = 7.07\ V\underline{/90°}$$

14-3.   IMPEDANCE

The three basic components are resistors, capacitors, and inductors. To account for the effect of these components on sinusoidal currents and voltages, we need a new concept called *impedance*. The key questions are

> *How is impedance defined?*
> *What is the impedance of the basic components?*

Definition

Impedance is defined as the phasor voltage across a load divided by the phasor current into the load. In symbols,

$$\mathbf{Z} = \frac{\mathbf{V}}{\mathbf{I}} \tag{14-5}$$

where $\mathbf{Z}$ = impedance
  $\mathbf{V}$ = phasor voltage
  $\mathbf{I}$ = phasor current

Since $\mathbf{V}$ and $\mathbf{I}$ are complex numbers, $\mathbf{Z}$ is a complex number; this is why impedance is printed as a boldface $\mathbf{Z}$.

Impedance of resistor

Look at Fig. 14-4a. If this resistor is ideal (negligible lead inductance and stray capacitance), a sinusoidal voltage across the resistor produces an in-phase sinusoidal

Figure 14-4. Impedance of an ideal resistor.

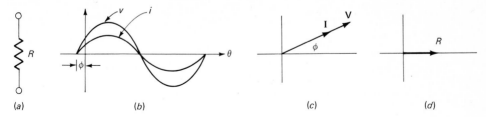

(a)   (b)   (c)   (d)

current as shown in Fig. 14-4*b*. Since both waves have the same phase angle, the vectors for phasor voltage and current likewise have the same phase angle (Fig. 14-4*c*). Regardless of the value of ϕ, the vectors are always superimposed.

With defining formula (14-5), the impedance of an ideal resistor is

$$\mathbf{Z}_R = \frac{\mathbf{V}}{\mathbf{I}} = \frac{V\;\underline{/\phi}}{I\;\underline{/\phi}} = \frac{V}{I}$$

or $\mathbf{Z}_R = R$ (ideal resistor)   (14-6)

This says the impedance of any ideal resistor equals the resistance. For instance, an ideal 1-kΩ resistor has an impedance of

$$\mathbf{Z}_R = 1 \text{ k}\Omega$$

An ideal 47-kΩ resistor has an impedance of

$$\mathbf{Z}_R = 47 \text{ k}\Omega$$

Because the impedance of an ideal resistor is always a real number, we can represent $\mathbf{Z}_R$ as a vector pointing at R (see Fig. 14-4*d*). As long as lead inductance and stray capacitance are negligible, the impedance vector of a resistor always is drawn as shown; a horizontal vector pointing to the right.

Impedance of inductor

If the inductor of Fig. 14-5*a* is ideal (negligible resistance and stray capacitance), a sinusoidal current through the inductor induces a sinusoidal voltage across the inductor. From previous discussions, the voltage leads the current by 90° as shown in Fig. 14-5*b*. Because of this, the phasor current is

$$\mathbf{I} = I\;\underline{/\phi}$$

and the phasor voltage is

$$\mathbf{V} = V\;\underline{/\phi + 90°}$$

Figure 14-5*c* illustrates the vectors for phasor voltage and current for any value of ϕ; the vectors are always perpendicular.

Figure 14-5. Impedance of an ideal inductor.

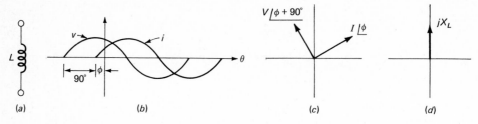

(a)   (b)   (c)   (d)

With the defining formula for impedance,

$$\mathbf{Z}_L = \frac{\mathbf{V}}{\mathbf{I}} = \frac{V \;\underline{/\phi + 90°}}{I \;\underline{/\phi}} = \frac{V}{I}\;\underline{/90°}$$

Recall from Chap. 12 that rms voltage divided by rms current equals the inductive reactance; so the foregoing equation becomes

$$\mathbf{Z}_L = X_L \;\underline{/90°} \quad \text{(polar)} \tag{14-7a}$$

which is equivalent to

$$\mathbf{Z}_L = jX_L \quad \text{(rectangular)} \tag{14-7b}$$

If an ideal inductor has a reactance of 2 kΩ, its impedance is

$$\mathbf{Z}_L = X_L \;\underline{/90°} = 2\text{ kΩ} \;\underline{/90°}$$

equivalent to

$$\mathbf{Z}_L = j2\text{ kΩ}$$

Figure 14-5*d* shows how to visualize the impedance of any ideal inductor; it's a vector pointing at jX_L. In polar terms, it's a vector with a magnitude of X_L and an angle of 90°.

Impedance of capacitor

If the capacitor of Fig. 14-6*a* is ideal, a sinusoidal voltage across it results in a sinusoidal current. Furthermore, the current leads the voltage by 90°, as shown in Fig. 14-6*b*. Because of this, phasor voltage and current are

$$\mathbf{V} = V \;\underline{/\phi}$$

$$\mathbf{I} = I \;\underline{/\phi + 90°}$$

Figure 14-6*c* shows this phasor voltage and phasor current.

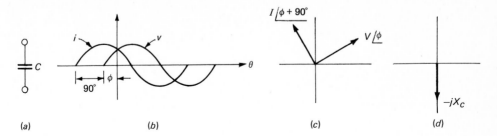

Figure 14-6. Impedance of an ideal capacitor.

(a) (b) (c) (d)

With defining formula (14-5),

$$\mathbf{Z}_C = \frac{\mathbf{V}}{\mathbf{I}} = \frac{V\ \underline{/\phi}}{I\ \underline{/\phi + 90°}} = \frac{V}{I}\ \underline{/-90°}$$

Chapter 12 showed that rms voltage divided by rms current equals capacitive reactance; therefore, the foregoing equation may be written as

$$\mathbf{Z}_C = X_C\ \underline{/-90°} \qquad \text{(polar)} \tag{14-8a}$$

equivalent to

$$\mathbf{Z}_C = -jX_C \qquad \text{(rectangular)} \tag{14-8b}$$

Figure 14-6d shows how to visualize this impedance as a vector. For an ideal capacitor, the vector always points down to $-jX_C$. In polar terms, the vector always has a magnitude of X_C and an angle of $-90°$.

EXAMPLE 14-2.
What is the impedance of an ideal 50-mH inductor at 10 kHz?

SOLUTION.
First, get the reactance:

$$X_L = 2\pi f L = 6.28(10^4)50(10^{-3})$$

$$= 3.14\ \text{k}\Omega$$

Then you can express the impedance either as

$$\mathbf{Z}_L = jX_L = j3.14\ \text{k}\Omega$$

or as

$$\mathbf{Z}_L = X_L\ \underline{/90°} = 3.14\ \text{k}\Omega\ \underline{/90°}$$

EXAMPLE 14-3.
What is the impedance of a 50-pF capacitor at 20 MHz?

SOLUTION.
First, calculate the reactance:

$$X_C = \frac{1}{2\pi fC} = \frac{1}{6.28(20)10^6(50)10^{-12}}$$
$$= 159 \ \Omega$$

Next, express the impedance either as

$$\mathbf{Z}_C = -jX_C = -j159 \ \Omega$$

or as

$$\mathbf{Z}_C = X_C \ \underline{/-90°} = 159 \ \Omega \ \underline{/-90°}$$

Test 14-1

1. Which of the following belongs least?
   (*a*) real number (*b*) polar number (*c*) impedance
   (*d*) rectangular number .... ( )
2. For the case of an ideal resistor, impedance is
   (*a*) a real number (*b*) an imaginary number (*c*) a vector pointing up
   (*d*) a vector pointing down .... ( )
3. The impedance vector of an ideal capacitor always points
   (*a*) up (*b*) down (*c*) right (*d*) left .... ( )
4. The impedance vector of an ideal inductor always points
   (*a*) up (*b*) down (*c*) right (*d*) left .... ( )
5. The phasor current into an ideal capacitor always leads the phasor voltage by
   (*a*) 20° (*b*) −90° (*c*) 90° (*d*) 180° .... ( )
6. The vectors representing the impedance of an ideal inductor and an ideal capacitor are always how far apart?
   (*a*) 0° (*b*) 90° (*c*) −90° (*d*) 180° .... ( )

14-4. OHM'S AC LAWS

Steinmetz (1865–1923) was the one who worked out the complete solution of ac circuits using complex numbers. He did not revise existing theory; he created a whole new world of it. Among his towering contributions is the extension of Ohm's law to ac circuits. The things to learn in this section are

What is Ohm's phasor law?
What is Ohm's rms law?

Phasor voltage and current

Steinmetz proved Ohm's law can be extended to ac loads, provided phasor voltage and phasor current are used. In other words, he showed the impedance of any combination of linear R's, L's, and C's remains constant at a given frequency. We will refer to this extension of Ohm's law to ac circuits as *Ohm's phasor law*.

With the understanding that $\mathbf{Z}$ is constant for a given load, the three ways of expressing Ohm's phasor law are

$$\mathbf{I} = \frac{\mathbf{V}}{\mathbf{Z}} \qquad (14\text{-}9a)$$

$$\mathbf{V} = \mathbf{Z}\mathbf{I} \qquad (14\text{-}9b)$$

$$\mathbf{Z} = \frac{\mathbf{V}}{\mathbf{I}} \qquad (14\text{-}9c)$$

Given two of the three quantities, you can calculate the third using one of the foregoing equations.

For example, given an ac load with a phasor voltage of 10 V $\underline{/30°}$ and an impedance of 2 kΩ $\underline{/70°}$, the phasor current into this load is

$$\mathbf{I} = \frac{\mathbf{V}}{\mathbf{Z}} = \frac{10 \text{ V } \underline{/30°}}{2 \text{ kΩ } \underline{/70°}} = 5 \text{ mA } \underline{/-40°}$$

So you see, the use of Ohm's phasor law is similar to the use of Ohm's dc law; the difference is that complex numbers are involved instead of real numbers.

Rms voltage and current

Ohm's phasor law says

$$\mathbf{I} = \frac{\mathbf{V}}{\mathbf{Z}}$$

or expressed in polar numbers,

$$I \underline{/\phi_1} = \frac{V \underline{/\phi_2}}{Z \underline{/\phi_3}} = \frac{V}{Z} \underline{/\phi_2 - \phi_3}$$

The magnitude of the left side is I, and the magnitude of the right side is V/Z; therefore, we can write

$$I = \frac{V}{Z} \qquad (14\text{-}10a)$$

This says rms current equals rms voltage divided by the *magnitude* of impedance. Especially important, notice that the magnitude of impedance is printed as an italic Z to distinguish it from impedance which is always printed as a boldface $\mathbf{Z}$. Equation (14-10a) will be called *Ohm's rms law*.

Here's an example. If the phasor voltage and impedance of a load are

$$\mathbf{V} = 50 \text{ V} \angle 30°$$

$$\mathbf{Z} = 10 \text{ k}\Omega \angle{-45°}$$

the rms voltage and magnitude of impedance are

$$V = 50 \text{ V}$$

$$Z = 10 \text{ k}\Omega$$

Then, with Ohm's rms law the rms current equals

$$I = \frac{V}{Z} = \frac{50 \text{ V}}{10 \text{ k}\Omega} = 5 \text{ mA}$$

Equation (14-10a) may be written in the alternative form

$$V = ZI \qquad\qquad (14\text{-}10b)$$

which says rms voltage equals magnitude of impedance times rms current. Another form is

$$Z = \frac{V}{I} \qquad\qquad (14\text{-}10c)$$

which tells us magnitude of impedance equals rms voltage divided by rms current.

Ohm's phasor law and Ohm's rms law are both important in ac analysis. The phasor law is useful as a starting point for deriving new formulas; the rms law is more helpful in practical work, because we can measure rms voltage with moving-coil ac voltmeters, digital ac voltmeters, etc.

EXAMPLE 14-4.
Calculate the phasor current in Fig. 14-7a.

SOLUTION.
First, get the capacitive reactance:

$$X_C = \frac{1}{2\pi fC} = \frac{1}{6.28(5)10^3(0.01)10^{-6}} = 3.18 \text{ k}\Omega$$

Second, write the impedance of the capacitor in polar form:

$$\mathbf{Z}_C = X_C \angle{-90°} = 3.18 \text{ k}\Omega \angle{-90°}$$

Third, use Ohm's phasor law to get the phasor current:

$$\mathbf{I} = \frac{\mathbf{V}}{\mathbf{Z}} = \frac{10 \text{ mV} \angle 0°}{3.18 \text{ k}\Omega \angle{-90°}} = 3.14 \text{ }\mu\text{A} \angle 90°$$

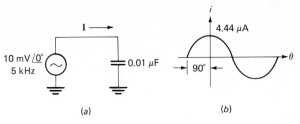

Figure 14-7. Calculating phasor
current.

The answer says the phasor current equals 3.14 μA at an angle of 90°. This means the rms current equals 3.14 μA and the phase angle is 90°. The corresponding sinusoidal current has a peak value of

$$I_p = 1.414I = 1.414(3.14\ \mu\text{A}) = 4.44\ \mu\text{A}$$

Figure 14-7b shows the sinusoidal current.

EXAMPLE 14-5.
If an ac voltmeter were connected across the capacitor of Fig. 14-8a, what rms voltage would it read? What would V_{pp} equal?

SOLUTION.
The source frequency is 5 kHz and the capacitance is 0.01 μF, the same as in the preceding example; therefore,

$$X_C = 3.18\ \text{k}\Omega$$

With Ohm's rms law,

$$V = ZI = X_C I = 3.18\ \text{k}\Omega \times 2\ \text{mA} = 6.36\ \text{V}$$

This is what an ac voltmeter would read.
   If you connected an oscilloscope across the capacitor, you would see a sinusoidal voltage with a peak value of

$$V_p = 1.414V = 1.414(6.36\ \text{V}) = 9\ \text{V}$$

Figure 14-8. Calculating rms
voltage.

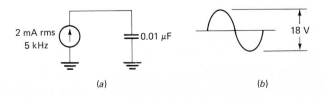

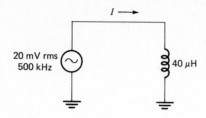

Figure 14-9. Calculating rms current.

or a peak-to-peak value of

$$V_{pp} = 2V_p = 2(9 \text{ V}) = 18 \text{ V}$$

(See Fig. 14-8*b*).

EXAMPLE 14-6.
Treat the inductor of Fig. 14-9 as ideal, and calculate the rms current.

SOLUTION.
First, get the inductive reactance:

$$X_L = 2\pi fL = 6.28(500)10^3(40)10^{-6} = 126 \ \Omega$$

Second, apply Ohm's rms law to get

$$I = \frac{V}{Z} = \frac{V}{X_L} = \frac{20 \text{ mV}}{126 \ \Omega} = 0.159 \text{ mA}$$

Test 14-2

1. Ohm's phasor law uses
    (*a*) real numbers   (*b*) complex numbers   (*c*) positive numbers
    (*d*) negative numbers .... ( )
2. Ohm's rms law involves
    (*a*) real numbers   (*b*) complex numbers   (*c*) logarithms
    (*d*) polar numbers .... ( )
3. Magnitude of impedance is to impedance as length of a vector is to
    (*a*) angle   (*b*) rectangular number   (*c*) vector   (*d*) real number ...... ( )
4. The magnitude of impedance gives no information about which of these?
    (*a*) ratio of rms voltage to rms current
    (*b*) length of impedance vector   (*c*) angle of impedance vector .... ( )

14-5.   KIRCHHOFF'S PHASOR LAWS

Another important thing Steinmetz did was to prove Kirchhoff's laws are also true when phasor current and phasor voltage are used.

The *Kirchhoff phasor-current law* says the sum of all phasor currents into a point equals the sum of all phasor currents out of the point. The *Kirchhoff phasor-voltage law* says the sum of all phasor voltages around a loop is zero. You apply these laws the same way you would in a dc circuit, except that you have to work with complex numbers instead of real numbers.

In the preceding section, we were able to start with Ohm's phasor law and derive Ohm's rms law. Unfortunately, it's impossible to do something similar with Kirchhoff's phasor laws. The laws simply do *not* hold true for rms voltage and rms current. In general, the sum of rms currents into a point does *not* equal the sum of rms currents out of the point, and the sum of rms voltages around a loop does *not* equal zero.

Now, we have these key ideas about ac circuits with linear R's, L's, and C's:

1. Ohm's law is valid for phasor voltage and phasor current.
2. Ohm's law is valid for rms voltage and rms current.
3. Kirchhoff's laws are valid for phasor voltage and phasor current.

Table 14-1 summarizes these crucial ideas, along with the validity of Ohm's and Kirchhoff's laws for instantaneous voltage and current.

TABLE 14-1. LINEAR AC LAWS

	INSTANTANEOUS	RMS	PHASOR
Ohm's law	No	Yes	Yes
Kirchhoff's laws	Yes	No	Yes

Here's what Table 14-1 says. Be careful about applying Ohm's and Kirchhoff's laws to linear ac circuits. In general, Ohm's law is not true for instantaneous values, but Kirchhoff's laws are always true for instantaneous values. On the other hand, if you're dealing with rms values, Ohm's law is always true, but Kirchhoff's laws are not. Finally, if phasor quantities are involved, Ohm's and Kirchhoff's laws are always true.

The only exception worth mentioning is *resistive circuits* (Chap. 8). When no inductance or capacitance is in a circuit, all sinusoidal voltages and currents are in phase. For this special case, it can be shown that Ohm's law is true for instantaneous voltage and current, and that Kirchhoff's laws are true for rms voltage and current. In other words, if a circuit is purely resistive, every entry in Table 14-1 will be *yes*.

14-6. EQUIVALENT IMPEDANCE OF SERIES CIRCUITS

The *equivalent impedance* between two points in a series circuit equals the sum of the impedances between the points. As a formula,

$$\mathbf{Z} = \mathbf{Z}_1 + \mathbf{Z}_2 + \mathbf{Z}_3 + \cdots \qquad (14\text{-}11)$$

Here's the proof for two impedances in series as shown in Fig. 14-10. Kirchhoff's

Figure 14-10. Impedances in series.

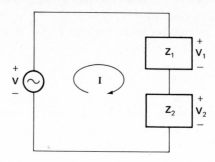

phasor-voltage law gives

$$\mathbf{V}_1 + \mathbf{V}_2 - \mathbf{V} = 0$$

or

$$\mathbf{V} = \mathbf{V}_1 + \mathbf{V}_2$$

Applying Ohm's phasor law to each load, we can rewrite the foregoing equation as

$$\mathbf{V} = \mathbf{Z}_1\mathbf{I} + \mathbf{Z}_2\mathbf{I} = (\mathbf{Z}_1 + \mathbf{Z}_2)\mathbf{I}$$

or

$$\frac{\mathbf{V}}{\mathbf{I}} = \mathbf{Z}_1 + \mathbf{Z}_2$$

The ratio $\mathbf{V}/\mathbf{I}$ is the same as the equivalent or total impedance; therefore, two impedances in series have an equivalent impedance of

$$\mathbf{Z} = \mathbf{Z}_1 + \mathbf{Z}_2$$

If you review Sec. 3-5, you will realize the derivation for equivalent $\mathbf{Z}$ is the same as for equivalent R, except for using $\mathbf{Z}$ in the place of R, and phasor voltage and current instead of dc voltage and current. Because of this, many ac formulas have the same form as dc formulas, except that

$\mathbf{V}$	is used instead of	V
$\mathbf{I}$	is used instead of	I
$\mathbf{Z}$	is used instead of	R

From now on, when the steps in deriving an ac formula are similar to those given earlier for a dc formula, we will state the ac formula without repeating the derivation.

14-7.   LEAD NETWORK

A *lead network* is one whose output voltage leads the input voltage. The most widely used lead network in electronics is the series RC circuit of Fig. 14-11. In what follows, be sure to learn the answers to

What is the equivalent impedance?
What is the rms output voltage?
What is the output phase angle?

Equivalent impedance

The total or equivalent impedance of a series ac circuit equals the sum of the impedances. In Fig. 14-11, this means

$$\mathbf{Z} = \mathbf{Z}_R + \mathbf{Z}_C = R - jX_C$$

Expressed in polar form,

$$\mathbf{Z} = \sqrt{R^2 + X_C{}^2} \; \underline{/-\arctan\,(X_C/R)} \qquad (14\text{-}12)$$

The magnitude of impedance is

$$Z = \sqrt{R^2 + X_C{}^2} \qquad (14\text{-}13)$$

This says the magnitude of impedance is the quadratic sum of resistance and reactance. The same result applies to any one-loop *RC* circuit, provided you use loop resistance for R and loop reactance for X_C.

Rms output voltage

Applying Ohm's rms law to Fig. 14-11 gives

$$I = \frac{V}{Z} = \frac{V_{\text{in}}}{\sqrt{R^2 + X_C{}^2}}$$

The rms output voltage is

$$V_{\text{out}} = RI$$

or

$$V_{\text{out}} = \frac{R}{\sqrt{R^2 + X_C{}^2}} \, V_{\text{in}} \qquad (14\text{-}14)$$

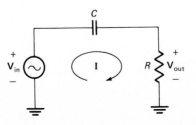

Figure 14-11. Lead network.

For the special case of X_C much smaller than R, the foregoing equation simplifies to

$$V_{out} = V_{in} \qquad (X_C \ll R)$$

In this case, the capacitor acts like an ac short and is called a coupling capacitor. As you recall, a typical rule for good coupling action is to make X_C less than one-tenth of R at the lowest frequency to be coupled. When this condition is satisfied, the output voltage is within 1 percent of the input voltage.

Output phase angle

In Fig. 14-11, let the input voltage have a phase angle of 0°; then when we calculate other phase angles, they will be with respect to the input voltage. The phasor input voltage therefore equals

$$\mathbf{V}_{in} = V_{in} \; \underline{/0°}$$

By applying Ohm's phasor law to Fig. 14-11, we can get the phasor current:

$$\mathbf{I} = \frac{\mathbf{V}_{in}}{\mathbf{Z}} = \frac{V_{in} \; \underline{/0°}}{Z \; \underline{/-\arctan \; (X_C/R)}}$$

or
$$\mathbf{I} = \frac{V_{in}}{Z} \; \underline{/\arctan \; (X_C/R)} \qquad (14\text{-}15)$$

The output phasor voltage equals

$$\mathbf{V}_{out} = R\mathbf{I} = \frac{RV_{in}}{Z} \; \underline{/\arctan \; (X_C/R)}$$

Therefore, the phase angle of output voltage is

$$\phi = \arctan \frac{X_C}{R} \qquad (14\text{-}16)$$

Since X_C is between 0 and ∞,

$$0° < \phi < 90°$$

In other words, the output voltage of any lead network like Fig. 14-11 always leads the input voltage by an angle between 0° and 90°. Figure 14-12a summarizes this idea in terms of vectors.

Recall how X_C varies with f; it's inversely proportional. So when f varies from 0 to ∞, X_C varies from ∞ to 0. With Eq. (14-16), we conclude ϕ varies from 90° to 0° when f increases from 0 to ∞. Figure 14-12b shows this variation in phase angle with frequency for any lead network like Fig. 14-11.

Here's what Fig. 14-12b says. At low frequencies the output phase angle is approximately 90°, and at high frequencies it's approximately 0°. In between, we get intermediate values of phase angle, depending on the particular frequency. For this reason, a

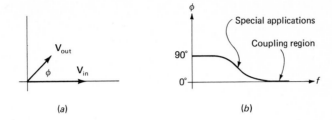

(a)

(b)

lead network can be used in special applications where you want to advance the phase angle of a signal.

For the condition X_C much smaller than R, good coupling action results and the phase angle is approximately $0°$ (coupling region of Fig. 14-12b).

EXAMPLE 14-7.
Calculate the rms output voltage for the lead network of Fig. 14-13a.

SOLUTION.
The magnitude of impedance equals the quadratic sum of the resistance and reactance:

$$Z = \sqrt{R^2 + X_C^2} = \sqrt{3000^2 + 2000^2}$$
$$= 3.61 \text{ k}\Omega$$

The rms current is

$$I = \frac{V_{in}}{Z} = \frac{10 \text{ V}}{3.61 \text{ k}\Omega} = 2.77 \text{ mA}$$

The rms output voltage is

$$V_{out} = RI = 3 \text{ k}\Omega \times 2.77 \text{ mA} = 8.31 \text{ V}$$

Figure 14-13. Example of lead net-work. (a) *Circuit.* (b) *Phasor volt-ages.*

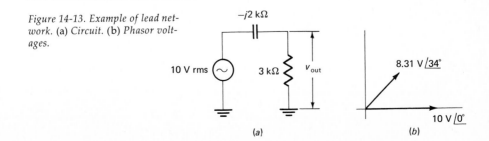

(a)

(b)

EXAMPLE 14-8.

Calculate the output phase angle in Fig. 14-13a.

SOLUTION.

With Eq. (14-16),

$$\phi = \arctan \frac{X_C}{R} = \arctan \frac{2000}{3000} = 34°$$

Figure 14-13b shows the phasor input and output voltages; the output leads the input by 34°.

EXAMPLE 14-9.

What does the rms voltage of Fig. 14-14a equal when reactance equals resistance?

SOLUTION.

This is a special case. We're given

$$X_C = R$$

Therefore, the magnitude of impedance is

$$Z = \sqrt{R^2 + X_C{}^2} = \sqrt{R^2 + R^2} = 1.414R$$

The rms current for this special case is

$$I = \frac{V_{in}}{Z} = \frac{V_{in}}{1.414R} = 0.707 \frac{V_{in}}{R}$$

and the rms output voltage is

$$V_{out} = RI = R \times 0.707 \frac{V_{in}}{R}$$

or $\qquad V_{out} = 0.707 V_{in} \qquad$ (for $X_C = R$) $\qquad\qquad$ (14-17a)

So whenever the reactance equals the resistance in a lead network, the rms output voltage equals 70.7 percent of the rms input voltage.

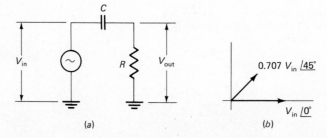

Figure 14-14. Special case: $X_C = R$.

(a)

(b)

EXAMPLE 14-10.
What is the output phase angle in the preceding example?

SOLUTION.
Again we are dealing with the special case of

$$X_C = R$$

Because of this, the output phase angle is

$$\phi = \arctan \frac{X_C}{R} = \arctan \frac{R}{R} = \arctan 1$$

or $\qquad \phi = 45° \qquad$ (for $X_C = R$) $\qquad\qquad$ (14-17b)

Figure 14-14b summarizes the results of this and the preceding example. For a phasor input voltage of

$$\mathbf{V}_{in} = V_{in} \angle 0°$$

the phasor output voltage is

$$\mathbf{V}_{out} = 0.707 V_{in} \angle 45°$$

for the special condition

$$X_C = R$$

EXAMPLE 14-11.
Derive a formula for the frequency at which X_C equals R in a lead network.

SOLUTION.
The derivation starts with

$$X_C = R$$

which may be written

$$\frac{1}{2\pi fC} = R$$

Solving for f gives

$$f = \frac{1}{2\pi RC}$$

This is a useful formula in electronics, because it pins down the frequency for which the rms output voltage equals 0.707 times the rms input voltage in a lead network like Fig. 14-14a.

The frequency where $X_C = R$ is so important in electronics that it's given names

Figure 14-15. Calculating critical frequency.

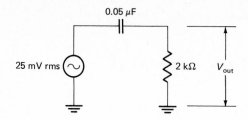

such as *critical* frequency, *cutoff* frequency, *corner* frequency, and others. We will refer to it as the critical frequency and use f_c to symbolize it. Because of this, the formula for critical frequency is written

$$f_c = \frac{1}{2\pi RC} \tag{14-18}$$

EXAMPLE 14-12.
Calculate the critical frequency for the lead network of Fig. 14-15.

SOLUTION.

$$f_c = \frac{1}{2\pi RC} = \frac{1}{6.28(2)10^3(0.05)10^{-6}}$$

$$= 1.59 \text{ kHz}$$

This means that a source frequency of 1.59 kHz results in an rms output voltage of

$$V_{out} = 0.707 V_{in} = 0.707(25 \text{ mV}) = 17.7 \text{ mV}$$

and an output phase angle of

$$\phi = 45°$$

Test 14-3

1. In a series RC circuit, the magnitude of impedance equals
    (a) the sum of R and X_C (b) the algebraic sum of R and X_C
    (c) the quadratic sum of R and X_C .... ( )
2. For the special case X_C much smaller than R, the output voltage of a lead network
    (a) leads the input by approximately 90°
    (b) is much smaller than the input voltage
    (c) lags the input by about 90°
    (d) approximately equals the input voltage .... ( )
3. When the capacitor acts like a coupling capacitor, the output phase angle of a lead network is approximately
    (a) 0°   (b) 45°   (c) 90°   (d) 180° .... ( )

4. Which does not belong?
   (a) coupling capacitor   (b) ac open   (c) ac short   (d) dc open .... ( )
5. At very low frequencies the output phase angle of a lead network approaches
   (a) 0°   (b) 45°   (c) 90°   (d) 180° .... ( )
6. Which does not belong?
   (a) $\phi = 0°$ (b) $X_C = R$ (c) $f_c = 1/2\pi RC$ (d) $V_{out} = 0.707V_{in}$ .... ( )

14-8. LAG NETWORK

A *lag network* is one whose output voltage lags the input voltage. The most widely used lag network in electronics is the series *RC* circuit of Fig. 14-16. The key questions are

What is the equivalent impedance?
What is rms output voltage?
What is the output phase angle?

Equivalent impedance

The total or equivalent impedance of a series ac circuit equals the sum of the impedances. In Fig. 14-16, this means

$$\mathbf{Z} = \mathbf{Z}_R + \mathbf{Z}_C = R - jX_C$$

which converts into a polar form

$$\mathbf{Z} = \sqrt{R^2 + X_C^2} \ \underline{/-\arctan (X_C/R)} \qquad (14\text{-}19)$$

The magnitude of this impedance is

$$Z = \sqrt{R^2 + X_C^2} \qquad (14\text{-}20a)$$

This says the magnitude of impedance in the lag network of Fig. 14-16 equals the quadratic sum of the resistance and reactance.

Rms output voltage

Applying Ohm's rms law to Fig. 14-16 gives

$$I = \frac{V}{Z} = \frac{V_{in}}{\sqrt{R^2 + X_C^2}}$$

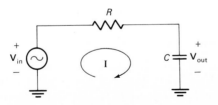

Figure 14-16. Lag network.

The rms output voltage is

$$V_{out} = X_C I$$

or
$$V_{out} = \frac{X_C}{\sqrt{R^2 + X_C^2}} V_{in} \qquad (14\text{-}20b)$$

At low frequencies the capacitor appears open, and the rms output voltage approximately equals the rms input voltage. At high frequencies the capacitor appears like a short, and the rms output voltage approaches zero.

Output phase angle

In Fig. 14-16, let the phasor input voltage have a phase angle of 0°. Then,

$$\mathbf{V}_{in} = V_{in} \,\angle\underline{0°}$$

Applying Ohm's phasor law to Fig. 14-16 gives

$$\mathbf{I} = \frac{\mathbf{V}_{in}}{\mathbf{Z}} = \frac{V_{in} \,\angle\underline{0°}}{Z \,\angle\underline{-\arctan\ (X_C/R)}}$$

or
$$\mathbf{I} = \frac{V_{in}}{Z} \,\angle\underline{\arctan\ (X_C/R)} \qquad (14\text{-}21)$$

The phasor output voltage equals

$$\mathbf{V}_{out} = \mathbf{Z}_C \mathbf{I} = X_C \,\angle\underline{-90°}\, \frac{V_{in}}{Z} \,\angle\underline{\arctan\ (X_C/R)}$$

$$= \frac{X_C V_{in}}{Z} \,\angle\underline{-90° + \arctan\ (X_C/R)}$$

Therefore, the output phase angle is

$$\phi = -90° + \arctan \frac{X_C}{R} \qquad (14\text{-}22)$$

Since X_C is between 0 and ∞,

$$-90° < \phi < 0°$$

This says the output voltage of a lag network in the form of Fig. 14-16 always lags the input voltage by an angle between 0° and −90°. Figure 14-17a summarizes this idea in terms of vectors.

When f varies from 0 to ∞, X_C varies from ∞ to 0. With Eq. (14-22), we conclude ϕ varies from 0° to −90° when f increases from 0 to ∞. Figure 14-17b shows this variation in phase angle with frequency for a lag network like Fig. 14-16.

Figure 14-17b says the following. At low frequencies the output phase angle is approximately 0°, and at high frequencies it's approximately −90°. In between, we get

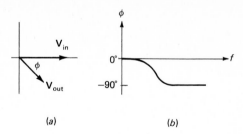

Figure 14-17. Lag network. (a)
Phasor voltages. (b) Angle versus
frequency.

intermediate values of phase angle, depending on the particular frequency. For this reason, a lag network can be used in special applications where you want to retard the phase angle of a signal.

EXAMPLE 14-13.
Calculate the rms output voltage for the lag network of Fig. 14-18a.

SOLUTION.
The magnitude of impedance is

$$Z = \sqrt{R^2 + X_C^2} = \sqrt{1000^2 + 2000^2} = 2.24 \text{ k}\Omega$$

The rms current is

$$I = \frac{V}{Z} = \frac{10 \text{ V}}{2.24 \text{ k}\Omega} = 4.46 \text{ mA}$$

and the rms output voltage is

$$V_{\text{out}} = X_C I = 2 \text{ k}\Omega \times 4.46 \text{ mA} = 8.92 \text{ V}$$

EXAMPLE 14-14.
What is the output phase angle in Fig. 14-18a?

Figure 14-18. Rms output voltage
of lag network.

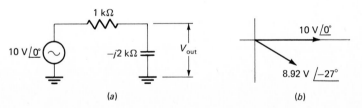

SOLUTION.
With Eq. (14-22),

$$\phi = -90° + \arctan \frac{X_C}{R} = -90° + \arctan \frac{2000}{1000}$$
$$= -90° + 63° = -27°$$

Figure 14-18b shows the phasor input and output voltages; the output lags the input by 27°.

EXAMPLE 14-15.
Calculate the rms output voltage and the output phase angle in Fig. 14-19a for the special case $X_C = R$.

SOLUTION.
The magnitude of impedance is

$$Z = \sqrt{R^2 + X_C^2} = \sqrt{R^2 + R^2} = 1.414R$$

The rms current is

$$I = \frac{V}{Z} = \frac{V_{in}}{1.414R} = 0.707 \frac{V_{in}}{R}$$

And the rms output voltage is

$$V_{out} = X_C I = RI = R \times 0.707 \frac{V_{in}}{R} = 0.707 V_{in}$$

With Eq. (14-22),

$$\phi = -90° + \arctan \frac{X_C}{R} = -90° + \arctan \frac{R}{R}$$

$$= -90° + 45° = -45°$$

Figure 14-19b summarizes these results. For a phasor input voltage of

$$\mathbf{V}_{in} = V_{in} \angle 0°$$

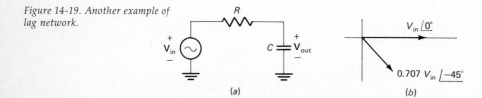

Figure 14-19. Another example of lag network.

(a)

(b)

the phasor output voltage is

$$\mathbf{V}_{out} = 0.707V_{in} \; \underline{/-45°}$$

for the special condition

$$X_C = R$$

By a derivation similar to that given in Example 14-11, the critical frequency is

$$f_c = \frac{1}{2\pi RC}$$

At this frequency, $X_C = R$, $V_{out} = 0.707V_{in}$, and $\phi = -45°$ for the lag network of Fig. 14-19a.

Test 14-4

1. Which of these is not the magnitude of impedance in a lag network?
   (a) Z   (b) $\sqrt{R^2 + X_C^2}$
   (c) quadratic sum of resistance and reactance
   (d) algebraic sum of resistance and reactance .... ( )
2. At lower frequencies the capacitor appears like an open in a lag network. Because of this, the rms output voltage approximately equals
   (a) 0   (b) 0.707V_{in} (c) V_{in} (d) 2V_{in} .... ( )
3. The smallest possible value of magnitude of impedance in a lag network is
   (a) 0   (b) R   (c) 1.414R   (d) X_C .... ( )
4. At very low frequencies the output phase angle of a lag network approaches
   (a) 0°   (b) 45°   (c) −45°   (d) −90° .... ( )
5. Which does not belong?
   (a) critical frequency   (b) $V_{out} = 0.707V_{in}$
   (c) capacitor appears shorted   (d) $\phi = -45°$ .... ( )
6. Lead network is to lag network as positive phase angle is to
   (a) lead angle   (b) negative phase angle   (c) lag angle of −45°
   (d) zero phase angle .... ( )

14-9. CHOKE FILTER CIRCUIT

Figure 14-20 shows a choke filter circuit, first discussed in Chap. 12. As you read this new material, dig out the answers to

What is the equivalent impedance?
What is rms output voltage?

Figure 14-20. Choke filter circuit.

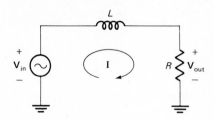

Equivalent impedance

The equivalent impedance of Fig. 14-20 is

$$\mathbf{Z} = \mathbf{Z}_R + \mathbf{Z}_L = R + jX_L$$

Expressed in polar form,

$$\mathbf{Z} = \sqrt{R^2 + X_L^2} \ \underline{/\arctan (X_L/R)} \qquad (14\text{-}23)$$

The magnitude of impedance is

$$Z = \sqrt{R^2 + X_L^2} \qquad (14\text{-}24)$$

This says the magnitude of impedance is the quadratic sum of the resistance and reactance. The same result applies to any one-loop *RL* circuit, provided you use loop resistance and loop reactance.

Rms output voltage

Applying Ohm's rms law to Fig. 14-20 gives

$$I = \frac{V}{Z} = \frac{V_{\text{in}}}{\sqrt{R^2 + X_L^2}}$$

The rms output voltage is

$$V_{\text{out}} = RI$$

or

$$V_{\text{out}} = \frac{R}{\sqrt{R^2 + X_L^2}} V_{\text{in}} \qquad (14\text{-}25)$$

For the special case $X_L = 10R$, the foregoing equation simplifies to

$$V_{\text{out}} = 0.1V_{\text{in}} \qquad (X_L = 10R) \qquad (14\text{-}26)$$

This says the output voltage is only 10 percent of the input voltage when the inductive reactance is 10 times the resistance. In this case, the inductor effectively blocks the ac signal from the output. When used in this way, the inductor is called a *choke*.

It's possible to derive a formula for the output phase angle, but we won't bother because a choke filter circuit like Fig. 14-20 is seldom used for shifting the phase of a signal.

EXAMPLE 14-16.
Calculate the output voltage in Fig. 14-21a.

SOLUTION.
This is a two-source circuit, so we can solve the problem with the superposition theorem. Figure 14-21b shows the *ac equivalent circuit,* the one that results when the dc source is reduced to zero. The inductive reactance at 120 Hz is

$$X_L = 2\pi fL = 6.28(120)20 = 15 \text{ k}\Omega$$

The rms current in this ac equivalent circuit is

$$I = \frac{V_{in}}{\sqrt{R^2 + X_L{}^2}} = \frac{5 \text{ V}}{\sqrt{1000^2 + 15{,}000^2}} = 0.333 \text{ mA}$$

and the rms output voltage is

$$V_{out} = RI = 1 \text{ k}\Omega \times 0.333 \text{ mA} = 0.333 \text{ V}$$

Figure 14-21c shows the *dc equivalent circuit,* the one you get when the ac source is reduced to zero. If the winding resistance is negligible, the inductor appears as a dc

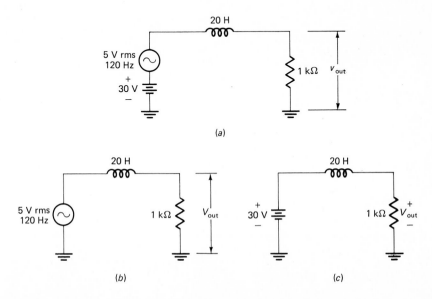

Figure 14-21. Applying superposition theorem to choke filter circuit.

(a)

(b)   (c)

short and

$$V_{out} = 30 \text{ V}$$

If the winding resistance is not negligible, you can use the voltage-divider theorem to calculate the output dc voltage.

Ideally, the output voltage has a dc component of 30 V and an ac component of 0.333 V rms. The main idea is that the choke passes the dc voltage but blocks the ac voltage.

Test 14-5

1. Which comes closest in meaning to choke?
   (*a*) dc open   (*b*) ac open   (*c*) high resistance   (*d*) ac short .... ( )
2. Choke is to coupling capacitor as ac open is to
   (*a*) ac short   (*b*) dc short   (*c*) ac open   (*d*) high resistance .... ( )
3. Which does not belong?
   (*a*) choke filter circuit   (*b*) small rms current
   (*c*) large rms output voltage   (*d*) small rms output voltage .... ( )
4. The magnitude of impedance in a choke filter circuit does not equal
   (*a*) the algebraic sum of the resistance and reactance
   (*b*) the quadratic sum of the resistance and reactance
   (*c*) the length of the impedance vector   (*d*) $\sqrt{R^2 + X_L^2}$ .... ( )

14-10. PARALLEL *RC* CIRCUIT

Figure 14-22*a* shows a parallel *RC* circuit often encountered in electronics. The easiest way to find the output voltage is to Thevenize the circuit to the left of the capacitor. This results in the circuit of Fig. 14-22*b*, where

$$\mathbf{V}_{in} = R\mathbf{I}_{in}$$

Figure 14-22. (a) Current-driven parallel RC circuit. (b) Equivalent circuit.

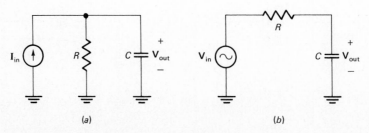

(a)   (b)

Figure 14-23. Example of converting to series RC circuit.

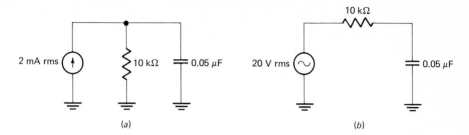

(a)　　　　　　　　　　(b)

The circuit of Fig. 14-22b is the lag network analyzed earlier. Therefore, the rms output voltage is

$$V_{\text{out}} = \frac{X_C}{\sqrt{R^2 + X_C^2}} V_{\text{in}}$$

and the phase angle is

$$\phi = -90° + \arctan \frac{X_C}{R}$$

For instance, given a parallel *RC* circuit like Fig. 14-23a, you can Thevenize the current source and resistance to get Fig. 14-23b. With the circuit in this form, you can analyze the lag network as done earlier. This is easiest way to find the rms output and phase angle.

Circuits like Fig. 14-23a are common in electronics, because a transistor acts like a current source. Often, the load on a transistor is a parallel *RC* circuit, discussed further in electronics courses.[1]

SUMMARY OF FORMULAS

DEFINING

$$\mathbf{I} = I \ \underline{/\phi} \tag{14-2}$$

$$\mathbf{V} = V \ \underline{/\phi} \tag{14-4}$$

$$\mathbf{Z} = \frac{\mathbf{V}}{\mathbf{I}} \tag{14-5}$$

[1] See Malvino, A. P.: *Electronic Principles,* McGraw-Hill Book Company, New York, 1973, Chap. 16.

DERIVED

$$\mathbf{Z}_R = R \tag{14-6}$$

$$\mathbf{Z}_L = X_L \ \underline{/90^\circ} \tag{14-7a}$$

$$\mathbf{Z}_L = jX_L \tag{14-7b}$$

$$\mathbf{Z}_C = X_C \ \underline{/-90^\circ} \tag{14-8a}$$

$$\mathbf{Z}_C = -jX_C \tag{14-8b}$$

$$\mathbf{I} = \frac{\mathbf{V}}{\mathbf{Z}} \tag{14-9a}$$

$$I = \frac{V}{Z} \tag{14-10a}$$

$$\mathbf{Z} = \mathbf{Z}_1 + \mathbf{Z}_2 + \mathbf{Z}_3 + \cdots \tag{14-11}$$

$$Z = \sqrt{R^2 + X_C{}^2} \tag{14-12}$$

$$f_c = \frac{1}{2\pi RC} \tag{14-18}$$

$$Z = \sqrt{R^2 + X_L{}^2} \tag{14-24}$$

Problems

14-1.   A sinusoidal current has a peak value of 20 mA and a phase angle of 35°. What is the phasor current?

14-2.   What is the phasor current for each of these sinusoidal currents:
    a. $i = 0.2 \sin (\theta + 75^\circ)$
    b. $i = 0.005 \sin (\theta - 50^\circ)$
    c. $i = 5 \sin (\theta + 150^\circ)$
    d. $i = 0.03 \sin (\theta - 80^\circ)$

Figure 14-24.

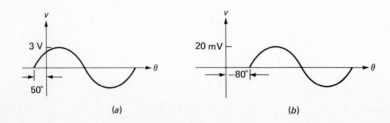

(a)   (b)

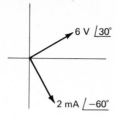

Figure 14-25.

14-3.   What is the phasor voltage for each of these sinusoidal voltages:

    a. $v = 50 \sin (\theta + 35°)$

    b. $v = 2 \sin (\theta - 120°)$

    c. $v = 0.4 \sin (\theta + 25°)$

    d. $v = 0.005 \sin (\theta - 45°)$

14-4.   An ac voltmeter measures an rms voltage of 50 mV. If the phase angle of this sinusoidal voltage is $-35°$, what is the phasor voltage?

14-5.   If an oscilloscope displays a sinusoidal voltage with a peak-to-peak value of 50 mV and a phase angle of 30°, what is the phasor voltage?

14-6.   What is the phasor voltage for Fig. 14-24a? Fig. 14-24b?

14-7.   What is the impedance of a 500-Ω resistor, assuming negligible lead inductance and stray capacitance?

14-8.   An inductor has a reactance of 350 Ω. If it's ideal, what is its impedance in polar form? In rectangular form?

14-9.   An inductor has an rms voltage across it of 50 V and an rms current through it of 20 mA. Ideally, what is its impedance?

14-10. Figure 14-25 shows the phasor voltage and phasor current for an ideal inductor. What is its impedance?

14-11. A capacitor has an rms voltage of 20 V and an rms current of 40 μA. What is its impedance?

14-12. A capacitor has a reactance of 6 kΩ. Ideally, what is its impedance in polar form? In rectangular form?

14-13. What is the impedance of a 1000-pF capacitor at 2 MHz?

14-14. What is the impedance of a 20-μF capacitor at

    a.   60 Hz

    b.   120 Hz

    c.   240 Hz

14-15. An ac load has a phasor voltage of 20 mV $\angle 30°$ and a phasor current of 5 mA $\angle -45°$. What is the impedance of this load?

14-16. An ac load has an impedance of 3 kΩ $\angle 50°$ and a phasor voltage of 75 mV $\angle 120°$. What is the phasor current into this load?

14-17. If the impedance of an ac load is 10 kΩ $\angle -45°$ and the phasor current into this load is 2 mA $\angle 30°$, what is the phasor voltage across the load?

Figure 14-26.

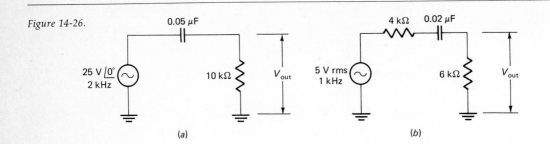

(a)　　　　　(b)

Figure 14-27.

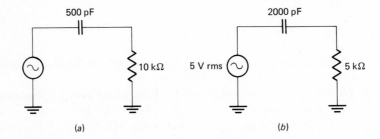

(a)　　　　　(b)

14-18.　In Problem 14-15, what is the magnitude of impedance?

14-19.　In Problem 14-16, what is the rms current?

14-20.　In Problem 14-17, what is the rms voltage?

14-21.　Calculate the rms output voltage in Fig. 14-26a.

14-22.　What is the output phase angle in Fig. 14-26a?

14-23.　If the source frequency of Fig. 14-26a is doubled, what is the rms output voltage? The phase angle?

14-24.　What does the rms current equal in Fig. 14-26b? The rms output voltage?

14-25.　To get good coupling action in Fig. 14-26a, what size should the capacitor be changed to?

14-26.　What is the critical frequency in Fig. 14-27a?

14-27.　If the source frequency of Fig. 14-27b equals the critical frequency, what does the rms output voltage equal?

Figure 14-28.

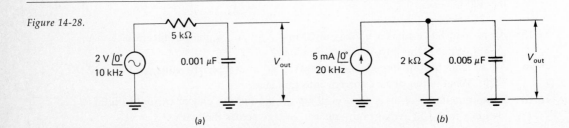

(a)　　　　　(b)

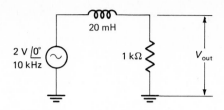

Figure 14-29.

14-28. Calculate the rms output voltage in Fig. 14-28a.

14-29. What is the output phase angle in Fig. 14-28a?

14-30. If the source frequency of Fig. 14-28a is doubled, what would an ac voltmeter read when connected across the output?

14-31. If the capacitor of Fig. 14-28a is changed to 0.002 μF, what would the rms output voltage equal?

14-32. Thevenize the circuit driving the capacitor of Fig. 14-28b. Then calculate the rms output voltage.

14-33. A series RC circuit has an rms source voltage of 2 V. If the resistance is 10 kΩ and the reactance is 15 kΩ, what does the rms current equal?

14-34. What does the rms output voltage equal in Fig. 14-29?

14-35. Calculate the rms output voltage in Fig. 14-29 for a source frequency of 50 kHz.

ANSWERS TO TESTS

14-1. *a, a, b, a, c, d*

14-2. *b, a, c, c*

14-3. *c, d, a, b, c, a*

14-4. *d, c, b, a, c, b*

14-5. *b, a, c, a*

15. Advanced AC Topics

This chapter continues our discussion of ac analysis. We will concentrate on those ideas that are most useful in electronics. Otherwise, we'd be caught up in complicated theoretical ac problems that only experts need to solve. In other words, this chapter emphasizes topics needed for practical electronics work.

15-1. CONDUCTANCE, SUSCEPTANCE, AND ADMITTANCE

The previous chapter restricted the analysis to series ac circuits, except for the special case of a current source driving a parallel RC circuit. Parallel-circuit analysis requires new concepts which will be discussed in this section. In particular, the first things to learn are

> *What is conductance?*
> *What is susceptance?*
> *What is admittance?*

Conductance

Chapter 5 defined *conductance.* As you recall, the defining formula is

$$G = \frac{1}{R} \qquad\qquad (15\text{-}1)$$

where G is the conductance. Furthermore, the reciprocal of ohms results in a unit of measure known as mhos. So whenever an answer is in mhos, you automatically know it represents conductance and not resistance.

For example, if a resistor has a resistance of 1 kΩ, the conductance of this resistor is

$$G = \frac{1}{1 \text{ k}\Omega} = 0.001 \text{ mho}$$

Sometimes, measuring instruments read conductance rather than resistance, and this is one reason for the use of conductance. If an instrument indicates a conductance of 0.02 mho, you can calculate the resistance as

$$R = \frac{1}{G} = \frac{1}{0.02 \text{ mho}} = 50 \ \Omega$$

Susceptance

Susceptance is the reciprocal of reactance. Since there are two kinds of reactance (inductive and capacitive), there are two kinds of susceptance.

Inductive susceptance is defined by the formula

$$B_L = \frac{1}{X_L} \tag{15-2}$$

where B_L is the inductive susceptance in mhos. For instance, if an inductor has an inductive reactance of 500 Ω, it has an inductive susceptance of

$$B_L = \frac{1}{X_L} = \frac{1}{500 \ \Omega} = 0.002 \text{ mho}$$

Capacitive susceptance is the reciprocal of capacitive reactance, and is defined by the formula

$$B_C = \frac{1}{X_C} \tag{15-3}$$

where B_C is the capacitive susceptance in mhos. So if a capacitor has a capacitive reactance of 40 Ω, its capacitive susceptance is

$$B_C = \frac{1}{X_C} = \frac{1}{40 \ \Omega} = 0.025 \text{ mho}$$

Admittance

Admittance is defined as the reciprocal of impedance. In symbols,

$$\mathbf{Y} = \frac{1}{\mathbf{Z}} \tag{15-4}$$

where **Y** is the admittance. Since there are three basic components, it's important to know the admittance of each.

The admittance of an ideal resistor is

$$\mathbf{Y}_R = \frac{1}{\mathbf{Z}_R} = \frac{1}{R}$$

or $\qquad\qquad \mathbf{Y}_R = G \qquad\qquad$ (15-5)

This says the admittance of an ideal resistor is a real number equal to the conductance. A 1-kΩ resistor, for instance, has an admittance of

$$\mathbf{Y}_R = 0.001 \text{ mho}$$

The admittance of an ideal inductor is

$$\mathbf{Y}_L = \frac{1}{\mathbf{Z}_L} = \frac{1}{jX_L} = -j\frac{1}{X_L}$$

or $\qquad\qquad \mathbf{Y}_L = -jB_L \qquad\qquad$ (15-6)

This says the admittance of an ideal inductor is a negative imaginary number, exactly opposite the impedance of an ideal inductor. The magnitude of the admittance equals B_L, the inductive susceptance. As an example, if an ideal inductor has an X_L of 1 kΩ, then its impedance is

$$\mathbf{Z}_L = jX_L = j1 \text{ kΩ}$$

and its admittance is

$$\mathbf{Y}_L = -jB_L = -j0.001 \text{ mho}$$

The admittance of an ideal capacitor is

$$\mathbf{Y}_C = \frac{1}{\mathbf{Z}_C} = \frac{1}{-jX_C} = j\frac{1}{X_C}$$

or $\qquad\qquad \mathbf{Y}_C = jB_C \qquad\qquad$ (15-7)

This tells us the admittance of an ideal capacitor is always a positive imaginary number, the exact opposite of the impedance of the ideal capacitor. The magnitude of the admittance equals B_C, the capacitive susceptance. For instance, if an ideal capacitor has an X_C of 5 kΩ, it has an impedance of

$$\mathbf{Z}_C = -jX_C = -j5 \text{ kΩ}$$

and it has an admittance of

$$\mathbf{Y}_C = jB_C = j0.0002 \text{ mho}$$

Memorize defining formulas (15-1) through (15-4) and derived formulas (15-5) through (15-7). Conductance, susceptance, and admittance are extremely useful for dealing with parallel connections of basic components.

EXAMPLE 15-1.
Figure 15-1a shows a parallel *RL* circuit. Label each basic component with its admittance.

SOLUTION.
The circuit shows the impedance of each component. With the basic formulas just derived, the admittance of the resistor is

$$\mathbf{Y}_R = \frac{1}{20 \ \Omega} = 0.05 \text{ mho}$$

and the admittance of the inductor is

$$\mathbf{Y}_L = \frac{1}{j40 \ \Omega} = -j0.025 \text{ mho}$$

Figure 15-1b shows the original circuit with each component relabeled with its admittance.

EXAMPLE 15-2.
Relabel each component of Fig. 15-1c with its admittance.

SOLUTION.
The admittance of the resistor is

$$\mathbf{Y}_R = \frac{1}{50 \ \Omega} = 0.02 \text{ mho}$$

Figure 15-1. Admittance.

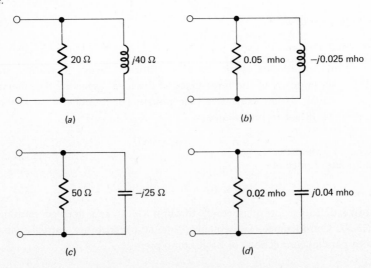

(a) (b)

(c) (d)

and the admittance of the capacitor is

$$\mathbf{Y}_C = \frac{1}{-j25\ \Omega} = j0.04 \text{ mho}$$

So in the equivalent circuit of Fig. 15-1d, we see an admittance of 0.02 mho in parallel with another admittance of $j0.04$ mho.

Test 15-1

1. How many kinds of susceptance are there?
   (a) 0   (b) 1   (c) 2   (d) 3 .... ( )
2. Inductive susceptance is the reciprocal of
   (a) resistance   (b) inductive reactance   (c) capacitive reactance
   (d) impedance .... ( )
3. Resistance is to conductance as impedance is to
   (a) susceptance   (b) admittance   (c) inductive susceptance
   (d) capacitive susceptance .... ( )
4. Which of the following does not belong?
   (a) positive imaginary number   (b) admittance of ideal capacitor
   (c) admittance of ideal inductor   (d) j times capacitive susceptance ...... ( )
5. If the admittance of an ideal inductor were expressed in polar form, it would have an angle of
   (a) 0°   (b) −90°   (c) 90°   (d) 180° .... ( )
6. If the reactance of an ideal capacitor is 100 Ω, the admittance is
   (a) 0.01 mho   (b) −j0.01 mho   (c) j0.02 mho   (d) 0.01 mho $\angle 90°$
   .... ( )

15-2. PARALLEL CIRCUITS

When we analyze parallel circuits, we can find either the equivalent impedance or the equivalent admittance. This section answers

What is the equivalent impedance of a parallel circuit?
What is the equivalent admittance of a parallel circuit?

Equivalent impedance

Suppose two impedances are in parallel, as shown in Fig. 15-2a. Ohm's and Kirchhoff's phasor laws apply to this circuit. Because of this, the equivalent impedance turns out to be

$$\mathbf{Z} = \frac{\mathbf{Z}_1\mathbf{Z}_2}{\mathbf{Z}_1 + \mathbf{Z}_2} \tag{15-8}$$

Figure 15-2. Parallel circuits. (a)
Impedances combine by product-
over-sum. (b) Admittances add.

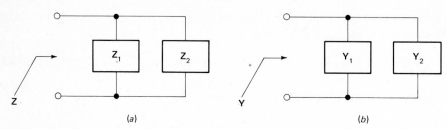

(a)   (b)

The steps in the derivation of this formula are the same as given earlier for resistances (Sec. 3-6), except for using phasor voltage, phasor current, and impedance in place of dc voltage, dc current, and resistance.

When more than two impedances are in parallel, the equivalent impedance is

$$Z = \frac{1}{1/Z_1 + 1/Z_2 + 1/Z_3 + \cdots} \qquad (15\text{-}9)$$

Because complex numbers are involved, both of these equations are difficult to use except for very simple parallel circuits.

Parallel admittance

Since admittance is defined as

$$Y = \frac{1}{Z}$$

we can replace each impedance in Eq. (15-8) by its admittance. After simplifying as much as possible, we get

$$Y = Y_1 + Y_2 \qquad (15\text{-}10)$$

This says the equivalent admittance of two parallel admittances equals the sum of the admittances.

To put it another way, suppose you have a circuit like Fig. 15-2a, where each load is specified by its impedance. Using Eq. (15-8) may be difficult because of the complex numbers. For this reason, you can relabel each component with its admittance, as in Examples 15-1 and 15-2; this results in an equivalent circuit like Fig. 15-2b. Then it's easy to get the equivalent admittance because all you have to do is add the parallel admittances. This may seem roundabout, but it's usually easier than working with Eq. (15-8).

As an example, given a circuit like Fig. 15-3a, convert each impedance to its admittance:

$$\mathbf{Y}_R = \frac{1}{100} = 0.01 \text{ mho}$$

$$\mathbf{Y}_C = \frac{1}{-j200} = j0.005 \text{ mho}$$

Relabeling each component with its admittance results in Fig. 15-3b. Then the equivalent admittance is

$$\mathbf{Y} = \mathbf{Y}_R + \mathbf{Y}_C = 0.01 \text{ mho} + j0.005 \text{ mho}$$

To simplify the appearance, this may also be written as

$$\mathbf{Y} = 0.01 + j0.005$$

where the unit of mho is understood to be part of the answer.

When more than two components are parallel, the equivalent admittance equals the sum of the individual admittances. This can be seen as follows. Replacing each impedance of Eq. (15-9) by its admittance gives

$$\frac{1}{\mathbf{Y}} = \frac{1}{\mathbf{Y}_1 + \mathbf{Y}_2 + \mathbf{Y}_3 + \cdot \cdot \cdot}$$

By inverting both sides,

$$\mathbf{Y} = \mathbf{Y}_1 + \mathbf{Y}_2 + \mathbf{Y}_3 + \cdot \cdot \cdot \tag{15-11}$$

So if you have a parallel circuit with several components, each labeled with its admittance, add up the admittances to get the equivalent admittance.

Special cases

There are three special cases of parallel circuits that occur frequently in electronics. Figure 15-4a shows a parallel RC circuit. With each component described by its admittance as shown, the equivalent admittance is

$$\mathbf{Y} = G + jB_C \tag{15-12}$$

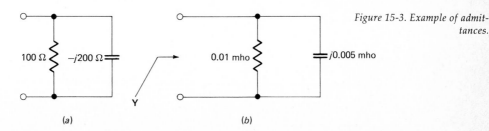

Figure 15-3. Example of admittances.

100 Ω   −j200 Ω

Y

0.01 mho   j0.005 mho

(a)   (b)

Figure 15-4. *Admittance of basic circuits. (a) Parallel RC. (b) Parallel RL. (c) Parallel RLC.*

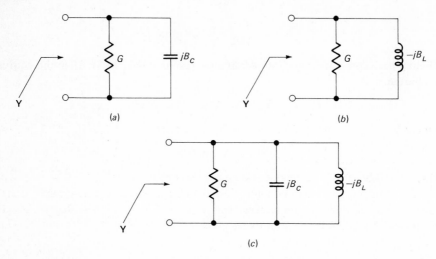

(a)

(b)

(c)

Similarly, the parallel *RL* circuit of Fig. 15-4*b* has an equivalent admittance of

$$\mathbf{Y} = G - jB_L \tag{15-13}$$

And finally, the parallel *RLC* circuit of Fig. 15-4*c* has an equivalent admittance of

$$\mathbf{Y} = G + jB_C - jB_L \tag{15-14}$$

EXAMPLE 15-3.
What is the equivalent admittance of Fig. 15-5*a* at 2 MHz?

SOLUTION.
First, get the capacitive reactance:

$$X_C = \frac{1}{2\pi f C} = \frac{1}{6.28(2)10^6(500)10^{-12}} = 159 \ \Omega$$

Then calculate the admittance of the resistor and the capacitor:

$$\mathbf{Y}_R = \frac{1}{400 \ \Omega} = 0.0025 \ \text{mho}$$

$$\mathbf{Y}_C = \frac{1}{-j159 \ \Omega} = j0.0063 \ \text{mho}$$

Figure 15-5*b* shows the circuit labeled with its admittances at 2 MHz. The equiva-

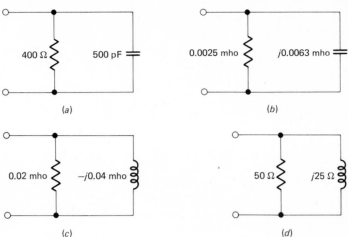

Figure 15-5. Examples of admittance.

lent admittance is

$$\mathbf{Y} = 0.0025 + j0.0063$$

EXAMPLE 15-4.

The equivalent admittance of a circuit is

$$\mathbf{Y} = 0.02 - j0.04$$

Show the simplest circuit that has this admittance.

SOLUTION.

The admittance is the sum of 0.02 mho and $-j0.04$ mho; therefore, the circuit must be a parallel circuit as shown in Fig. 15-5c. If you prefer working with ohms, you can take the reciprocal of each admittance to get the equivalent circuit of Fig. 15-5d.

Test 15-2

1. Which does not belong?
   (a) $\mathbf{Z}_1 + \mathbf{Z}_2$ (b) product-over-sum of impedances
   (c) two parallel impedances   (d) $\mathbf{Z}_1\mathbf{Z}_2/(\mathbf{Z}_1 + \mathbf{Z}_2)$ .... ( )
2. If parallel loads are labeled with their admittances, the equivalent admittance is
   (a) the same as the equivalent impedance
   (b) equal to the sum of the admittances
   (c) the product-over-sum of the impedances
   (d) the product-over-sum of the admittances .... ( )

3. If $\mathbf{Y} = 0.01 + j0.02$, the circuit with this admittance is a
    (a) resistor in series with a capacitor
    (b) resistor in parallel with an inductor
    (c) resistor in parallel with a capacitor
    (d) resistor in series with an inductor .... ( )
4. If a resistance of 1 Ω is in parallel with an inductive reactance of 1 Ω, the equivalent admittance of this parallel circuit is
    (a) $1 - j1$ (b) $1 + j1$ (c) $j1/(1 + j1)$ (d) $j1/(1 - j1)$ .... ( )
5. A parallel RLC circuit has $R = 1$ Ω, $X_L = 2$ Ω, and $X_C = 4$ Ω. The admittance of this parallel circuit is
    (a) $1 - j0.25 + j0.5$ (b) $1 + j0.25 - j0.8$ (c) $1 - j0.1 + j0.2$
    (d) $1 - j0.5 + j0.25$ .... ( )

15-3.   SECOND APPROXIMATION OF AN INDUCTOR

Up to now, we have neglected the resistance of an inductor. In this section, we take resistance into account. Look for the answers to

What is the Q of an inductor?
What is the dissipation factor?

Q of an inductor

An inductor dissipates some power because of its winding resistance and core losses. To account for these power losses, visualize a resistance R_s in series with the inductance L, as shown in Fig. 15-6. This equivalent circuit for an inductor is called the *second approximation*. The impedance in this case equals

$$Z = R_s + jX_L \tag{15-15}$$

The *quality factor* Q of an inductor is defined as the inductive reactance divided by

---

Figure 15-6. Second approximation of an inductor.

the series resistance:

$$Q = \frac{X_L}{R_s} \qquad (15\text{-}16)$$

For instance, if $X_L = 100 \ \Omega$ and $R_s = 2 \ \Omega$,

$$Q = \frac{100 \ \Omega}{2 \ \Omega} = 50$$

Note that the unit of ohms cancels.

Q is called the quality factor because the higher its value, the more closely the inductor approaches an ideal inductor. If R_s were zero, the inductor would be ideal and its Q would be infinite. The inductors used in electronics typically have Q's from 50 to 150 when used at the frequencies for which they are designed.

It's important to remember defining formula (15-16). Among other things, measuring instruments often indicate the value of Q rather than R_s. For this reason, inductors are usually specified in terms of their inductance and their Q, rather than their inductance and series resistance. You can measure Q with an instrument such as a Q meter, an RLC bridge, etc.

Dissipation factor

Closely related to Q is the *dissipation factor D*. It's defined by

$$D = \frac{1}{Q} \qquad (15\text{-}17)$$

An inductor with a Q of 50 has a dissipation factor of

$$D = \frac{1}{Q} = \frac{1}{50} = 0.02$$

The dissipation factor is important with certain measuring instruments. In some cases, it's easier to design instruments that measure D rather than Q. So if you measure the D of an inductor as 0.05, you can find the Q by

$$Q = \frac{1}{D} = \frac{1}{0.05} = 20$$

EXAMPLE 15-5.
What is the Q of a 50-μH inductor at 1 MHz, given a series resistance of 5 Ω? What is the dissipation factor?

SOLUTION.
First, calculate inductive reactance:

$$X_L = 2\pi f L = 6.28(10^6)50(10^{-6}) = 314 \ \Omega$$

Next,

$$Q = \frac{X_L}{R_s} = \frac{314 \ \Omega}{5 \ \Omega} \cong 63$$

Finally,

$$D = \frac{1}{Q} = \frac{1}{63} \cong 0.016$$

Test 15-3

1. Which does not belong?
   (a) large Q (b) small X_L/R_s (c) almost ideal inductor
   (d) very small R_s .... ( )
2. Q always has which of these units?
   (a) ohms   (b) mhos   (c) volts   (d) no units .... ( )
3. An ideal inductor has a D of
   (a) 0   (b) 1   (c) ∞ .... ( )

15-4.   SERIES-TO-PARALLEL CONVERSION

Sometimes it's easier to work with a parallel circuit than with a series circuit. For this reason, it helps to know how to convert series RL circuits to equivalent parallel RL circuits, and how to convert series RC circuits to equivalent parallel RC circuits.

Given a series RL or RC circuit, the Q of the circuit is defined as the reactance divided by the resistance:

$$Q = \frac{X}{R} \qquad \text{(series)} \tag{15-18}$$

When applying this formula, use X_L for series RL circuits and X_C for series RC circuits. For example, if a series RL circuit has $X_L = 1 \ \text{k}\Omega$ and $R = 50 \ \Omega$,

$$Q = \frac{X_L}{R} = \frac{1000 \ \Omega}{50 \ \Omega} = 20$$

If a series RC circuit has $X_C = 400 \ \Omega$ and $R = 25 \ \Omega$,

$$Q = \frac{X_C}{R} = \frac{400 \ \Omega}{25 \ \Omega} = 16$$

Here's how to convert series circuits to equivalent parallel circuits:

1. If you have a series RL circuit like Fig. 15-7a, you can replace it with the equivalent parallel circuit shown in Fig. 15-7a. The key quantity is Q; once it's known, the formulas shown in Fig. 15-7a automatically give the values of the resistance and inductance in the equivalent parallel circuit.

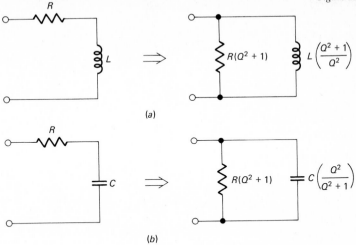

Figure 15-7. Series-to-parallel conversion.

(a)

(b)

2. Similarly, a series RC circuit like Fig. 15-7b can be replaced by the equivalent parallel RC circuit shown in the same figure. Again, Q is the key to the calculation of the parallel resistance and capacitance.

   The proof of the equivalence between the series and parallel circuits of Fig. 15-7 is too complicated to go into here.[1]

EXAMPLE 15-6.
If the frequency is 1 kHz, what is the equivalent parallel circuit for Fig. 15-8a?

SOLUTION.
First, calculate X_L:

$$X_L = 2\pi fL = 6.28(10^3)5(10^{-3}) = 31.4 \ \Omega$$

Next, get Q:

$$Q = \frac{X_L}{R} = \frac{31.4 \ \Omega}{10 \ \Omega} = 3.14$$

Now use the formulas given in Fig. 15-7a:

$$R(Q^2 + 1) = 10(3.14^2 + 1) = 109 \ \Omega$$

and

$$L\left(\frac{Q^2 + 1}{Q^2}\right) = 0.005 \left(\frac{3.14^2 + 1}{3.14^2}\right) = 5.51 \ \text{mH}$$

[1] If you are interested in the proof, see Malvino, A. P.: *Electronic Instrumentation Fundamentals*, McGraw-Hill Book Company, New York, 1967, pp. 134–138.

Figure 15-8. Examples of series-to-parallel conversion.

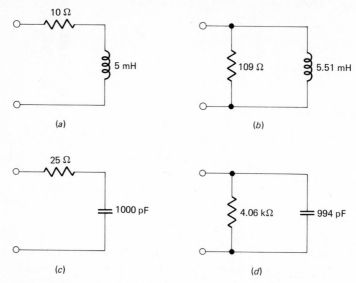

(a)

(b)

(c)

(d)

Figure 15-8b shows the equivalent parallel circuit. This circuit acts the same as the original circuit of Fig. 15-8a. A word of caution: Q changes with frequency. Since the calculations we just went through depend on the value of Q, Fig. 15-8b is equivalent to Fig. 15-8a only at 1 kHz. If the frequency changes, you have to recalculate Q and the values of the parallel resistance and inductance.

EXAMPLE 15-7.
Work out the equivalent parallel circuit for Fig. 15-8c for a frequency of 500 kHz.

SOLUTION.
The capacitive reactance in Fig. 15-8c at 500 kHz is

$$X_C = \frac{1}{2\pi fC} = \frac{1}{6.28(500)10^3(10^{-9})} = 318 \ \Omega$$

and the Q is

$$Q = \frac{X_C}{R} = \frac{318 \ \Omega}{25 \ \Omega} = 12.7$$

With the formulas of Fig. 15-7b,

$$R(Q^2 + 1) = 25(12.7^2 + 1) = 4.06 \ \text{k}\Omega$$

and

$$C\left(\frac{Q^2}{Q^2 + 1}\right) = 1000 \ \text{pF}\left(\frac{12.7^2}{12.7^2 + 1}\right) = 994 \ \text{pF}$$

Figure 15-8d shows the equivalent parallel circuit. As long as the frequency is 500 kHz, you can replace the series RC circuit of Fig. 15-8c with the parallel circuit of Fig. 15-8d.

EXAMPLE 15-8.
When Q is much greater than unity, the conversion formulas of Fig. 15-7 can be simplified. What are the approximate conversion formulas for large Q?

SOLUTION.
For large Q, the parallel resistance simplifies to

$$R(Q^2 + 1) \cong RQ^2 = Q^2R$$

So for either equivalent parallel circuit of Fig. 15-7, resistance is approximately Q^2 times the series resistance R.

The parallel inductor of Fig. 15-7a reduces as follows for large Q:

$$L\left(\frac{Q^2 + 1}{Q^2}\right) \cong L$$

Similarly, the parallel capacitor of Fig. 15-7b reduces as follows for large Q:

$$C\left(\frac{Q^2}{Q^2 + 1}\right) \cong C$$

Figure 15-9 summarizes these approximations. The parallel equivalent resistance

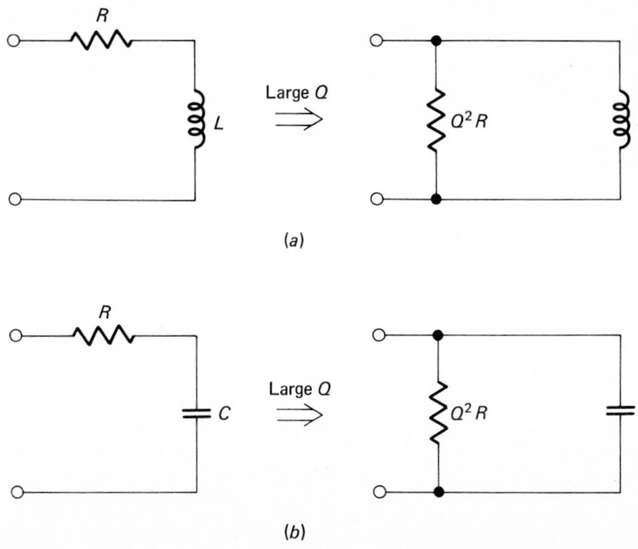

Figure 15-9. Series-to-parallel conversion for large Q.

(a)

(b)

Figure 15-10. (a) Second approxi-
mation of an inductor. (b) Equiva-
lent parallel circuit for inductor.

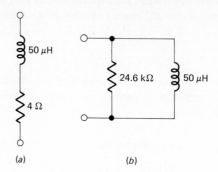

(a)   (b)

equals Q^2R in either case; furthermore, the L and C values remain unchanged when converting from series to parallel. These approximations for large Q are important, because the Q of typical electronic circuits is greater than 10. Since Q is squared in all conversion formulas given earlier, less than 1 percent error results with the approximations of Fig. 15-9 when Q is greater than 10.

EXAMPLE 15-9.

Figure 15-10*a* shows the second approximation of an inductor at 1 MHz. Work out the equivalent parallel circuit for this inductor.

SOLUTION.

First, get the reactance:

$$X_L = 2\pi fL = 6.28(10^6)50(10^{-6}) = 314 \ \Omega$$

Second, get the Q:

$$Q = \frac{X_L}{R} = \frac{314 \ \Omega}{4 \ \Omega} = 78.5$$

Third, since Q is greater than 10, the parallel R is approximately

$$Q^2R = 78.5^2 \times 4 \ \Omega = 24.6 \ k\Omega$$

and the parallel L is approximately the same as the series L:

$$L = 50 \ \mu H$$

Figure 15-10*b* shows the equivalent parallel circuit for the inductor.

Test 15-4

1. The values of the equivalent parallel circuit are affected by changes in
    (*a*) Q (*b*) frequency   (*c*) R (*d*) all of these .... ( )

2. If $Q = 10$ and $R = 50 \ \Omega$, the equivalent parallel resistance equals
    (a) 50 Ω (b) 500 Ω (c) 5 kΩ (d) 50 kΩ .... ( )
3. Which of the following is true for large Q and a series RL circuit?
    (a) parallel resistance approximately the same as series resistance
    (b) parallel capacitance approximately the same as series capacitance
    (c) parallel inductance approximately the same as series inductance
    (d) parallel resistance much smaller than series resistance .... ( )
4. Which does not belong?
    (a) series RC circuit   (b) large Q
    (c) parallel C approximately the same as series C
    (d) parallel R almost the same as series R .... ( )
5. An inductor has an L of 10 mH and a Q of 10. Which of the following is true?
    (a) parallel R is 100 times series R
    (b) parallel L is approximately equal to 10 mH   (c) both are true .... ( )

15-5. PARALLEL-TO-SERIES CONVERSION

In the preceding section, series circuits were converted to equivalent parallel circuits. It's also possible to convert the other way, from parallel to series.

To begin with, the Q of a parallel RL or RC circuit is defined as

$$Q = \frac{R}{X} \quad \text{(parallel)} \tag{15-19}$$

When applying this formula, use X_L for parallel RL circuits and X_C for parallel RC circuits. So if a parallel RL circuit has $R = 10$ kΩ and $X_L = 250 \ \Omega$, the Q of the parallel circuit is

$$Q = \frac{R}{X_L} = \frac{10,000 \ \Omega}{250 \ \Omega} = 40$$

Similarly, if a parallel RC circuit has an R of 5 kΩ and an X_C of 200 Ω, Q is

$$Q = \frac{R}{X_C} = \frac{5000 \ \Omega}{200 \ \Omega} = 25$$

Figure 15-11a gives the conversion formulas for a parallel RL circuit. For the important special case of large Q, the inductance in the series equivalent circuit has approximately the same value as the original parallel circuit.

Likewise, Fig. 15-11b shows how to convert parallel RC circuits to their equivalent series circuits. Again notice that the large-Q case results in a series C approximately equal to the original parallel C.

EXAMPLE 15-10.
Convert the parallel RL circuit of Fig. 15-12a to its equivalent series form at 265 kHz.

Figure 15-11. Parallel-to-series
conversion.

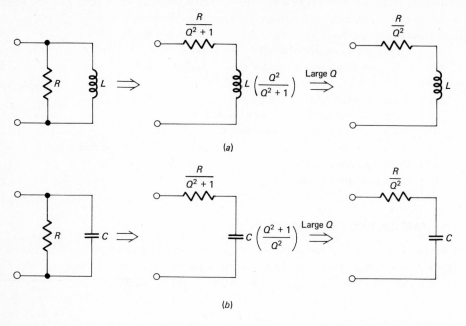

(a)

(b)

Figure 15-12. Examples of parallel-
to-series conversion.

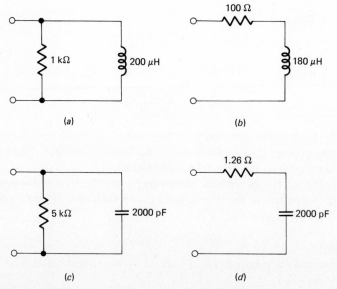

(a)

(b)

(c)

(d)

SOLUTION.

The inductive reactance is

$$X_L = 2\pi fL = 6.28(265)10^3(200)10^{-6} = 333 \ \Omega$$

and Q is

$$Q = \frac{R}{X_L} = \frac{1000 \ \Omega}{333 \ \Omega} = 3$$

Since Q is less than 10, use the exact formulas of Fig. 15-11a:

$$\frac{R}{Q^2 + 1} = \frac{1000 \ \Omega}{3^2 + 1} = 100 \ \Omega$$

and

$$L\left(\frac{Q^2}{Q^2 + 1}\right) = 200 \ \mu\text{H} \left(\frac{3^2}{3^2 + 1}\right) = 180 \ \mu\text{H}$$

Figure 15-12b summarizes the results. This circuit is equivalent to the original circuit of Fig. 15-12a when the frequency is 265 kHz.

EXAMPLE 15-11.

Convert the parallel RC circuit of Fig. 15-12c to its equivalent series circuit at 1 MHz.

SOLUTION.

The capacitive reactance is

$$X_C = \frac{1}{2\pi fC} = \frac{1}{6.28(10^6)2000(10^{-12})} = 79.6 \ \Omega$$

and Q is

$$Q = \frac{R}{X_C} = \frac{5000 \ \Omega}{79.6 \ \Omega} \cong 63$$

Since Q is greater than 10, the approximate conversion formulas of Fig. 15-11b can be used:

$$\frac{R}{Q^2} = \frac{5000 \ \Omega}{63^2} = 1.26 \ \Omega$$

$$C = 2000 \ \text{pF}$$

Figure 15-12d shows the equivalent series circuit. Remember: this is a valid equivalent circuit only at the frequency where X_C and Q have been calculated, in this case, 1 MHz.

Test 15-5

1. If a parallel RC circuit has an R of 4 kΩ and an X_C of 125 Ω, what is the Q?
    (a) 16   (b) 32   (c) 48   (d) 64 .... ( )

2. A parallel RC circuit has an R of 10 kΩ, a C of 500 pF, and a Q of 25. The equivalent series resistance is closest to
    (a) 5   (b) 10   (c) 15   (d) 20 .... (   )
3. A parallel RL circuit has an inductance of 20 mH, a Q of 50, and an R of 7 kΩ. In the equivalent series circuit the inductance is approximately
    (a) 50 μH   (b) 20 mH   (c) 2.8 Ω (d) 7 kΩ .... (   )
4. Which of these does not belong?
    (a) parallel RC circuit   (b) large Q
    (c) series R much smaller than parallel R
    (d) series C much smaller than parallel C .... (   )

15-6.   AC WHEATSTONE BRIDGES

Figure 15-13 shows an ac Wheatstone bridge, similar to the dc Wheatstone bridge of Sec. 4-5. Because an ac source drives the bridge, we must use impedances for each branch as shown. In this section, here are the key questions:

> *When is an ac Wheatstone bridge balanced?*
> *What frequency balances a Wien bridge?*
> *What are other practical bridge types?*

Balance condition

Section 4-5 proved that a dc Wheatstone bridge balances when

$$\frac{R_1}{R_2} = \frac{R_3}{R_4}$$

By using phasor voltage, phasor current, and impedance in place of dc voltage, dc current, and resistance, we can prove the balance condition for an ac Wheatstone bridge is

$$\frac{\mathbf{Z}_1}{\mathbf{Z}_2} = \frac{\mathbf{Z}_3}{\mathbf{Z}_4} \tag{15-20}$$

Figure 15-13. Ac Wheatstone bridge.

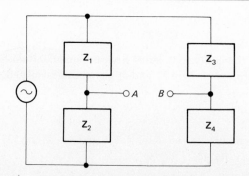

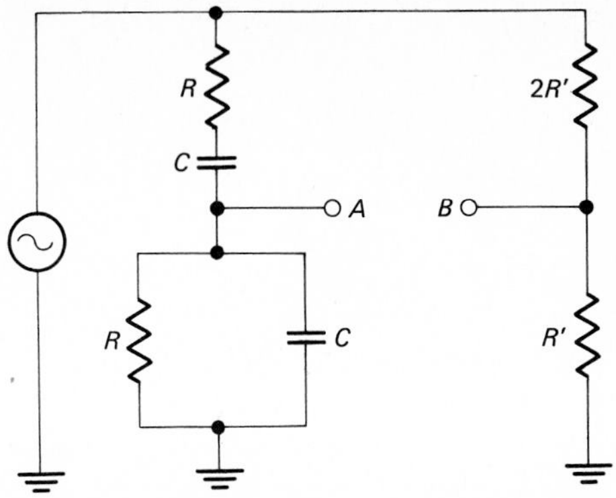

Figure 15-14. Wien bridge.

Wien bridge

One of the most important ac bridges you encounter in electronics is the *Wien bridge* of Fig. 15-14; this bridge is used in almost all commercial audio generators.

We are particularly interested in the frequency that produces bridge balance. To get a formula for this frequency, we begin by noting

$$\mathbf{Z}_1 = R - jX_C$$

$$\mathbf{Z}_2 = R \parallel (-jX_C)$$

$$\mathbf{Z}_3 = 2R'$$

$$\mathbf{Z}_4 = R'$$

Substituting into Eq. (15-20) gives

$$\frac{R - jX_C}{R \parallel (-jX_C)} = \frac{2R'}{R'} = 2$$

Expanding the denominator by the product-over-sum rule,

$$\frac{R - jX_C}{-jRX_C/(R - jX_C)} = \frac{(R - jX_C)^2}{-jRX_C} = 2$$

Then, expanding the numerator gives

$$\frac{R^2 - j2RX_C - X_C^2}{-jRX_C} = 2$$

After cross-multiplying,

$$R^2 - j2RX_C - X_C^2 = -j2RX_C$$

The imaginary terms drop out, leaving

$$R^2 - X_C^2 = 0$$

or
$$R^2 = X_C{}^2$$

Taking the square root of both sides,

$$R = X_C$$

or
$$R = \frac{1}{2\pi fC}$$

Finally, solving for f gives

$$f = \frac{1}{2\pi RC} \qquad \text{(balance)} \tag{15-21}$$

This is an important result. It says the Wien bridge of Fig. 15-14 balances when the source has a frequency of 1 divided by 2π times RC. This is the key formula used in the design of most *audio oscillators* (electronic circuits that generate frequencies from less than 20 Hz to more than 20 kHz).

Other ac bridges

An *RLC bridge* is a commercial instrument for measuring R, L, C, Q, and D. A typical *RLC* bridge uses several kinds of ac bridges to measure a wide range of inductance, capacitance, etc. Figure 15-15 shows six of the most common ac bridges used in commercial instruments. Each bridge has certain properties that make it better suited than others for particular measurements. For instance, the *series-capacitance comparison* bridge of Fig. 15-15a is useful for measuring C and D (the dissipation factor); the *Maxwell bridge* of Fig. 15-15c is ideal for measuring L and Q; etc. All six bridges are analyzed and discussed in advanced books.[2]

EXAMPLE 15-12.
Figure 15-16 shows a Wien bridge with variable capacitors. These capacitors are *ganged*, meaning they are turned by a common shaft. In this way, they have the same capacitance value throughout the variable range.

Calculate the frequency that balances the bridge for the maximum capacitance value. For the minimum capacitance value.

SOLUTION.
For maximum capacitance, the key quantities are

$$R = 100 \text{ k}\Omega$$

$$C = 500 \text{ pF}$$

---

[2] Among others, see Malvino, A. P.: *Electronic Instrumentation Fundamentals*, McGraw-Hill Book Company, New York, 1967, Chap. 6.

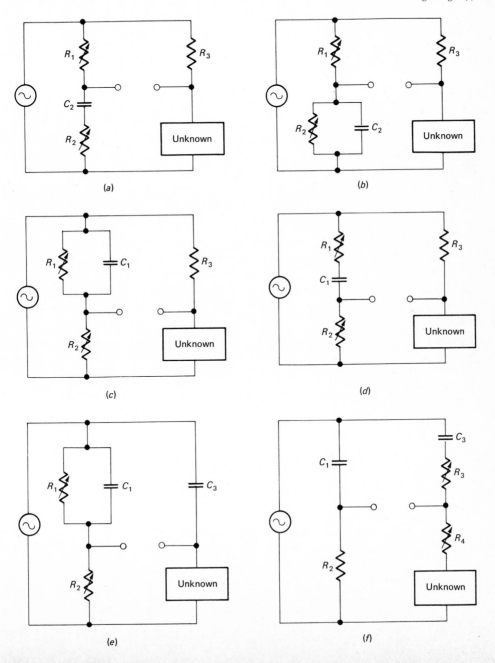

Figure 15-15. Ac bridges. (a) *Series-capacitance comparison bridge.* (b) *Parallel-capacitance comparison bridge.* (c) *Maxwell bridge.* (d) *Hay bridge.* (e) *Schering bridge.* (f) *Owen bridge.*

(a)

(b)

(c)

(d)

(e)

(f)

Figure 15-16. *Tunable Wien bridge.*

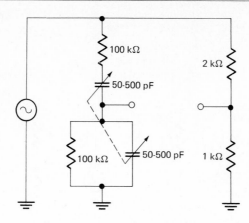

With Eq. (15-21),

$$f = \frac{1}{2\pi RC} = \frac{1}{6.28(100)10^3(500)10^{-12}}$$
$$= 3.18 \text{ kHz}$$

This is the source frequency that produces bridge balance.

At the other extreme, $C = 50$ pF and

$$f = \frac{1}{2\pi RC} = \frac{1}{6.28(100)10^3(50)10^{-12}}$$
$$= 31.8 \text{ kHz}$$

So when the capacitors are adjusted to their minimum value, it takes a source frequency of 31.8 kHz to balance the Wien bridge.

Test 15-6

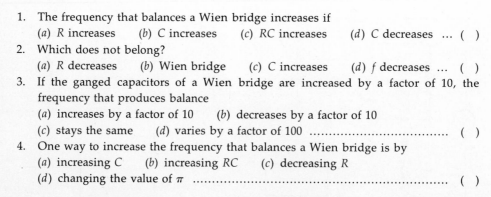

1. The frequency that balances a Wien bridge increases if
    (*a*) R increases   (*b*) C increases   (*c*) RC increases   (*d*) C decreases ... (   )
2. Which does not belong?
    (*a*) R decreases   (*b*) Wien bridge   (*c*) C increases   (*d*) f decreases ... (   )
3. If the ganged capacitors of a Wien bridge are increased by a factor of 10, the frequency that produces balance
    (*a*) increases by a factor of 10   (*b*) decreases by a factor of 10
    (*c*) stays the same   (*d*) varies by a factor of 100 .... (   )
4. One way to increase the frequency that balances a Wien bridge is by
    (*a*) increasing C (*b*) increasing RC (*c*) decreasing R
    (*d*) changing the value of π .... (   )

15-7.   POWER IN A RESISTOR

The key questions in this section are

> *What is instantaneous power?*
> *What is average power?*

Instantaneous power

Chapter 8 discussed *instantaneous power;* it equals

$$p = vi \qquad\qquad (15\text{-}22)$$

where v and i are the instantaneous voltage and current. In ac circuits, v and i are sinusoidal; therefore, the instantaneous power is continuously changing during the cycle. To calculate the instantaneous power, multiply the voltage by the current at the instant you're interested in.

Average Power

Average power is defined as the average of the instantaneous power over one cycle. Average power is more useful than instantaneous power, because average power is easier to measure; wattmeters normally indicate average power.

When a load is purely resistive, the sinusoidal voltage and current are in phase as shown in Fig. 15-17a. The instantaneous power in this case equals

$$p = vi = (V_p \sin \theta) \ (I_p \sin \theta)$$
$$= V_p I_p \sin^2 \theta$$

Using the trigonometric identity

$$\sin^2\theta = \tfrac{1}{2} - \tfrac{1}{2} \cos 2\theta$$

Figure 15-17. (a) *In-phase sinusoids.* (b) *Average and instantaneous power.*

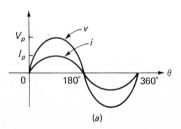

(a)

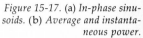
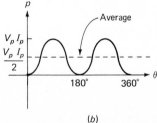

(b)

the instantaneous power becomes

$$p = \frac{V_p I_p}{2} (1 - \cos 2\theta) \tag{15-23}$$

Graph Eq. (15-23), and Fig. 15-17b results. As shown, instantaneous power varies sinusoidally from a minimum of 0 to a maximum of $V_p I_p$. Because of this, the average power P over one cycle is

$$P = \frac{V_p I_p}{2} \quad \text{(resistor)} \tag{15-24}$$

This applies to any ideal and linear resistance; it says the average power equals half the peak voltage times the peak current.

EXAMPLE 15-13.

A 4-kΩ resistor has a sinusoidal voltage across it with a peak of 20 V. What is the average power?

SOLUTION.

The current is also sinusoidal and has a peak of

$$I_p = \frac{V_p}{R} = \frac{20 \text{ V}}{4 \text{ k}\Omega} = 5 \text{ mA}$$

The average power is

$$P = \frac{V_p I_p}{2} = \frac{20 \text{ V} \times 5 \text{ mA}}{2} = 50 \text{ mW}$$

Test 15-7

1. The instantaneous power in a resistor
    (a) varies during the cycle   (b) is constant during the cycle
    (c) equals $V_p I_p/2$ (d) is negative during the cycle .... ( )
2. Instantaneous power is to average power as variable is to
    (a) constant   (b) variable   (c) $V_p I_p$ (d) cycle .... ( )
3. Which of the following does not belong?
    (a) instantaneous power   (b) average power   (c) $V_p I_p/2$
    (d) average over one cycle .... ( )
4. A resistor has a peak voltage of 5 V across it and a peak current of 2 mA through it.
    The average power is
    (a) 10 μW   (b) 5 mW   (c) 10 mW   (d) 20 mW .... ( )

15-8.   RMS VALUES

This section derives the formulas for the rms values of a sine wave; it also derives a formula for the average power.

Deriving rms values for a sine wave

To begin with, *rms voltage* is defined as the dc voltage that produces the same average power. For instance, the average power in the resistor of Fig. 15-18a is

$$P = \frac{V_p I_p}{2}$$

Since $I_p = V_p/R$, the average power may be rewritten as

$$P = \frac{V_p^2}{2R} \qquad (15\text{-}25a)$$

The dc source of Fig. 15-18b produces a power of

$$P_{dc} = \frac{V^2}{R}$$

If this dc power equals the average power produced by the sinusoidal source, the dc voltage is called the rms voltage and labeled V_{rms} as shown in Fig. 15-18c. In this case,

$$P_{dc} = \frac{V_{rms}^2}{R} \qquad (15\text{-}25b)$$

To get a useful formula, equate the right members of Eqs. (15-25a) and (15-25b):

$$\frac{V_{rms}^2}{R} = \frac{V_p^2}{2R}$$

which simplifies to

$$V_{rms}^2 = \frac{V_p^2}{2}$$

Taking the square root of both sides gives

$$V_{rms} = \frac{V_p}{\sqrt{2}}$$

or $\qquad\qquad V_{rms} = 0.707 V_p \qquad$ (sine wave)

In a similar way, *rms current* is defined as the dc current producing the same average

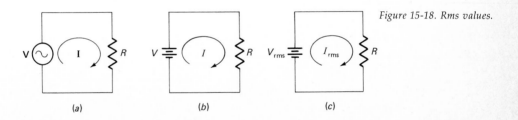

Figure 15-18. Rms values.

(a)   (b)   (c)

power. By a derivation like the one for voltage, we can prove

$$I_{\text{rms}} = 0.707I_p \quad \text{(sine wave)}$$

Average power in terms of rms values

For sinusoidal current and voltage, the average power in a resistor is

$$P = \frac{V_p I_p}{2}$$

which may be written as

$$P = \frac{V_p}{\sqrt{2}} \frac{I_p}{\sqrt{2}}$$

or
$$P = V_{\text{rms}} I_{\text{rms}} \quad \text{(resistor)} \tag{15-26a}$$

Two alternative forms of this useful equation are

$$P = R I_{\text{rms}}^2 \tag{15-26b}$$

$$P = \frac{V_{\text{rms}}^2}{R} \tag{15-26c}$$

EXAMPLE 15-14.

The rms voltage across a resistor is 25 V, and the rms current is 30 mA. What is the average power in the resistor?

SOLUTION.

$$P = V_{\text{rms}} I_{\text{rms}} = 25 \text{ V} \times 30 \text{ mA} = 750 \text{ mW}$$

EXAMPLE 15-15.

An ac voltage of 20 V rms is across a 5-kΩ resistor. What does the average power equal?

SOLUTION.

$$P = \frac{V_{\text{rms}}^2}{R} = \frac{20^2}{5000} = 0.08 \text{ W} = 80 \text{ mW}$$

EXAMPLE 15-16.

A 2-kΩ resistor has an rms current of 10 mA. What is the average power?

SOLUTION.

$$P = R I_{\text{rms}}^2 = 2000 \times 0.01^2 = 0.2 \text{ W} = 200 \text{ mW}$$

Test 15-8

1. Rms voltage produces the same
   (*a*) instantaneous power   (*b*) average power   (*c*) peak power .... (   )
2. Rms voltage
   (*a*) changes during the cycle   (*b*) is a fixed value
   (*c*) is time-varying   (*d*) equals V_p .... (   )
3. Which belongs least?
   (*a*) sawtooth   (*b*) sine wave   (*c*) $0.707V_p$ (*d*) rms voltage .... (   )

15-9. AC LOAD POWER

When the load is not purely resistive, the voltage and current are no longer in phase. Because of this, the average power depends on the phase angle between the sinusoidal voltage and current. As you read, find the answers to

> *What is the average power in an ideal capacitor?*
> *What is the average power in an ideal inductor?*
> *What is the power in any ac load?*

Capacitor

The instantaneous power is

$$p = vi$$

In an ideal capacitor, the current leads the voltage by 90° as shown in Fig. 15-19*a*. Since

$$v = V_p \sin \theta$$

and

$$i = I_p \cos \theta$$

the instantaneous power equals

$$p = (V_p \sin \theta)\,(I_p \cos \theta)$$
$$= V_p I_p \sin \theta \cos \theta$$

Using the trigonometric identity

$$\sin \theta \cos \theta = \frac{\sin 2\theta}{2}$$

the instantaneous power may be written as

$$p = \frac{V_p I_p}{2} \sin 2\theta \tag{15-27}$$

Graph Eq. (15-27), and Fig. 15-19*b* results. The instantaneous power varies sinusoi-

Figure 15-19. (a) *Sinusoids for an ideal capacitor. (b) Average power is zero.*

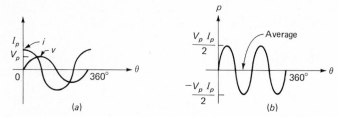

(a)   (b)

dally from a minimum of $-V_pI_p/2$ to a maximum of $V_pI_p/2$. Because of this, the average power is *zero*. This means the capacitor absorbs energy during the part of the cycle it's charging, and releases energy during the part of the cycle it's discharging. Therefore, an ideal capacitor does not dissipate power like a resistor. In symbols,

$$P = 0 \qquad \text{(ideal capacitor)} \qquad (15\text{-}28)$$

Inductor power

In an ideal inductor, the sinusoidal voltage leads the sinusoidal current by 90°. With a derivation like the one just given, we can prove the average power in an inductor is zero. During the part of the cycle that the magnetic field is increasing, the inductor absorbs energy; but during the part of the cycle when the magnetic field decreases, the inductor returns energy to the circuit. Neglecting any resistance in the inductor, therefore, the average power is zero:

$$P = 0 \qquad \text{(ideal inductor)} \qquad (15\text{-}29)$$

Any ac load

By an advanced derivation, the average power for any ac load is

$$P = VI \cos \phi \qquad (15\text{-}30)$$

where V and I are rms values and ϕ is the angle between the sinusoidal voltage and sinusoidal current (the same as the angle of $\mathbf{Z}$).

The quantity $\cos \phi$ is called the *power factor*. For a purely resistive load, the angle of the impedance is 0°; therefore, $\cos \phi = 1$ for a resistor. On the other hand, for a purely reactive load, ϕ is $\pm90°$ and $\cos \phi = 0$. For all other loads, $\cos \phi$ has a value between 0 and 1.

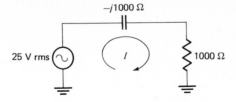

Figure 15-20.

EXAMPLE 15-17.
What average power does the source deliver to the series RC circuit of Fig. 15-20.

SOLUTION.
The magnitude of equivalent impedance is

$$Z = \sqrt{R^2 + X_C^2} = \sqrt{1000^2 + 1000^2} = 1.41 \text{ k}\Omega$$

The rms current is

$$I = \frac{V}{Z} = \frac{25 \text{ V}}{1.41 \text{ k}\Omega} = 17.7 \text{ mA}$$

The angle of the impedance is

$$\phi = -\arctan \frac{X_C}{R} = -\arctan \frac{1000}{1000} = -45°$$

So the average power to the RC circuit is

$$P = VI \cos \phi = (25 \text{ V})(17.7 \text{ mA})(\cos -45°)$$
$$= 313 \text{ mW}$$

Test 15-9

1. For the average power to be zero in an inductor, the
   (a) inductor must be lossy
   (b) sinusoidal current must be in phase with the sinusoidal voltage
   (c) inductor must be ideal   (d) windings must dissipate power .... ( )
2. Which of the following belongs least?
   (a) small ϕ (b) large power factor
   (c) phasor current and phasor voltage almost have same angle
   (d) impedance is mostly reactive .... ( )
3. If $\phi = 60°$, the power factor is
   (a) 0   (b) 0.5   (c) 0.707   (d) 1 .... ( )
4. For a resistor, the power factor equals
   (a) 0°   (b) 90°   (c) 1   (d) $\cos 90°$ .... ( )

SUMMARY OF FORMULAS

DEFINED

$$B_L = \frac{1}{X_L} \qquad (15\text{-}2)$$

$$B_C = \frac{1}{X_C} \qquad (15\text{-}3)$$

$$\mathbf{Y} = \frac{1}{\mathbf{Z}} \qquad (15\text{-}4)$$

$$Q = \frac{X_L}{R_s} \qquad (15\text{-}16)$$

$$D = \frac{1}{Q} \qquad (15\text{-}17)$$

$$Q = \frac{X}{R} \qquad \text{(series)} \qquad (15\text{-}18)$$

$$Q = \frac{R}{X} \qquad \text{(parallel)} \qquad (15\text{-}19)$$

DERIVED

$$\mathbf{Y}_R = G \qquad (15\text{-}5)$$

$$\mathbf{Y}_L = -jB_L \qquad (15\text{-}6)$$

$$\mathbf{Y}_C = jB_C \qquad (15\text{-}7)$$

$$\mathbf{Z} = \frac{\mathbf{Z}_1\mathbf{Z}_2}{\mathbf{Z}_1 + \mathbf{Z}_2} \qquad (15\text{-}8)$$

$$\mathbf{Y} = \mathbf{Y}_1 + \mathbf{Y}_2 \qquad (15\text{-}10)$$

$$\frac{\mathbf{Z}_1}{\mathbf{Z}_2} = \frac{\mathbf{Z}_3}{\mathbf{Z}_4} \qquad (15\text{-}20)$$

$$f = \frac{1}{2\pi RC} \qquad (15\text{-}21)$$

$$P = VI \cos \phi \qquad (15\text{-}30)$$

Problems

15-1. What is the conductance of a 5-kΩ resistor?

15-2. Calculate the conductance of a 300-Ω resistor.

15-3. Work out the inductive susceptance for each of these:
   a. $X_L = 8$ kΩ
   b. $L = 100$ μH and $f = 1$ MHz
   c. $L = 25$ mH and $f = 200$ kHz

15-4. Calculate the capacitive susceptance for each of these:
   a. $X_C = 2$ kΩ
   b. $C = 10$ μF and $f = 120$ Hz
   c. $C = 500$ μF and $f = 2$ MHz

15-5. What is the admittance for each of these:
   a. $R = 1$ kΩ
   b. $X_L = 20$ Ω
   c. $C = 400$ pF and $f = 5$ MHz
   d. $L = 250$ μH and $f = 450$ kHz

15-6. A 2-kΩ resistor is in parallel with a capacitive reactance of 3 kΩ. What is the equivalent admittance?

15-7. A 1-kΩ resistor is in parallel with a 200-pF capacitor. What is the equivalent admittance at 1 MHz?

15-8. The equivalent admittance of a circuit is

$$\mathbf{Y} = 0.1 + j0.25$$

What is the parallel circuit with this admittance.

15-9. A parallel circuit has an admittance of

$$\mathbf{Y} = 0.002 - j0.001$$

What is the parallel resistance? The parallel reactance?

15-10. An inductor has an X_L of 2500 Ω and an R_s of 125 Ω. What is the value of Q? The value of D?

15-11. An inductor has an X_L of 10 kΩ and a Q of 125. What is the value of R_s? The value of D?

15-12. What is the Q of a 200-μH inductor at 1 MHz, given an R_s of 10 Ω?

15-13. A series RC circuit has an X_C of 2 kΩ and an R of 100 Ω. What is the Q?

15-14. A series RL circuit has an R of 25 Ω and an L of 10 mH. What is the equivalent parallel circuit at 1 kHz?

15-15. A series RL circuit has an R of 100 Ω and an L of 25 mH. What is the equivalent parallel circuit at 500 Hz?

15-16. $R = 50$ Ω and $C = 500$ pF in a series RC circuit. What is the equivalent parallel RC circuit at 450 kHz?

15-17.   An inductor has a Q of 100 and an L of 200 μH. What is the equivalent parallel circuit for the inductor at 1 MHz?

15-18.   A series RL circuit has $R = 50$ Ω and $L = 1$ mH. If Q equals 75, what is the equivalent parallel circuit?

15-19.   The Q of Fig. 15-21a equals 100. What is the equivalent parallel circuit?

15-20.   Suppose the Q of Fig. 15-21b equals 150. What is the equivalent parallel circuit?

15-21.   A parallel RL circuit has a resistance of 2 kΩ and an inductance of 150 μH. Convert this circuit to its equivalent series form at 250 kHz.

15-22.   A parallel RC circuit has an R of 10 kΩ and a C of 1000 pF. What is the equivalent series circuit at 1 MHz?

15-23.   A measuring instrument indicates that an inductor has an equivalent parallel resistance of 4 kΩ. If Q is 100, what is the equivalent series resistance of the inductor?

15-24.   At what frequency does the Wien bridge of Fig. 15-22 balance?

15-25.   A Wien bridge has an R of 10 kΩ and a C of 100 pF. What frequency balances the bridge? If R is increased by a factor of 10, what is the frequency that balances the bridge?

15-26.   A sinusoidal voltage with a peak of 10 V is across a 500-Ω resistor. What is the average power in the resistor?

15-27.   The rms voltage across a 470-Ω resistor is 100 mV. What is the average power?

15-28.   A 50-kΩ resistor has 20 mA rms through it. What is the average power?

15-29.   The voltage across a resistor is 250 mV rms, and the current through it is 0.5 A rms. What is the average power?

15-30.   A series RC circuit has an impedance of

$$\mathbf{Z} = 2 \text{ k}\Omega - j3 \text{ k}\Omega$$

If the rms voltage across it is 20 V rms, what is the average power into the circuit.

15-31.   The voltage across a series RC circuit is 25 V rms, and the current is 50 mA rms. What is the average power for each of these:
   a. $\phi = 30°$
   b.   A circuit where the capacitor acts like a coupling capacitor
   c.   A lead network operating at the critical frequency

Figure 15-21.

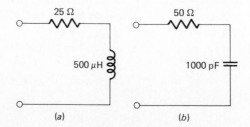

(a)   (b)

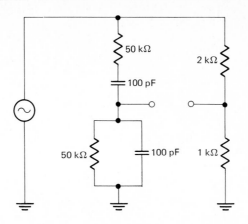

Figure 15-22.

ANSWERS TO TESTS

15-1. *c, b, b, c, b, d*
15-2. *a, b, c, a, d*
15-3. *b, d, a*
15-4. *d, c, c, d, c*
15-5. *b, c, b, d*
15-6. *d, a, b, c*
15-7. *a, a, a, b,*
15-8. *b, b, a*
15-9. *c, d, b, c*

16. Resonance

Resonance occurs when X_L equals X_C in a series or parallel *RLC* circuit. The two opposite types of reactance cancel, leaving only the resistance. Because resonance occurs at only one frequency, resonant circuits can filter out one frequency from all others. In fact, this is how a radio or TV receiver is able to tune to one particular station, even though the antenna simultaneously picks up all signals.

16-1. SERIES RESONANT CIRCUITS

Figure 16-1*a* shows a series *RLC* circuit. The current passes through the basic components in series. Here are the main questions:

> *What is the rms current in the circuit?*
> *What is the resonant frequency of the circuit?*

Basic idea

At very low frequencies, the inductor of Fig. 16-1*a* appears shorted (ideally), and the circuit acts like an equivalent series *RC* circuit (Fig. 16-1*b*). On the other hand, at very high frequencies the capacitor appears shorted, and the circuit acts like a series *RL* circuit (Fig. 16-1*c*). Because of the opposite nature of inductive and capacitive reactance, there's an intermediate frequency where the effects of the inductor cancel the effects of the capacitor. At this frequency the circuit acts like a purely resistive circuit (Fig. 16-1*d*).

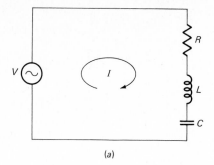

Figure 16-1. Series resonant circuit. (a) Original circuit. (b) Below the resonant frequency. (c) Above the resonant frequency. (d) At the resonant frequency.

(a)

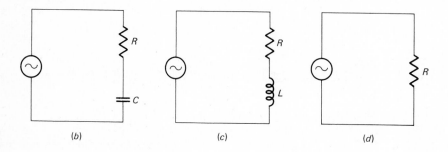

(b)   (c)   (d)

Rms current

In Fig. 16-1a, the equivalent impedance is

$$\mathbf{Z} = R + jX_L - jX_C$$

or

$$\mathbf{Z} = R + j(X_L - X_C) \tag{16-1}$$

The magnitude of this impedance is

$$Z = \sqrt{R^2 + (X_L - X_C)^2} \tag{16-2}$$

The *net reactance* in a series RLC circuit is defined as the difference $X_L - X_C$. The magnitude of the impedance therefore equals the quadratic sum of the resistance and net reactance.

With Ohm's rms law, the rms current in Fig. 16-1a is

$$I = \frac{V}{Z} = \frac{V}{\sqrt{R^2 + (X_L - X_C)^2}} \tag{16-3}$$

This says the rms current equals the rms source voltage divided by the quadratic sum of the resistance and net reactance.

Resonant frequency

Resonance is the condition of the circuit when

$$X_L = X_C \tag{16-4}$$

When the inductive and capacitive reactances are equal, the net reactance of the circuit equals zero and the rms current has a maximum value of

$$I = \frac{V}{R}$$

This makes sense, because at resonance the circuit acts purely resistive (Fig. 16-1*d*).

The *resonant frequency* f_r is the source frequency that makes inductive and capacitive reactances equal. In other words, if we graph X_L and X_C on the same pair of axes, Fig. 16-2 results. X_L increases with frequency, but X_C decreases with frequency; therefore, a single frequency exists where the two reactances are equal. By definition, we call this frequency the resonant frequency, and symbolize it as f_r.

Figure 16-3 shows the *resonance curve*, a graph of rms current versus source frequency. At very low frequencies the rms current approaches zero, because the capacitor appears open. Similarly, at very high frequencies the rms current approaches zero, because the inductor appears open. In between these extremes, the rms current increases and reaches a maximum value at the resonant frequency.

Here's how to derive an important formula for resonant frequency. When the source frequency equals f_r,

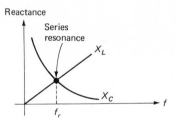

Figure 16-2. Reactance curves.

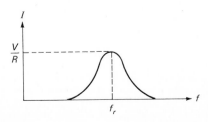

Figure 16-3. Resonance curve for series RLC circuit.

$$X_L = X_C$$

which may be written as

$$2\pi f_r L = \frac{1}{2\pi f_r C}$$

This rearranges into

$$f_r^2 = \frac{1}{(2\pi)^2 LC}$$

Taking the square root of both sides gives

$$f_r = \frac{1}{2\pi \sqrt{LC}} \tag{16-5}$$

This is an important formula. With it, you can calculate the resonant frequency, given the values of L and C on a schematic diagram. Also notice that resonant frequency is not related to resistance; it depends only on the values of L and C.

EXAMPLE 16-1.
Calculate the rms current in Fig. 16-4a for a source frequency of 1 MHz.

SOLUTION.

$$X_L = 2\pi fL = 6.28(10^6)20(10^{-6}) = 126 \ \Omega$$

$$X_C = \frac{1}{2\pi fC} = \frac{1}{6.28(10^6)500(10^{-12})} = 318 \ \Omega$$

Figure 16-4b shows the circuit with its individual impedances for a source frequency of 1 MHz. From Eq. (16-3), the rms current is

$$I = \frac{V}{Z} = \frac{10}{\sqrt{10^2 + (126 - 318)^2}} \cong 52 \ \text{mA}$$

EXAMPLE 16-2.
Work out the rms current in Fig. 16-4a for a source frequency of 2 MHz.

SOLUTION.
In the preceding example, the source frequency is 1 MHz; in this example, it's 2 MHz. Since the source frequency has doubled, X_L doubles and X_C halves. With the values calculated in the preceding example, the reactances at 2 MHz are

$$X_L = 2(126 \ \Omega) = 252 \ \Omega$$

$$X_C = \frac{318 \ \Omega}{2} = 159 \ \Omega$$

Figure 16-4c shows the circuit with its impedances at 2 MHz. The rms current is

$$I = \frac{V}{Z} = \frac{10}{\sqrt{10^2 + (252 - 159)^2}} = 107 \text{ mA}$$

EXAMPLE 16-3.
What is the resonant frequency of Fig. 16-4a?

SOLUTION.
Use Eq. (16-5):

$$f_r = \frac{1}{2\pi \sqrt{LC}} = \frac{1}{2\pi \sqrt{20(10^{-6})500(10^{-12})}}$$
$$= 1.59 \text{ MHz}$$

When the source frequency equals 1.59 MHz, X_L equals X_C and the circuit acts like Fig. 16-4d. In this case, the rms current is 1 A.

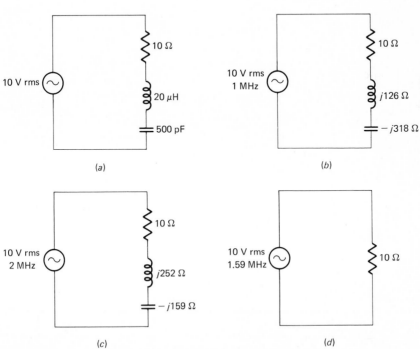

Figure 16-4. Example of series RLC circuit.

(a)

(b)

(c)

(d)

EXAMPLE 16-4.

Sketch the resonance curve for Fig. 16-4a.

SOLUTION.

We just calculated a resonant frequency of 1.59 MHz. And in preceding examples, we have found

$$I = 52 \text{ mA} \quad \text{for} \quad f = 1 \text{ MHz}$$

$$I = 1 \text{ A} \quad \text{for} \quad f = 1.59 \text{ MHz}$$

$$I = 107 \text{ mA} \quad \text{for} \quad f = 2 \text{ MHz}$$

Figure 16-5 shows the resonance curve.

EXAMPLE 16-5.

Calculate the phasor voltage across the resistor, inductor, and capacitor of Fig. 16-6a.

SOLUTION.

The source frequency is 1.59 MHz, which is the resonant frequency of the circuit; therefore, the circuit acts purely resistive with an equivalent impedance of 10 Ω. As a result, the phasor current at resonance is

$$\mathbf{I} = \frac{\mathbf{V}}{\mathbf{Z}} = \frac{10 \text{ V } \underline{/0°}}{10 \text{ Ω}} = 1 \text{ A } \underline{/0°}$$

The phasor voltage across the resistor is

$$\mathbf{V}_R = R\mathbf{I} = 10 \text{ Ω} \times 1 \text{ A } \underline{/0°} = 10 \text{ V } \underline{/0°}$$

To get the phasor voltage across the inductor, first calculate the reactance:

$$X_L = 2\pi f_r L = 6.28(1.59)10^6(20)10^{-6} = 200 \text{ Ω}$$

For an ideal inductor,

$$\mathbf{Z}_L = X_L \underline{/90°} = 200 \text{ Ω} \underline{/90°}$$

Figure 16-5. Resonance curve.

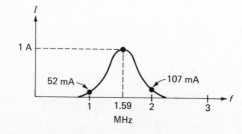

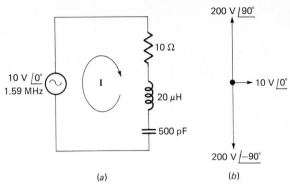

Figure 16-6. *Calculating phasor voltages across the components of a series RLC circuit.*

(a)   (b)

The phasor voltage across the inductor is

$$\mathbf{V}_L = \mathbf{Z}_L\mathbf{I} = 200\ \Omega\ \underline{/90°} \times 1\ \text{A}\ \underline{/0°} = 200\ \text{V}\ \underline{/90°}$$

Because of resonance,

$$X_C = X_L = 200\ \Omega$$

so that

$$\mathbf{Z}_C = 200\ \Omega\ \underline{/-90°}$$

The phasor voltage across the capacitor is

$$\mathbf{V}_C = \mathbf{Z}_C\mathbf{I} = 200\ \Omega\ \underline{/-90°} \times 1\ \text{A}\ \underline{/0°} = 200\ \text{V}\ \underline{/-90°}$$

Figure 16-6b summarizes the results. The phasor voltage across the inductor has an angle of 90°, but the phasor voltage across the capacitor has an angle of −90°. These two phasor voltages add to zero. In other words, if you connect an ac voltmeter across the inductor, it will read 200 V. Likewise, an ac voltmeter across the capacitor reads 200 V. But an ac voltmeter across the inductor and capacitor reads 0 V.

As a final point, the source voltage of Fig. 16-6a is 10 V rms. The voltage across the inductor (or capacitor) is 200 V rms, 20 times greater than the source voltage. This step-up in voltage across each reactive component of a series resonant circuit is called the *resonant rise of voltage.*

Test 16-1

1. When the source frequency approaches infinity, which of the following does not occur in a series resonant circuit?
    (a) X_L approaches infinity   (b) X_C approaches zero
    (c) Z approaches infinity   (d) I approaches infinity .... (   )

2. Which of these does not belong?
   (*a*) resonance of series circuit   (*b*) $X_L = X_C$
   (*c*) rms current equals rms source voltage divided by resistance
   (*d*) circuit looks like a *RC* circuit .... ( )
3. When the source frequency is greater than the resonant frequency, a series *RLC* circuit acts like a
   (*a*) resistive circuit   (*b*) series *RC* circuit
   (*c*) series *RL* circuit   (*d*) purely reactive circuit .... ( )
4. If you double the resistance in a series resonant circuit, the resonant frequency
   (*a*) stays the same   (*b*) doubles   (*c*) quadruples   (*d*) decreases ...... ( )
5. If the inductance is quadrupled in a series resonant circuit, the resonant frequency
   (*a*) stays the same   (*b*) doubles   (*c*) halves
   (*d*) decreases by a factor of 4 .... ( )

16-2. PARALLEL RESONANT CIRCUITS

Figure 16-7*a* shows a parallel *RLC* circuit driven by a current source (typically a transistor). To understand more about this important circuit, learn the answers to

> *What is the rms voltage across the circuit?*
> *What is the resonant frequency of the circuit?*

Basic idea

At very low frequencies the capacitor of Fig. 16-7*a* appears open, and the circuit acts like an equivalent parallel *RL* circuit (Fig. 16-17*b*). On the other hand, at very high

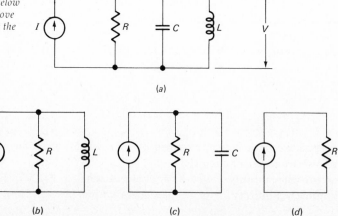

Figure 16-7. Parallel resonant circuit. (a) Original circuit. (b) Below the resonant frequency. (c) Above the resonant frequency. (d) At the resonant frequency.

frequencies the inductor appears open, and the circuit acts like a parallel RC circuit (Fig. 16-7c). Again, because of the opposite nature of the two types of reactances, there's an intermediate frequency where the effects of the inductor and capacitor cancel, leaving only the resistance (Fig. 16-7d).

Rms voltage

In Fig. 16-7a, the equivalent admittance is

$$\mathbf{Y} = G + jB_C - jB_L$$

or

$$\mathbf{Y} = G + j(B_C - B_L) \tag{16-6}$$

The magnitude of this admittance is

$$Y = \sqrt{G^2 + (B_C - B_L)^2} \tag{16-7}$$

Ohm's rms law applied to Fig. 16-7a gives

$$V = ZI = \frac{I}{Y} = \frac{I}{\sqrt{G^2 + (B_C - B_L)^2}} \tag{16-8}$$

This says the rms voltage across the parallel RLC circuit equals the rms source current divided by the magnitude of the admittance.

Resonant frequency

Resonance is still defined as the condition

$$X_L = X_C$$

equivalent to

$$B_L = B_C$$

At resonance the net susceptance $B_C - B_L$ is zero, and the rms voltage across the circuit has a maximum value of

$$V = \frac{I}{G} = RI$$

This makes sense, because at resonance the circuit is purely resistive (Fig. 16-7d).

The *resonant frequency* again is the source frequency that makes $X_L = X_C$, equivalent to $B_L = B_C$. If we graph B_L and B_C on the same set of axes, Fig. 16-8 results. B_C increases with frequency, and B_L decreases with frequency; therefore, there's a single frequency at which the two are equal. By definition, this is the resonant frequency f_r.

Figure 16-9 shows the *resonance curve* of a parallel RLC circuit; this graph indicates the voltage approaches zero at very low and very high frequencies. In between, the rms voltage increases and reaches a maximum value at the resonant frequency f_r.

Figure 16-8. Susceptance curves.

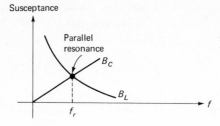

Susceptance

Parallel resonance

B_C

B_L

f

f_r

Again, the condition

$$X_L = X_C$$

leads to this formula for resonant frequency:

$$f_r = \frac{1}{2\pi \sqrt{LC}} \quad \text{(ideal)} \qquad (16\text{-}9)$$

EXAMPLE 16-6.

What is the resonant frequency in Fig. 16-10a? Sketch the resonance curve.

SOLUTION.

$$f_r = \frac{1}{2\pi \sqrt{LC}} = \frac{1}{6.28 \sqrt{50(10^{-6})2(10^{-9})}}$$
$$= 504 \text{ kHz}$$

At this frequency the parallel *RLC* circuit of Fig. 16-10a appears purely resistive to the source. Because of this, all the source current passes through the 5-kΩ resistor, and the rms voltage across the circuit is

$$V = RI = 5 \text{ k}\Omega \times 2 \text{ mA} = 10 \text{ V}$$

Figure 16-10b shows the resonance curve.

Figure 16-9. Resonance curve for parallel RLC circuit.

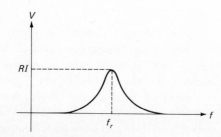

V

RI

f_r

f

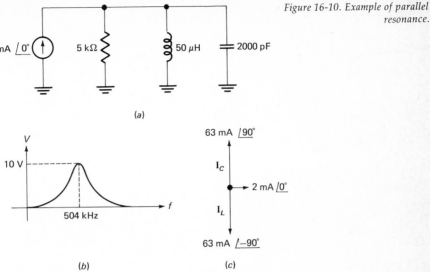

Figure 16-10. Example of parallel resonance.

(a)

(b)

(c)

EXAMPLE 16-7.
Calculate the phasor current in each component of Fig. 16-10a at resonance.

SOLUTION.
At resonance, the current in the resistor equals the source current:

$$\mathbf{I}_R = 2 \text{ mA } \underline{/0^\circ}$$

Because of this, the phasor voltage across the circuit is

$$\mathbf{V} = R\mathbf{I}_R = 5 \text{ k}\Omega \times 2 \text{ mA } \underline{/0^\circ} = 10 \text{ V } \underline{/0^\circ}$$

The inductive reactance at the resonant frequency (504 kHz) is

$$X_L = 2\pi fL = 6.28(504)10^3(50)10^{-6} = 158 \ \Omega$$

So the phasor current into the inductor is

$$\mathbf{I}_L = \frac{\mathbf{V}}{\mathbf{Z}_L} = \frac{10 \text{ V } \underline{/0^\circ}}{158 \ \Omega \ \underline{/90^\circ}} \cong 63 \text{ mA } \underline{/-90^\circ}$$

The capacitive reactance also equals 158 Ω at resonance; therefore, the phasor current into the capacitor is

$$\mathbf{I}_C = \frac{\mathbf{V}}{\mathbf{Z}_C} = \frac{10 \text{ V } \underline{/0^\circ}}{158 \ \Omega \ \underline{/-90^\circ}} \cong 63 \text{ mA } \underline{/90^\circ}$$

Figure 16-10c summarizes these phasor currents. The phasor current into the capacitor has an angle of 90°, but the phasor current into the inductor has an angle of −90°. These two phasor currents represent sinusoidal currents that are 180° apart; therefore, the sum of these two currents is zero. In other words, there's an rms current of 63 mA into the capacitor, and an rms current of 63 mA into the inductor, but the sum is zero because the instantaneous currents are always equal and opposite. (When inductor current is downward, capacitor current is upward in Fig. 16-10a, and vice versa.)

The source current in Fig. 16-10a, has a value of 2 mA rms, but the inductor current is 63 mA rms and the capacitor current is 63 mA rms. Each reactive current is approximately 30 times greater than the source current. This step-up in current is called the *resonant rise of the current.* Because the reactive currents are much larger than the source current, large amounts of energy are stored in the inductor and capacitor. This is why a parallel *RLC* circuit is often called a *tank circuit.*

Test 16-2

1. Which does not belong?
    (*a*) parallel resonance   (*b*) maximum equivalent impedance
    (*c*) minimum equivalent admittance   (*d*) minimum voltage across a circuit   ( )
2. Series resonant circuit is to parallel resonant circuit as maximum current is to
    (*a*) minimum voltage   (*b*) maximum voltage   (*c*) minimum current
    (*d*) maximum current .... ( )
3. The source driving a series resonant circuit is usually a voltage source. Likewise, the source driving a parallel resonant circuit is normally a
    (*a*) voltage source   (*b*) current source   (*c*) battery   (*d*) ac voltmeter   ( )
4. If the source frequency is less than the resonant frequency, a parallel *RLC* circuit acts like a
    (*a*) resistive circuit   (*b*) series *RC* circuit   (*c*) parallel *RL* circuit
    (*d*) parallel *RC* circuit .... ( )
5. At resonance, the parallel connection of the inductor and capacitor acts like an open because the sum of the two reactive currents is zero. This assumes
    (*a*) a large resistor   (*b*) ideal *L* and ideal *C*
    (*c*) a resonant rise in current   (*d*) the circuit is reactive .... ( )

16-3. PARALLEL RESONANCE WITH A LOSSY INDUCTOR

The preceding discussion of parallel resonance assumed an ideal inductor. This is all right for preliminary analysis, but there are many applications where we cannot neglect the resistance of the inductor. In particular, we now want to answer these questions:

> *What is the resonant frequency with a lossy inductor?*
> *What is the equivalent resistance?*

Resonant frequency

Figure 16-11a shows a tank circuit with the second approximation of an inductor (discussed in Sec. 15-3). The Q of the inductor equals

$$Q = \frac{X_L}{R_s}$$

Typically, the Q of an inductor is from 50 to 150; you can measure Q with an RLC bridge.

Section 15-4 explains how to convert a series RL circuit to an equivalent parallel RL circuit. Using the formulas of Fig. 15-7a, we can convert the series connection of L and R_s to an equivalent parallel circuit with a resistance of

$$R_s(Q^2 + 1)$$

and an inductance of

$$L\left(\frac{Q^2 + 1}{Q^2}\right)$$

Figure 16-11b shows the circuit after the lossy inductor has been converted to its equivalent parallel circuit. Resonance occurs when the reactance of the inductive branch equals the reactance of the capacitive branch. To get the formula for resonant frequency, therefore, we may proceed like this:

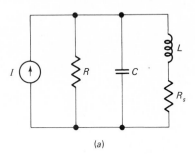

(a)

Figure 16-11. Resonance with a lossy inductor. (a) Original circuit. (b) Equivalent circuit.

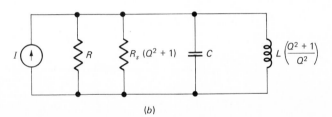

(b)

$$X_L = X_C$$

$$2\pi f_r L \left(\frac{Q^2 + 1}{Q^2}\right) = \frac{1}{2\pi f_r C}$$

or

$$f_r^2 = \frac{1}{(2\pi)^2 LC} \frac{Q^2}{Q^2 + 1}$$

After taking the square root of both sides,

$$f_r = \frac{1}{2\pi \sqrt{LC}} \sqrt{\frac{Q^2}{Q^2 + 1}} \qquad (16\text{-}10)$$

This tells us the resonant frequency of a circuit like Fig. 16-11a is always *lower* than the ideal resonant frequency given by Eq. (16-9). For instance, for a Q of 10 the actual resonant frequency is about ½ percent lower than the ideal resonant frequency. In practical electronics work, the Q of inductors is almost always greater than 10; therefore, the actual resonant frequency is within ½ percent of the ideal resonant frequency calculated by neglecting the R_s of the coil. For this reason, Eq. (16-10) is normally used only when Q is less than 10, or when exact answers are required in a particular application. Most of the time, Eq. (16-9) is accurate enough.

Equivalent resistance

The source of Fig. 16-11b sees a purely resistive circuit at resonance consisting of R in parallel with $R_s(Q^2 + 1)$, as shown in Fig. 16-12a. The case $Q > 10$ dominates in practical electronics. For this reason, we can use the approximate equivalent circuit of Fig. 16-12b for everyday work. With this equivalent circuit, the resonant frequency is

$$f_r = \frac{1}{2\pi \sqrt{LC}} \qquad (16\text{-}11)$$

and the rms voltage at resonance is

$$V = [R \parallel Q^2 R_s] \, I \qquad (16\text{-}12)$$

Often you will know the value of Q and X_L, but not R_s. For this reason, it helps to have an alternative formula for $Q^2 R_s$. Since $Q = X_L/R_s$, we can write

$$Q^2 R_s = Q \frac{X_L}{R_s} R_s$$

or

$$Q^2 R_s = QX_L \qquad (16\text{-}13)$$

This tells us the equivalent parallel resistance of the inductor equals QX_L. Because of this, the source of Fig. 16-12c looks into a purely resistive circuit at resonance whose

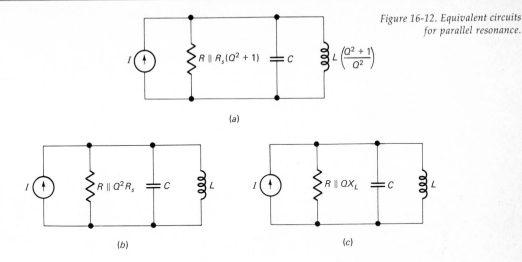

Figure 16-12. Equivalent circuits for parallel resonance.

(a)

(b)   (c)

resistance equals R in parallel with QX_L. The rms voltage across the circuit at resonance is

$$V = [R \parallel QX_L]I \qquad (16\text{-}14)$$

Equation (16-12) is helpful when the value of R_s is known. But usually, R_s is unknown and it's easier to work with Eq. (16-14).

EXAMPLE 16-8.
Calculate the rms voltage across the tank circuit of Fig. 16-13a at resonance.

SOLUTION.
First, get the resonant frequency. Since Q is large, Eq. (16-11) is adequate:

$$f_r = \frac{1}{2\pi\sqrt{LC}} = \frac{1}{6.28\sqrt{15(10^{-6})1000(10^{-12})}}$$
$$= 1.3 \text{ MHz}$$

At this frequency,

$$X_L = 2\pi f_r L = 6.28(1.3)10^6(15)10^{-6} = 122 \ \Omega$$

The equivalent parallel resistance of the inductor is

$$Q^2R_s = QX_L = 50(122 \ \Omega) = 6.1 \text{ k}\Omega$$

Therefore, we can visualize the circuit as shown in Fig. 16-13b.

At resonance, the source sees 5 kΩ in parallel with 6.1 kΩ. Therefore, the equiva-

Figure 16-13. Example of parallel resonance.

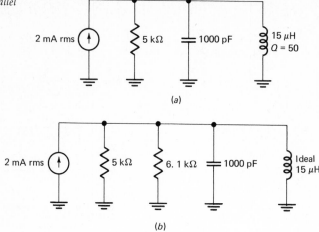

(a)

(b)

lent resistance the source works into is

$$R \parallel QX_L = 5 \text{ k}\Omega \parallel 6.1 \text{ k}\Omega = 2.75 \text{ k}\Omega$$

Because the rms source current is 2 mA, the rms voltage across the resonant circuit is

$$V = [R \parallel QX_L]I = 2.75 \text{ k}\Omega \times 2 \text{ mA} = 5.5 \text{ V}$$

Test 16-3

1. The exact resonant frequency of a parallel circuit with a lossy inductor is
   (*a*) the same as the ideal resonant frequency
   (*b*) lower than the ideal value   (*c*) higher than the ideal value .... ( )
2. For a Q of 10, the error in using the ideal resonant frequency for a tank circuit is
   (*a*) ½ percent   (*b*) 1 percent   (*c*) 2 percent   (*d*) 10 percent .... ( )
3. An inductor has a Q of 100 and an X_L of 200 Ω. Its equivalent parallel resistance is
   (*a*) 2 Ω   (*b*) 50 Ω   (*c*) 20 kΩ   (*d*) 50 kΩ .... ( )
4. Which does not belong?
   (*a*) rms voltage across a parallel resonant circuit
   (*b*) rms source current times $R \parallel QX_L$ (*c*) maximum rms voltage
   (*d*) minimum rms voltage .... ( )

16-4. Q_r OF PARALLEL RESONANT CIRCUITS

Parallel resonant circuits are much more important than series resonant circuits in practical electronics work. Typically, a transistor is connected to the tank circuit; this is why we keep showing a current source driving a parallel resonant circuit.

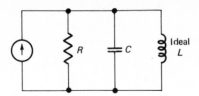

Figure 16-14. Q_r of parallel resonant circuit with ideal inductor.

In this section, find the answers to

> *How is the Q_r of an ideal parallel circuit defined?*
> *How is the Q_r with a lossy inductor defined?*

Ideal parallel RLC circuit

Figure 16-14 shows a parallel RLC circuit with an ideal inductor. The quality factor of this circuit is defined as

$$Q_r = \frac{R}{X_L} \qquad (16\text{-}15)$$

where Q_r = quality factor of tank circuit at resonance
R = parallel resistance
X_L = inductive reactance at resonance

An example, if $R = 10$ kΩ and $X_L = 125$ Ω,

$$Q_r = \frac{10 \text{ k}\Omega}{125 \text{ }\Omega} = 80$$

With lossy inductor

Figure 16-15a shows a parallel RLC circuit with a lossy inductor. With the second approximation, visualize this inductor as an ideal inductance L in series with a resistance R_s. The Q of the inductor equals

$$Q = \frac{X_L}{R_s}$$

As discussed in the preceding section, a large Q means we can replace the circuit by the equivalent circuit of Fig. 16-15b.

The Q_r of the parallel resonant circuit is still defined as the parallel resistance divided by the inductive reactance. In Fig. 16-15b, this means

$$Q_r = \frac{R \parallel Q^2 R_s}{X_L} \qquad (16\text{-}16)$$

*Figure 16-15. Q_r of parallel reso-
nant circuit with lossy inductor.*

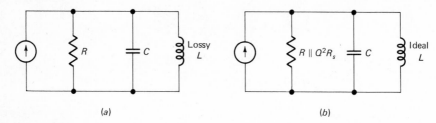

(a)   (b)

*Figure 16-15. Q_r of parallel reso-
nant circuit with lossy inductor.*

Because of Eq. (16-13), an alternative form is

$$Q_r = \frac{R \parallel QX_L}{X_L} \tag{16-17}$$

Don't get the Q's confused. In Eqs. (16-16) and (16-17), Q stands for the quality factor of the inductor, typically from 50 to 150. On the other hand, Q_r represents the quality factor of the parallel RLC circuit; because of this, it includes the effects of all the resistance in the circuit.

EXAMPLE 16-9.
What is the Q_r of the circuit shown in Fig. 16-16a?

SOLUTION.
Q_r is always calculated at the resonant frequency. Example 16-8 analyzed the circuit earlier and showed that

$$f_r = 1.3 \text{ MHz}$$

$$X_L = 122 \ \Omega$$

$$QX_L = 6.1 \text{ k}\Omega$$

*Figure 16-16. Calculating
Q_r.*

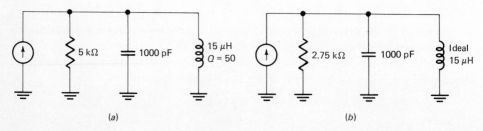

(a)   (b)

Therefore, the equivalent parallel resistance of the inductor (6.1 kΩ) may be lumped in with the 5-kΩ resistance of Fig. 16-16a to get an equivalent resistance of 2.75 kΩ, as shown in Fig. 16-16b.

With Eq. (16-17),

$$Q_r = \frac{R \parallel QX_L}{X_L} = \frac{2.75 \text{ k}\Omega}{122 \ \Omega} = 22.5$$

Notice that Q_r is lower than the Q of the inductor, given as 50 in Fig. 16-16a.

EXAMPLE 16-10.
Sometimes a current source drives a parallel resonant circuit which has no resistor (Fig. 16-17a). What is the Q_r of this circuit?

SOLUTION.
For the large-Q case, the lossy inductor can be converted into a parallel resistance of QX_L and an ideal inductor L, as shown in Fig. 16-17b. The Q_r is still defined as the equivalent parallel resistance divided by X_L:

$$Q_r = \frac{QX_L}{X_L} = Q$$

This says the Q_r of a parallel resonant circuit without a separate resistor equals the Q of the inductor. If the 5-kΩ resistor is opened in Fig. 16-16a, Q_r increases to 50.

You can get the same result with Eq. (16-17). If R goes to infinity, the numerator equals QX_L. After dividing by X_L, Q_r equals Q.

Remember this special case. It sets the upper limit on the value of Q_r. In other words, the Q_r of any parallel resonant circuit is always less than or equal to the Q of the inductor. If no resistor is used, or if the resistor has a very high resistance, Q_r approaches the Q of the inductor.

Figure 16-17. Another example of calculating Q_r.

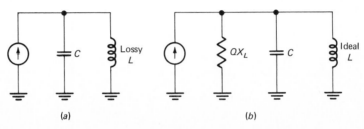

(a)   (b)

Test 16-4

1. Q includes the resistance of the
    (*a*) inductor only   (*b*) entire circuit   (*c*) capacitor only .... ( )
2. Q_r includes the resistance of the
    (*a*) inductor only   (*b*) entire circuit   (*c*) capacitor only .... ( )
3. For a parallel *RLC* circuit, Q_r always equals the equivalent parallel resistance divided by the inductive reactance at
    (*a*) 0   (*b*) f_r (*c*) $2f_r$ (*d*) any frequency .... ( )
4. An inductor has a Q of 125. A separate resistor is across a parallel resonant circuit. Q_r must be
    (*a*) equal to 125   (*b*) less than 125   (*c*) more than 125 .... ( )
5. The Q of an inductor equals 100. The QX_L of the inductor equals 10 kΩ. If a separate resistor with a resistance of 10 kΩ is connected across the tank circuit, Q_r equals
    (*a*) 25   (*b*) 50   (*c*) 75   (*d*) 100 .... ( )

16-5.   BANDWIDTH

Figure 16-18 shows the resonance curve for a parallel *RLC* circuit. The rms voltage reaches a maximum value of V_{max}. For very low and very high frequencies, the rms voltage approaches zero. As mentioned earlier, a resonant circuit can filter out a desired frequency from others. But the ability to do this depends on how fast the resonance curve decreases on either side of the resonant frequency.

As you read more about the filtering action of a parallel resonant circuit, find the answers to

> *How are cutoff frequencies defined?*
> *How is bandwidth defined?*
> *How is bandwidth related to Q_r?*

Figure 16-18. Cutoff frequencies.

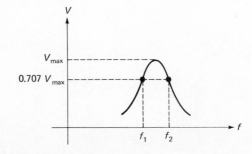

Cutoff frequencies

If you build a parallel resonant circuit, you can measure the rms voltage across it with an ac voltmeter. By varying the source frequency, you can find two frequencies, f_1 and f_2, where the rms voltage equals 70.7 percent of the maximum voltage (see Fig. 16-18). These frequencies are called the *cutoff frequencies.*

The average power equals the square of the rms voltage divided by the resistance. If the resistance is constant between f_1 and f_2, the average power at the cutoff frequencies equals half of the power at the resonant frequency. This is why cutoff frequencies are sometimes called the *half-power frequencies.* (Note: $0.707^2 = 0.5$; so if you calculate the power into a fixed resistance, you get half the power that occurs at the resonant frequency.)

Definition of bandwidth

The bandwidth B of a resonant circuit is defined as the upper cutoff frequency minus the lower cutoff frequency. In symbols,

$$B = f_2 - f_1 \qquad\qquad (16\text{-}18)$$

For instance, suppose you build a parallel resonant circuit and measure these frequencies:

$$f_1 = 985 \text{ kHz}$$

$$f_r = 1000 \text{ kHz}$$

$$f_2 = 1015 \text{ kHz}$$

Then, the bandwidth equals

$$B = f_2 - f_1 = 1015 \text{ kHz} - 985 \text{ kHz}$$
$$= 30 \text{ kHz}$$

Look at Fig. 16-18. The closer f_1 and f_2 are, the smaller the bandwidth and the better the filtering action. Ideally, if all you want is the resonant frequency, a circuit with a bandwidth near zero will be best. In some applications, however, it's important to pass the resonant frequency and frequencies near it. For instance, broadcast receivers use resonant circuits with bandwidths of 10 kHz; this is narrow enough to separate one station from another; at the same time it's wide enough to pass the resonant frequency and frequencies to within 5 kHz on each side.[1]

---

[1] The importance of this is discussed in Malvino, A. P.: *Electronic Principles*, McGraw-Hill Book Company, New York 1973, pp. 680–683.

Bandwidth and Q_r

By an advanced derivation, we can prove this important formula:

$$B = \frac{f_r}{Q_r} \qquad (16\text{-}19)$$

This says the bandwidth equals the resonant frequency divided by the Q_r of the circuit. As an example, if $f_r = 1000$ kHz and $Q_r = 50$,

$$B = \frac{f_r}{Q_r} = \frac{1000 \text{ kHz}}{50} = 20 \text{ kHz}$$

Note that we can control the bandwidth by varying the Q_r of a resonant circuit. As discussed in the preceding section, Q_r depends on the size of the separate resistor across the resonant circuit, as well as on the inductor. By varying this resistor, we can change Q_r which in turn changes the value of B. This is important because resonant circuits such as used in radio, TV, etc. must have the correct bandwidth for the kinds of signals being received.

EXAMPLE 16-11.
What is the bandwidth for the resonant circuit of Fig. 16-19?

SOLUTION.
We analyzed this circuit in Example 16-9 and found that

$$f_r = 1.3 \text{ MHz}$$

$$Q_r = 22.5$$

Therefore, the circuit has a bandwidth of

$$B = \frac{f_r}{Q_r} = \frac{1.3 \text{ MHz}}{22.5} \cong 58 \text{ kHz}$$

EXAMPLE 16-12.
If the 5-kΩ resistor of Fig. 16-19 is removed, what is the new value of the bandwidth?

Figure 16-19. Example of calculating bandwidth.

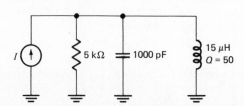

SOLUTION.

With no separate resistor, the Q_r of the circuit equals the Q of the inductor. In this case,

$$B = \frac{f_r}{Q_r} = \frac{1.3 \text{ MHz}}{50} = 26 \text{ kHz}$$

This is the smallest bandwidth we can get with Fig. 16-19, because the maximum Q_r occurs when there's no separate resistor.

Test 16-5

1. Which does not belong?
    (a) $0.5V_{max}$ (b) cutoff frequencies (c) $0.707V_{max}$
    (d) half-power frequencies .... ( )
2. The bandwidth always has the units:
    (a) amperes (b) volts (c) hertz (d) no units .... ( )
3. If the Q_r of a circuit is doubled, the bandwidth
    (a) halves (b) doubles (c) quadruples (d) stays the same .... ( )
4. Which belongs least?
    (a) high Q_r (b) small bandwidth (c) f_1 and f_2 close together
    (d) large bandwidth .... ( )
5. A tank circuit has a resonant frequency of 1 MHz. To get a bandwidth of 20 kHz, Q_r must equal
    (a) 20 (b) 50 (c) 100 (d) 500 .... ( )

16-6. AIR-CORE TRANSFORMERS

Inductors L_1 and L_2 of Fig. 16-20 are an air-core transformer. As you may recall, the coefficient of coupling k is very small with an air-core transformer, because the core is nonmagnetic. The analysis of the currents and voltages in Fig. 16-20 is too complicated to go into here; but there are a few general ideas you should know about.

Because k is so low with an air-core transformer, you almost always will see tuning

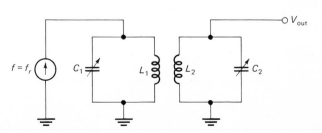

Figure 16-20. Tuned air-core transformer.

capacitors used in parallel with L_1 and L_2, as shown in Fig. 16-20. The idea is to adjust the capacitors until both parallel circuits are resonant at the source frequency. The reason for doing this is to get a *resonant rise in current* (see Example 16-7). The increased current results in a larger signal voltage being induced in the secondary winding. This helps to offset the low value of k. In other words, because of the resonant rise in current, a larger output voltage results than would be possible without the tuning capacitors.

SUMMARY OF FORMULAS

DEFINED

$$X_L = X_C \qquad \text{(resonance)} \tag{16-4}$$

$$Q_r = \frac{R \parallel QX_L}{X_L} \qquad \text{(lossy } L) \tag{16-17}$$

$$B = f_2 - f_1 \tag{16-18}$$

DERIVED

$$I = \frac{V}{\sqrt{R^2 + (X_L - X_C)^2}} \tag{16-3}$$

$$f_r = \frac{1}{2\pi\sqrt{LC}} \tag{16-5}$$

$$V = \frac{I}{\sqrt{G^2 + (B_C - B_L)^2}} \tag{16-8}$$

$$f_r = \frac{1}{2\pi\sqrt{LC}} \sqrt{\frac{Q^2}{Q^2 + 1}} \tag{16-10}$$

$$V = [R \parallel QX_L]I \tag{16-14}$$

$$B = \frac{f_r}{Q_r} \tag{16-19}$$

Problems

16-1.   If the source of Fig. 16-21a has a frequency of 500 kHz, what is the rms current?

16-2.   What is the rms current in Fig. 16-21a if the source frequency equals 1 MHz?

16-3.   Calculate the rms current in Fig. 16-21b. The given reactances are for a source

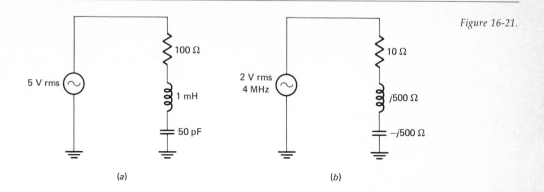

Figure 16-21.

(a)   (b)

frequency of 4 MHz; if the source frequency is doubled, what is the rms current?

16-4.   What is the resonant frequency of Fig. 16-21a? If the source frequency equals the resonant frequency, what does the rms current equal?

16-5.   What is the rms voltage across the capacitor of Fig. 16-21a at resonance?

16-6.   The capacitor of Fig. 16-22a is variable from 50 to 500 pF. What is the minimum resonant frequency? The maximum resonant frequency?

16-7.   You can combine similar components in the series circuit of Fig. 16-22b to get a single R, L, and C. What is the resonant frequency of this circuit?

16-8.   The circuit of Fig. 16-23a is resonant. What is the rms voltage across the circuit? What is the rms current in each component?

16-9.   What is the minimum resonant frequency in Fig. 16-23b? The maximum? The rms voltage across the circuit at resonance?

16-10. If you want to double the resonant frequency, by what factor should you decrease capacitance?

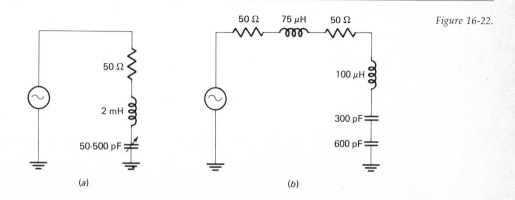

Figure 16-22.

(a)   (b)

Figure 16-23.

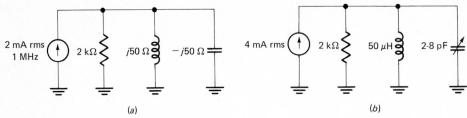

(a)   (b)

16-11. An inductor has an L of 50 μH and an R_s of 5 Ω. If R_s remains fixed even though frequency changes, what is the value of Q for each of these: 200 kHz, 400 kHz, 1 MHz, and 2 MHz.

(Actually, R_s increases with frequency in a lossy inductor, because power losses increase with frequency.)

16-12. A parallel *RLC* circuit has a Q of 5. If $L = 100$ μH and $C = 3000$ pF, what is the resonant frequency?

16-13. Calculate the rms voltage across the circuit of Fig. 16-24 at resonance.

16-14. If the inductor of Fig. 16-24 is replaced with an inductor with an L of 10 μH and a Q of 50, what will the new rms voltage across the circuit at resonance be?

16-15. What is the Q_r of the circuit in Fig. 16-24?

16-16. If the 10-kΩ resistor is replaced by a 5-kΩ resistor, what is the new Q_r for the circuit of Fig. 16-24?

16-17. If the 10-kΩ resistor of Fig. 16-24 is removed, what is the rms voltage across the circuit at resonance? The new value of Q_r?

16-18. What is the bandwidth for the circuit of Fig. 16-24?

16-19. If the 10-kΩ resistor of Fig. 16-24 is removed, what is the bandwidth of the circuit?

16-20. A circuit has a resonant frequency of 2 MHz and a bandwidth of 10 kHz. What is its Q_r?

16-21. If a resonant circuit has an f_r of 1 MHz and you want a bandwidth of 12.5 kHz, what Q_r should the circuit have?

Figure 16-24.

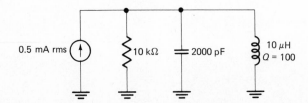

ANSWERS TO TESTS

16-1.   *d, d, c, a, c*
16-2.   *d, b, b, c, b*
16-3.   *b, a, c, d*
16-4.   *a, b, b, b, b*
16-5.   *a, c, a, d, b*

17. Instantaneous AC Analysis

This chapter is about calculating the instantaneous value of sinusoidal currents and voltages.

17-1. PHASORS

A *phasor* is a special kind of vector. As you read more about it, find the answers to

What is a phasor?
What is a radian?
What is the angular velocity of a phasor?

Basic idea

A phasor is a vector that rotates counterclockwise around the origin at a constant speed (see Fig. 17-1a). The magnitude of the phasor is M. As the phasor moves, its tip traces out a circle of radius M. Furthermore, the angle θ increases linearly with time.

Angle in radians

Degrees are not the only unit of measure for an angle. Look at Fig. 17-1b. The tip of the phasor has moved through a circumferential distance of d; the larger d is, the greater the value of θ. The defining formula for an angle in radians is

$$\theta = \frac{d}{M} \tag{17-1}$$

Figure 17-1. (a) Phasor. (b) Angle in radians. (c) 1 rad.

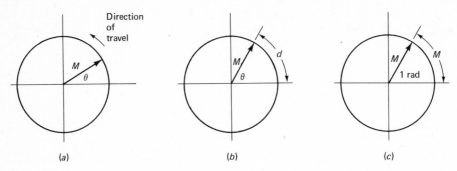

(a)   (b)   (c)

If $d = 3$ and $M = 2$, the angle in radians is

$$\theta = \frac{3}{2} = 1.5$$

Both d and M are length quantities. Because of this, d/M is a pure number. Nevertheless, it's customary to attach the label *radians* (abbreviated rad) to any answer in radians; this prevents confusing an angle in radians with an angle in degrees. For this reason, the answer in the foregoing example is normally written as

$$\theta = 1.5 \text{ rad}$$

Figure 17-1c shows the special case of 1 rad, the angle when the tip of the phasor moves through a distance equal to the radius of the circle.

Radians and degrees

When the phasor rotates through an angle of 360°, the tip of the phasor moves through a distance equal to the circumference of the circle. In this case,

$$d = 2\pi M$$

and the angle in radians is

$$\theta = \frac{d}{M} = \frac{2\pi M}{M} = 2\pi \text{ rad} = 6.28 \text{ rad}$$

For a half revolution, the phasor moves through an angle of 180°. In this case,

$$d = \pi M$$

and

$$\theta = \frac{d}{M} = \frac{\pi M}{M} = \pi \text{ rad} = 3.14 \text{ rad}$$

Therefore,

$$180° = \pi \text{ rad} = 3.14 \text{ rad}$$

It's often convenient to keep the factor π in answers, because π rad represents a half revolution. In other words, don't replace π by 3.14; then, all answers come out as a number times π rad, where the number stands for the number of half revolutions. For instance, given

$$\theta = 5\pi \text{ rad}$$

the angle equals five half revolutions, equivalent to 2.5 complete revolutions. Or given

$$\theta = \frac{\pi}{3} \text{ rad}$$

the angle equals one-third of a half revolution, equivalent to 60°.

Table 17-1 shows some useful degree-radian relations. Bearing in mind that π rad is a half revolution (180°), you can see why each pair is correct. For example, $\pi/6$ rad represents one-sixth of a half revolution, which equals 180°/6 or 30°.

TABLE 17-1. DEGREES AND RADIANS

Degrees	0	30	60	90	180	360
Radians	0	$\pi/6$	$\pi/3$	$\pi/2$	π	2π

In general, you can convert any angle in radians to its equivalent angle in degrees with

$$\theta_d = \frac{180°}{\pi \text{ rad}} \, \theta_r \tag{17-2}$$

where θ_d is the angle in degrees, and θ_r is the angle in radians. To convert from degrees to radians, use

$$\theta_r = \frac{\pi \text{ rad}}{180°} \, \theta_d \tag{17-3}$$

Angular velocity

The *angular velocity* of a phasor is defined as the angle it has traveled through divided by the time of travel. In symbols,

$$\omega = \frac{\theta}{t} \tag{17-4}$$

where ω = angular velocity
  θ = angle in radians
  t = time

Because the angle is in radians, ω has units of radians per second. Incidentally, ω is lower-case omega, whereas Ω is capital omega.

Here's an example. If a phasor moves through 20π rad in 4 s, its angular velocity is

$$\omega = \frac{\theta}{t} = \frac{20\pi \text{ rad}}{4 \text{ s}} = 5\pi \text{ rad/s}$$

Or if a phasor makes 10 revolutions in 3 s,

$$\omega = \frac{10 \times 2\pi \text{ rad}}{3 \text{ s}} = 6.67\pi \text{ rad/s}$$

Test 17-1

1. Which does not belong?
   (a) phasor   (b) clockwise motion   (c) constant speed
   (d) tip traces circle .... ( )
2. An angle of π rad corresponds to
   (a) a half revolution   (b) 180°   (c) both of these .... ( )
3. An angle of 7π rad means the phasor has made
   (a) 7 half revolutions   (b) 7 revolutions   (c) 14 revolutions .... ( )
4. θ is to ω as radians is to
   (a) degrees   (b) radians per second   (c) angle
   (d) degrees per second .... ( )

17-2.   PHASORS AND SINUSOIDS

The main questions answered in this section are

> *How is a phasor related to a sinusoid?*
> *How is angular velocity related to frequency?*

Graphical construction of sine wave

Look at Fig. 17-2a. The phasor has a magnitude M and an angle θ. The vertical side opposite angle θ is labeled y. This side has a length of

$$y = M \sin \theta$$

As the phasor rotates in a counterclockwise direction, θ increases linearly with time. Because of this, the instantaneous values of y produce a graph like Fig. 17-2b. In other words, by projecting the value of y for each value of θ, we can construct a sine wave without calculations.

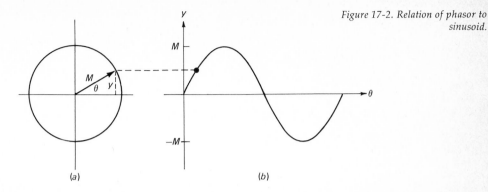

Figure 17-2. Relation of phasor to sinusoid.

(a)   (b)

Angular velocity and frequency

The sine wave of Fig. 17-2b has an equation of

$$y = M \sin \theta$$

By rearranging Eq. (17-4),

$$\theta = \omega t$$

Therefore,

$$y = M \sin \omega t \qquad (17\text{-}5)$$

This is a useful formula when M and ω are known. For instance, if $M = 50$ and $\omega = 2000\pi$ rad/s,

$$y = 50 \sin 2000\pi t$$

With an equation like this, we can calculate the value of y for different values of t.

The number of cycles per second equals the frequency of the sine wave in Fig. 17-2b. But each revolution of the phasor traces out one cycle. Since the period of a cycle equals T, the phasor moves through an angle of 2π rad in time T. Therefore, it has an angular velocity of

$$\omega = \frac{\theta}{t} = \frac{2\pi \text{ rad}}{T}$$

Since frequency is the reciprocal of period, this may be written as

$$\omega = 2\pi f \qquad (17\text{-}6)$$

As an example, if a sine wave has a frequency of 1 kHz, the angular velocity of the corresponding phasor is

$$\omega = 2\pi(1000) = 2000\pi \text{ rad/s}$$

General expression for a sinusoid

When there's a phase angle, the equation of a sinusoid is

$$y = M \sin (\theta + \phi)$$

which may be written in either of two equivalent ways:

$$y = M \sin (\omega t + \phi) \tag{17-7}$$

or
$$y = M \sin (2\pi ft + \phi) \tag{17-8}$$

Either of the last two equations is useful for instantaneous analysis, that is, calculating values of y for particular values of t.

EXAMPLE 17-1.
Sketch the waveform of

$$v = 50 \sin (\omega t + 30°)$$

SOLUTION.
The peak value is 50, and the phase angle is 30°. Figure 17-3a shows the waveform. Notice that ωt is used in place of θ.

EXAMPLE 17-2.
Write the equation for the sinusoidal current of Fig. 17-3b.

SOLUTION.
The peak value is 0.02 A, and the phase angle is −45°. Substituting these values into Eq. (17-7) gives

$$i = 0.02 \sin (\omega t - 45°)$$

Figure 17-3. Sinusoidal voltage and current.

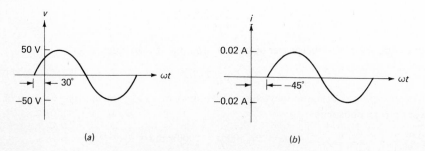

(a) (b)

EXAMPLE 17-3.

If ω equals 2000π rad/s, what does the sinusoidal current of the preceding example equal when $t = 1$ ms?

SOLUTION.

Substitute the given value of ω to get

$$i = 0.02 \sin (2000\pi t - 45°)$$

For $t = 1$ ms $= 0.001$ s,

$$i = 0.02 \sin (2\pi - 45°) = 0.02 \sin (360° - 45°)$$
$$= 0.02 \sin 315° = -0.01414 \text{ A} \cong -14.1 \text{ mA}$$

Test 17-2

1. The number of revolutions a phasor makes corresponds to
   (a) the number of radians per second   (b) the number of cycles of the sinusoid
   (c) the peak value of the sinusoid   (d) the phase angle .... (   )
2. When the phasor is above the horizontal axis, the sinusoid has a
   (a) negative value   (b) value of zero   (c) positive value
   (d) value equal to the peak value .... (   )
3. Which of these does not belong?
   (a) ω (b) $2\pi f$ (c) $2\pi/T$ (d) revolutions per second .... (   )
4. A sinusoid has a frequency of 10 Hz. The angular velocity of the corresponding phasor is
   (a) 10 rad/s   (b) 10π rad/s   (c) 20π rad/s   (d) 40π rad/s .... (   )

17-3. INSTANTANEOUS VOLTAGE AND CURRENT

After completing an ac analysis, you may have the phasor voltages and phasor currents of the circuit. If necessary, you can convert these phasor voltages and phasor currents back to their sinusoidal equations. This section tells you how.

To begin with, Eq. (17-7) may be written as follows for sinusoidal voltage:

$$v = V_p \sin (\omega t + \phi) \tag{17-9}$$

Similarly, for sinusoidal current, Eq. (17-7) gives

$$i = I_p \sin (\omega t + \phi) \tag{17-10}$$

Given a phasor voltage of

$$\mathbf{V} = V \angle \phi$$

or a phasor current of

$$\mathbf{I} = I \angle \phi$$

calculate the peak voltage or peak current. After substituting the peak value and phase angle into Eq. (17-9) or (17-10), you have the equation for the sinusoidal voltage or current. The following examples illustrate how it's done.

EXAMPLE 17-4.
Suppose an ac circuit has a source frequency of 1 kHz. What is the equation of the sinusoidal voltage whose phasor voltage equals

$$\mathbf{V} = 25 \text{ V } \underline{/30°}$$

SOLUTION.
The rms voltage is

$$V = 25 \text{ V}$$

which means the peak voltage is

$$V_p = 1.414V = 1.414(25 \text{ V}) = 35.4 \text{ V}$$

The phase angle is

$$\phi = 30°$$

and the angular velocity is

$$\omega = 2\pi f = 2\pi(1000) = 2000\pi \text{ rad/s}$$

When these values are substituted into Eq. (17-9),

$$v = 35.4 \sin (2000\pi t + 30°)$$

Once you've got an equation in this form, you can calculate the voltage at any instant in time by substituting the value of t and proceeding as shown earlier in Example 17-3.

EXAMPLE 17-5.
The phasor current in a circuit is

$$\mathbf{I} = 5 \text{ mA } \underline{/-60°}$$

If the frequency is 5 kHz, what is the equation for the sinusoidal current?

SOLUTION.
The peak current is

$$I_p = 1.414I = 1.414(5 \text{ mA}) = 7.07 \text{ mA}$$

The phase angle is

$$\phi = -60°$$

The angular velocity is

$$\omega = 2\pi f = 2\pi (5000) = 10{,}000\pi \text{ rad/s}$$

After substituting these values into Eq. (17-10),

$$i = 0.00707 \sin (10{,}000\pi t - 60°)$$

Test 17-3

1. The angle of a phasor voltage is $-45°$. Therefore, the value substituted into the equation for the sinusoidal voltage is
    (a) $-45°$ (b) $45°$ (c) $90°$ (d) $-90°$ .... ( )
2. An ac circuit has a source frequency of 40 kHz. The value of ω is
    (a) 40,000 rad/s   (b) 80,000π rad/s
    (c) 40,000 revolution per second   (d) 80,000 c/s .... ( )
3. The rms voltage across a load is 35 μV. The peak sinusoidal voltage is closest to
    (a) 35 μV   (b) 50 mV   (c) 75 mV   (d) 50 μV .... ( )

SUMMARY OF FORMULAS

DEFINING

$$\theta = \frac{d}{M} \tag{17-1}$$

$$\omega = \frac{\theta}{t} \tag{17-4}$$

DERIVED

$$\theta_d = \frac{180°}{\pi \text{ rad}} \theta_r \tag{17-2}$$

$$\theta_r = \frac{\pi \text{ rad}}{180°} \theta_d \tag{17-3}$$

$$v = V_p \sin (\omega t + \phi) \tag{17-9}$$

$$i = I_p \sin (\omega t + \phi) \tag{17-10}$$

Problems

17-1.   The tip of a phasor moves through a circumferential distance of 20 in. If the radius equals 3 in, what is the corresponding angle in radians?

17-2.   Convert each of the following to radians: 30°, 45°, 75°, 150°, 270°, and 560°.

17-3.   Convert each of these to degrees: 2 rad, 4π rad, 10 rad, and 35π rad.

17-4.   A phasor moves through an angle of 180° in 3 ms. What is the angular velocity?

17-5.   A phasor makes four revolutions in 50 μs. What is the angular velocity?

17-6.   A sinusoid has a period of 2 ms. What is the angular velocity of the corresponding phasor?

17-7.   Calculate the angular velocity for each of the following:
   a. $f = 20$ kHz
   b. $f = 1$ MHz
   c. $T = 20$ μs

17-8.   A sinusoidal voltage has this equation:

$$v = 20 \sin (2000\pi t + 45°)$$

   Calculate the instantaneous voltage for each of these:
   a. $t = 0$
   b. $t = 250$ μs
   c. $t = 0.5$ ms
   d. $t = 1$ ms

17-9.   The source frequency in an ac circuit is 2 kHz. What is the sinusoidal equation for each of these:
   a. $V = 10$ V $\underline{/0°}$
   b. $V = 25$ mV $\underline{/90°}$
   c. $I = 5$ mA $\underline{/-90°}$
   d. $I = 40$ mA $\underline{/75°}$

ANSWERS TO TESTS

17-1.   *b, c, a, b*
17-2.   *b, c, d, c*
17-3.   *a, b, b*

18. Switching Circuits

A *switching circuit* is one in which a voltage or current suddenly changes. The abrupt change may be caused by a switch, a source, or an electronic switching device. Regardless of the cause, the voltages and currents change throughout the circuit to satisfy Kirchhoff's laws.

Switching circuits are the backbone of pulse and digital electronics, one of the specialties in electronics. This chapter discusses the transients that occur in *RC* and *RL* switching circuits.

18-1. *R* SWITCHING CIRCUITS

Many circuits use only resistors for loads. If we neglect the stray capacitance and lead inductance, the analysis is ideal. In the discussion that follows, learn the answers to

> *What is the switching instant?*
> *What is the notation for this instant?*
> *What are step sources?*

The switching instant

Figure 18-1a shows a resistive (*R*) switching circuit. Before the switch is closed, there's one set of voltages and currents; after it's closed, another set. Unless otherwise indicated, $t = 0$ stands for the *switching instant,* the point in time when a switch or other devices cause an abrupt change in voltage or current, or both.

Notation at the switching instant

The switching instant is one of the most important points in time, because the circuit undergoes an abrupt change at this instant. Just before $t = 0$, there's one equivalent circuit; just after, another. To keep things straight, we will use this notation:

$t = 0^+$: this represents the *initial instant*, the point in time just after $t = 0$.

$t = 0^-$: this is the *preinitial instant*, the point in time just before $t = 0$.

v_i: this stands for the *initial voltage*, the voltage across a resistor or branch at $t = 0^+$.

v_p: this is the *preinitial voltage*, the voltage across a resistor or branch at $t = 0^-$.

For instance, suppose an oscilloscope (dc input) is across the 2-kΩ resistor of Fig. 18-1a. With the switch open, voltage-divider action produces 2 V across the 2-kΩ resistor. If the switch closes at $t = 0$, the voltage across this resistor suddenly jumps to 10 V, as shown in Fig. 18-1b. So the voltages across the 2-kΩ resistor are

$$v_p = 2 \text{ V}$$
$$v_i = 10 \text{ V}$$

Step sources

A waveform like Fig. 18-1b is called a *step*, because it steps abruptly from one value to another. When v_i is greater than v_p, the step is positive; when v_i is less than v_p, the step is negative.

Transistors and other devices can be used to build *step sources*, circuits that deliver a voltage or current that steps abruptly from one value to another. In other words, electronic circuits exist whose Thevenin circuit is approximately a voltage source that steps from V_1 to V_2, as shown in Fig. 18-2a; with another design, there are circuits whose Norton equivalent is a current source that steps from I_1 to I_2 (Fig. 18-2b).

Figure 18-1. (a) Switching circuit.
(b) Step.

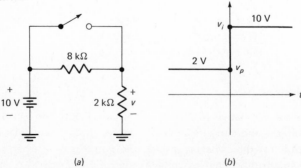

(a)   (b)

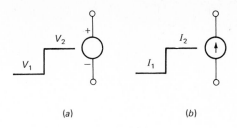

Figure 18-2. (a) *Step voltage source.* (b) *Step current source.*

(a)   (b)

EXAMPLE 18-1.
The switch of Fig. 18-3a closes at $t = 0$. What does the waveform of node voltage v look like?

SOLUTION.
With the switch open, there's no current through the 10-kΩ resistor; therefore, $v = 10$ V. After the switch closes, v drops to zero. So the waveform of v is a negative step like Fig. 18-3b, with $v_p = 10$ V and $v_i = 0$.

EXAMPLE 18-2.
What is the waveform of node voltage v in Fig. 18-4a?

SOLUTION.
At $t = 0^-$ the current source has a value of zero, equivalent to an open. Therefore, the circuit before the switching instant is equivalent to Fig. 18-4b. Clearly, $v = 10$ V.

At $t = 0^+$ the current source has a value of 10 mA, as shown in Fig. 18-4c. This 10

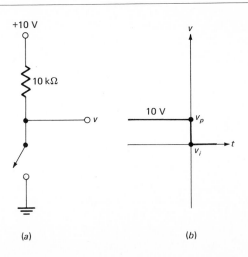

Figure 18-3. (a) *Switching circuit.* (b) *Negative step.*

(a)   (b)

Figure 18-4. Example of generating
a negative voltage step.

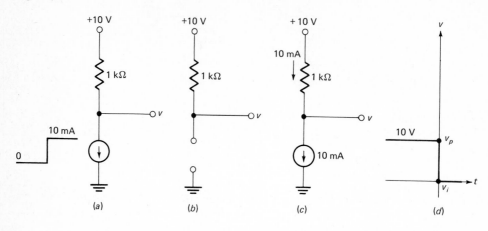

Figure 18-5. Example of generating
positive and negative steps.

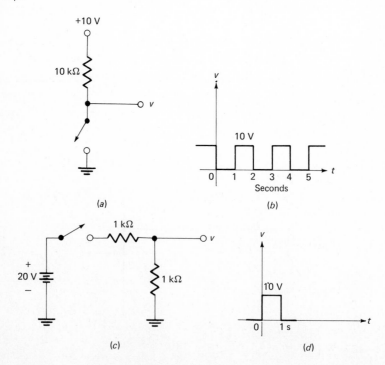

mA flows through the 1-kΩ resistor. Node voltage v drops to zero, because

$$v = 10 - Ri = 10 - 1000(0.01) = 0$$

Figure 18-4d shows the waveform of v, a negative step with $v_p = 10$ V and $v_i = 0$.

EXAMPLE 18-3.
The switch of Fig. 18-5a closes at $t = 0$, opens at $t = 1$ s, closes at $t = 2$ s, and so on. What does the waveform of v look like?

SOLUTION.
With the switch open, $v = 10$ V. With the switch closed, $v = 0$. Therefore, the waveform of v looks like Fig. 18-5b. A waveform like this is called a *square wave*.

EXAMPLE 18-4.
The switch of Fig. 18-5c closes at $t = 0$ and opens at $t = 1$ s. What does the waveform of v look like?

SOLUTION.
When the switch is open, v is zero. When the switch is closed, voltage-divider action produces a v of 10 V. This is why the waveform steps positively at $t = 0$ and negatively at $t = 1$ s (Fig. 18-5d). A waveform like Fig. 18-5d is called a *rectangular pulse*.

Test 18-1

1. Which of the following belongs least?
   (a) v_i (b) $t = 0^-$ (c) v_p (d) just before switching instant   .... ( )
2. Which of these does not belong?
   (a) $t = 0^+$ (b) v_p (c) v_i (d) just after switching instant   .... ( )
3. Initial voltage is to $t = 0^+$ as preinitial voltage is to
   (a) $t = 0$ (b) v_p (c) v_i (d) $t = 0^-$ .... ( )
4. In an R switching circuit with negligible stray capacitance and lead inductance, the changing voltages and current are
   (a) sine waves   (b) sawtooth waves   (c) exponentials   (d) steps   ... ( )
5. Which of these has an infinite Thevenin resistance?
   (a) step source   (b) step voltage source   (c) step current source
   (d) sinusoidal source   .... ( )

18-2.   CAPACITOR CURRENT AND VOLTAGE

Resistive-capacitive (RC) switching circuits use resistors and capacitors; the sources are either step sources or dc sources with switches. Ohm's law applies to each resistor, but not to the capacitors. In what follows, find the answers to

What is instantaneous current?
What is the capacitor formula?

Rate of voltage change

Figure 18-6a shows conventional flow into a capacitor. This results in an increasing voltage v, as shown in Fig. 18-6b. Since

$$q = Cv \qquad (18\text{-}1)$$

the waveform of q looks the same as the waveform of v (see Fig. 18-6c). As discussed in Sec. 10-8, dt is the special notation used for the time change between two nearby points in time. The corresponding voltage change is symbolized by dv, and the charge change by dq.

At $t = t_1$, Eq. (18-1) gives

$$q_1 = Cv_1$$

At a slightly later instant when $t = t_2$,

$$q_2 = Cv_2$$

The difference of these two equations is

$$q_2 - q_1 = Cv_2 - Cv_1 = C(v_2 - v_1)$$

which may be written

$$dq = C \, dv$$

Dividing by dt gives

$$\frac{dq}{dt} = C \frac{dv}{dt} \qquad (18\text{-}2)$$

Figure 18-6. Charging a capacitor.
(a) Circuit. (b) Voltage. (c) Charge
waveform.

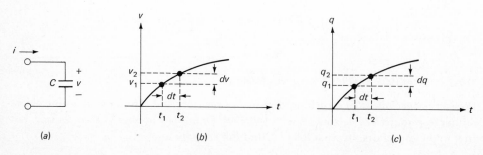

(a)   (b)   (c)

The meaning of dq/dt and dv/dt is similar to the meaning of $d\phi/dt$. As you recall, $d\phi/dt$ is the rate of flux change; it tells how fast the flux changes with time. In a similar way, dq/dt is the rate of charge change and dv/dt is the rate of voltage change.

The rate of charge change tells how fast the charge in Fig. 18-6c increases, equivalent to how steep the waveform is. The steeper the waveform, the larger the value of dq/dt. Similarly, dv/dt indicates how fast the voltage increases in Fig. 18-6b. The steeper the waveform, the larger the value of dv/dt.

Instantaneous current

When two points are extremely close together on a waveform, they appear as a single point. *Instantaneous current*, the current at a single point in time, is defined as the charge change divided by the time change between two infinitesimally close points in time. As a defining formula,

$$i = \frac{dq}{dt} \tag{18-3}$$

where i = instantaneous current
dq = charge change
dt = time change

In this defining formula, dt is infinitesimally small; as an approximation, this means the pair of points are so close together they appear as a single point on an oscillosope.
As an example, if $dq = 3\ \mu C$ and $dt = 1\ \mu s$, the instantaneous current is

$$i = \frac{dq}{dt} = \frac{3(10^{-6})\ C}{10^{-6}\ s} = 3\ C/s = 3\ A$$

The capacitor formula

Equation (18-3) allows us to rewrite Eq. (18-2) as

$$i = C\,\frac{dv}{dt} \tag{18-4}$$

This is an important formula. It says the instantaneous current into a capacitor equals the capacitance times the rate of voltage change. It applies to all capacitors, and is as basic for capacitors as Ohm's law is for resistors.

No matter what the waveform of capacitor voltage (sawtooth, sine wave, exponential, etc.), Eq. (18-4) applies. The steeper the voltage waveform, the larger the value of dv/dt and the greater the current.

EXAMPLE 18-5.
A 2-μF capacitor has a dv/dt of 100 V/s. What is the instantaneous current?

SOLUTION.

$$i = C\frac{dv}{dt} = 2(10^{-6})100 = 200 \ \mu A$$

EXAMPLE 18-6.
The instantaneous current into a 50-μF capacitor is 0.5 mA. What is the rate of voltage change?

SOLUTION.
Equation (18-4) rearranges to

$$\frac{dv}{dt} = \frac{i}{C}$$

Substituting the given values,

$$\frac{dv}{dt} = \frac{0.5(10^{-3})}{50(10^{-6})} = 10 \text{ V/s}$$

(Read the answer as ten volts per second.)

Test 18-2

1. Which of these belongs least?
   (a) instantaneous current   (b) rate of charge change   (c) dq/dt
   (d) dv/dt .... ( )
2. What do $d\phi/dt$, dq/dt, and dv/dt have in common?
   (a) units of measure   (b) the time change is extremely small
   (c) webers   (d) a large time change .... ( )
3. For a given capacitor, doubling the rate of voltage change
   (a) halves the current   (b) doubles the current
   (c) quadruples the current   (d) doesn't affect the current .... ( )
4. Given a dc voltage, the value of dv/dt is
   (a) 0   (b) infinite   (c) unity   (d) indeterminate .... ( )
5. A sinusoidal voltage has the greatest dv/dt at the
   (a) start of the sine wave   (b) positive peak   (c) negative peak
   (d) none of these .... ( )

18-3. THE CAPACITOR THEOREM

Analyzing *RC* switching circuits means finding the transient currents and voltages in the circuit. This is easier after you learn the answers to

Why can't capacitor voltage change instantaneously?
Why does a capacitor appear open when fully charged?

Figure 18-7. Voltage step.

Capacitor voltage at the switching instant

Chapter 11 mentioned that the voltage across a capacitor cannot change instantaneously. This is easy to prove as follows. The changing current in a capacitor is

$$i = C \frac{dv}{dt}$$

Assume the voltage across a capacitor does change instantaneously, as shown in Fig. 18-7. Then the rate of voltage change is infinite, because the voltage changes from V_1 to V_2 in zero time. Because of this,

$$i = C \frac{dv}{dt} = C \times \infty = \infty$$

But infinite current is impossible in any real circuit; therefore, the voltage across a capacitor *cannot* suddenly step as shown in Fig. 18-7.

To put it another way, the voltage across any capacitor in a switching circuit has the same value just after the switching instant as before. If there are 10 V across a capacitor just before a switch is thrown, there will be 10 V across the capacitor just after the switch is thrown. Or if 25 V are across a capacitor at $t = 0^-$, there will be 25 V across the capacitor at $t = 0^+$.

Long after switching instant

Given enough time, the capacitors in any RC switching circuit will be fully charged. When this happens, the charging current equals zero. This is equivalent to saying the capacitor appears *open* long after the switching instant.

Mathematically, when a capacitor is fully charged, its dv/dt equals zero. Since

$$i = C \frac{dv}{dt}$$

the charging current equals zero and the capacitor appears open.

Capacitor theorem

Let's summarize the two ideas just discussed:

1. At $t = 0^+$ each capacitor voltage in a switching circuit has exactly the same value as it had at $t = 0^-$.
2. Long after $t = 0^+$ each capacitor appears open.

These two ideas are essential in analyzing RC switching circuits. From now on, we refer to them as the *capacitor theorem*.

EXAMPLE 18-7.

The capacitor of Fig. 18-8a is uncharged. If the switch closes at $t = 0$, what is the initial voltage across the capacitor? The initial current? The final voltage? The final current?

SOLUTION.

Just before the switch closes, the voltage across the capacitor is zero; therefore, just after the switch closes, the voltage is still zero, which means the capacitor acts like a short at $t = 0^+$, as shown in Fig. 18-8b. Because of this, the initial current is

$$i_i = \frac{V}{R} = \frac{20}{1000} = 20 \text{ mA}$$

Long after the switch closes, the capacitor appears open, as shown in Fig. 18-8c. Because of this, the final voltage across the capacitor is

$$v_f = 20 \text{ V}$$

and the final current is

$$i_f = 0$$

Figure 18-8. *Calculating initial and final values.*

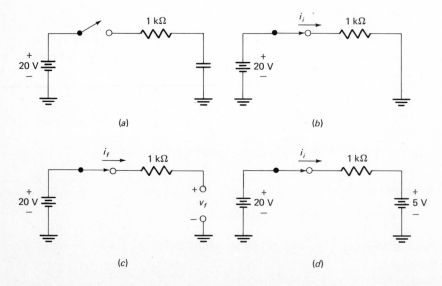

(a)   (b)

(c)   (d)

EXAMPLE 18-8.
Suppose the capacitor of Fig. 18-8a has a voltage of 5 V before the switch closes. What is the voltage across the capacitor at $t = 0^+$. The initial current? The final voltage across the capacitor? The final current?

SOLUTION.
At $t = 0^-$ the capacitor voltage is 5 V; therefore, at $t = 0^+$ the capacitor voltage is 5 V. You can visualize the capacitor as a 5-V battery at $t = 0^+$ (see Fig. 18-8d). In this case, the initial capacitor voltage is

$$v_i = 5 \text{ V}$$

and the initial capacitor current is

$$i_i = \frac{V}{R} = \frac{20 - 5}{1000} = 15 \text{ mA}$$

Given enough time, the capacitor eventually appears open, as shown in Fig. 18-8c. So the final voltage and final current are

$$v_f = 20 \text{ V}$$

$$i_f = 0$$

EXAMPLE 18-9.
The capacitor of Fig. 18-9a is uncharged before the switch is closed. What is the initial current? The final current?

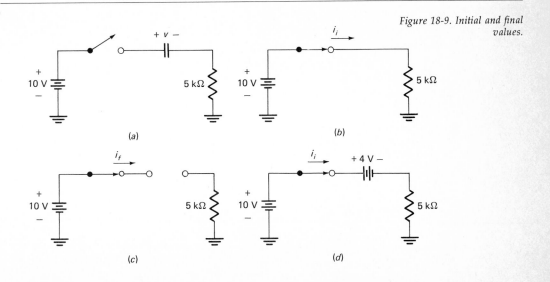

Figure 18-9. Initial and final values.

(a)

(b)

(c)

(d)

SOLUTION.

At $t = 0^+$ the capacitor appears like a short (Fig. 18-9b). For this reason, the current equals

$$i_i = \frac{V}{R} = \frac{10}{5000} = 2 \text{ mA}$$

Long after the switching instant, the capacitor appears open (Fig. 18-9c). So the final current is

$$i_f = 0$$

EXAMPLE 18-10.

If the capacitor of Fig. 18-9a has 4 V across it, what is the initial current? The final current?

SOLUTION.

At $t = 0^+$ you can visualize the capacitor as a 4-V battery (Fig. 18-9d). In this case,

$$i_i = \frac{V}{R} = \frac{10 - 4}{5000} = 1.2 \text{ mA}$$

After a long time, the capacitor appears open and

$$i_f = 0$$

Test 18-3

1. The voltage across a capacitor cannot change instantaneously because
   (*a*) current can't change instantaneously   (*b*) flux can't be zero
   (*c*) current can't be infinite   (*d*) C is constant   .... (   )
2. A capacitor eventually looks open because
   (*a*) current can't change instantaneously   (*b*) *dv/dt* can't be infinite
   (*c*) *dv/dt* goes to zero   (*d*) voltage can't change instantaneously   .... (   )
3. Which belongs least?
   (*a*) $t = 0^-$ (*b*) same voltage across capacitor   (*c*) $t = 0^+$
   (*d*) capacitor voltage changes   .... (   )
4. A capacitor that's uncharged at $t = 0^-$ looks like
   (*a*) an open at $t = 0^+$ (*b*) a short at $t = 0^+$
   (*c*) a short long after the switching instant   (*d*) a battery at $t = 0^+$ .... (   )

18-4. *RC* SWITCHING PROTOTYPE

Chapter 11 discussed the *RC* switching prototype of Fig. 18-10a for a special case of an initially uncharged capacitor. In general, the capacitor may have a voltage across it

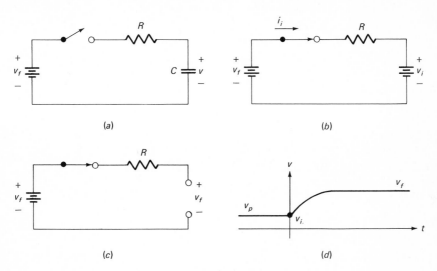

Figure 18-10. RC switching circuit.
(a) Before $t = 0$. (b) At $t = 0^+$. (c)
Long after $t = 0$. (d) Exponential
wave.

before the switching instant. If so, you can visualize the capacitor as a battery at $t = 0^+$ (Fig. 18-10b). The initial voltage across the capacitor is

$$v_i = v_p$$

where v_p is the voltage across the capacitor at $t = 0^-$.

Long after the switching instant, the capacitor looks like an open (Fig. 18-10c); therefore, the final voltage across it equals the source voltage.

Either experimentally or mathematically it can be shown that the capacitor voltage will be an exponential wave starting at v_i and finishing at v_f (Fig. 18-10d). The time constant for this waveform is still given by

$$\tau = RC$$

As before, the waveform is

63 percent finished after one time constant
99 percent finished after five time constants

EXAMPLE 18-11.
Suppose the capacitor of Fig. 18-11a has 5 V across it before the switch is thrown. What does the transient voltage across it look like after the switch is thrown?

Figure 18-11. Examples of exponential waves.

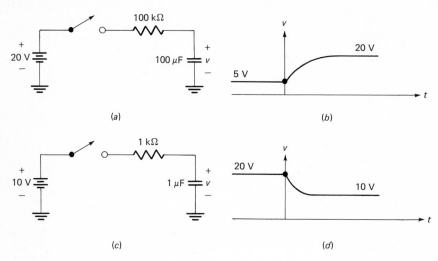

(a) (b)

(c) (d)

SOLUTION.

At $t = 0^+$ the initial voltage across the capacitor is

$$v_i = v_p = 5 \text{ V}$$

So the transient voltage across the capacitor is from 5 V to 20 V, as shown in Fig. 18-11b.

The time constant is

$$\tau = RC = 100(10^3) \times 100(10^{-6}) = 10 \text{ s}$$

Therefore, the transient is 99 percent finished after 50 s.

EXAMPLE 18-12.

The capacitor of Fig. 18-11c has 20 V across it before the switch is closed. What does the waveform of voltage across the capacitor look like after the switch is closed?

SOLUTION.

Here the initial voltage is

$$v_i = v_p = 20 \text{ V}$$

After the switch is closed, the capacitor will discharge back through the source because its voltage is greater than the source voltage. Eventually, it reaches a final voltage of 10 V.

Figure 18-11d shows the transient voltage. It starts at 20 V and decays to 10 V. The time constant is

$$\tau = RC = 10^3 \times 10^{-6} = 1 \text{ ms}$$

So the transient is 99 percent finished after 5 ms.

Test 18-4

1. In an RC switching prototype, the transient voltage across the capacitor starts at
    (a) v_i (b) v_f (c) source voltage .... ( )
2. Which of these does not belong?
    (a) final capacitor voltage   (b) source voltage   (c) one time constant
    (d) approximately five time constants .... ( )
3. The time constant does not depend on
    (a) resistance   (b) source voltage   (c) capacitance   (d) RC .... ( )
4. If the initial voltage across the capacitor of an RC switching prototype is greater
    than the source voltage, the transient voltage across the capacitor is
    (a) a decaying exponential   (b) an increasing exponential
    (c) finished after one time constant   (d) a negative step .... ( )

18-5. THE RC SWITCHING THEOREM

The solution of an RC switching prototype applies to more complicated switching circuits. The key things to learn in this section are

> What is a single time-constant circuit?
> How is its time constant defined?
> What is the RC switching theorem?

Single time constant circuit

Whenever a circuit can be reduced to a single loop, it's called a *single time-constant circuit*. For instance, Thevenin's theorem can reduce Fig. 18-12a to 18-12b. So the original circuit (Fig. 18-12a) is classified as a single time-constant circuit, because it is reducible to a single loop (Fig. 18-12b).

As another example, the 4-μF capacitors of Fig. 18-12c can be combined into a 2-μF capacitor. After Thevenizing the circuit facing this 2-μF capacitor, Fig. 18-12d results. The reduced circuit has a single loop; therefore, Fig. 18-12c is a single time-constant circuit.

One more example: in Fig. 18-12e the 2-kΩ resistors combine into a single 1-kΩ resistor. Also, the circuit left of the capacitor can be Thevenized. The resulting circuit is shown in Fig. 18-12f. Since this reduced circuit has a single loop, the original circuit is a single time-constant circuit.

Figure 18-12. Single time-constant circuits.

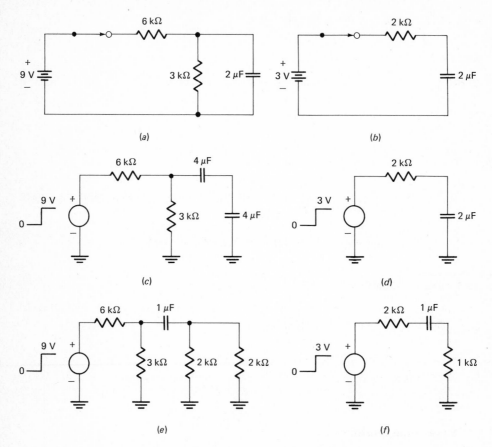

(a)

(b)

(c)

(d)

(e)

(f)

Time constant

A single time-constant circuit can be reduced to a single loop. The loop resistance is

$$R = R_1 + R_2 + R_3 + \cdots \qquad (18\text{-}5)$$

where R_1, R_2, R_3, etc., are the resistances in the loop. The loop capacitance is

$$C = \frac{1}{1/C_1 + 1/C_2 + 1/C_3 + \cdots}$$

where C_1, C_2, C_3, etc., are the capacitances in the loop. The time constant of the original

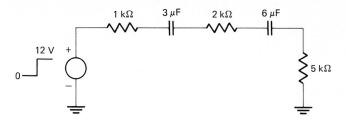

Figure 18-13. One-loop RC circuit.

circuit is defined as

$$\tau = RC \tag{18-7}$$

where R is the loop resistance and C is the loop capacitance.

As an example, suppose a complicated circuit has been reduced to Fig. 18-13. Then the loop resistance is

$$R = R_1 + R_2 + R_3 = 1\ k\Omega + 2\ k\Omega + 5\ k\Omega = 8\ k\Omega$$

and the loop capacitance is

$$C = \frac{C_1 C_2}{C_1 + C_2} = \frac{3\ \mu F \times 6\ \mu F}{3\ \mu F + 6\ \mu F} = 2\ \mu F$$

The time constant of Fig. 18-13 is

$$\tau = RC = 8(10^3) \times 2(10^{-6}) = 16\ ms$$

Figure 18-13 is a reduced form of some original circuit; so the original circuit is a single time-constant circuit with a time constant of 16 ms.

The theorem

Most of the switching circuits in everyday electronics are single time-constant circuits. Either experimentally or mathematically, we can prove the *RC switching theorem* of single time-constant circuits:

1. After $t = 0^+$ every voltage in the circuit is an exponential wave starting at v_i and finishing at v_f.
2. After $t = 0^+$ every current in the circuit is an exponential wave starting at i_i and finishing at i_f.
3. All exponential waves have the same time constant. You get this time constant by reducing the original circuit to a single loop and multiplying loop resistance by loop capacitance.

This powerful theorem is the key to analyzing almost all the *RC* switching circuits encountered in pulse and digital electronics. The following examples show how to apply it.

EXAMPLE 18-13.
What is the waveform of node voltage v in Fig. 18-14a if the switch opens at $t = 0$?

SOLUTION.
Before the switch opens, v is zero. Just after the switch opens,

$$v_i = v_p = 0$$

The time constant equals

$$\tau = RC = 10^3 \times 10^{-6} = 1 \text{ ms}$$

Figure 18-14. *Examples of increasing exponentials.*

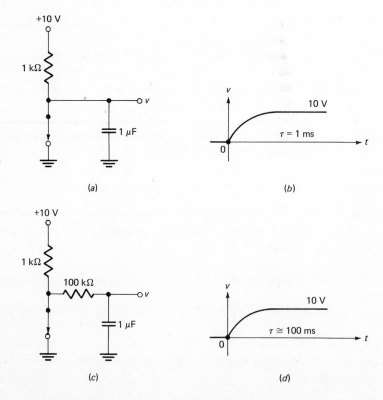

(a)

(b)

(c)

(d)

Given enough time, the capacitor fully charges and the final voltage is

$$v_f = 10 \text{ V}$$

Figure 18-14b shows the transient voltage, an exponential starting at 0 V and finishing at 10 V. After 1 ms, the waveform reaches the 63 percent point. After 5 ms it's at the 99 percent point, and is approximately finished.

EXAMPLE 18-14.
The switch of Fig. 18-14c has been closed for a long time so that the capacitor is uncharged. What is the waveform of v if the switch opens at $t = 0$?

SOLUTION.
At $t = 0^+$ the capacitor appears shorted, so

$$v_i = 0$$

The time constant is

$$\tau = RC = 101(10^3) \times 10^{-6} = 101 \text{ ms} \cong 100 \text{ ms}$$

Given enough time, the capacitor voltage reaches a final value of

$$v_f = 10 \text{ V}$$

Figure 18-14d shows the waveform, an exponential form 0 V to 10 V with a time constant of approximately 100 ms. After 500 ms, the transient is essentially over.

EXAMPLE 18-15.
The capacitor of Fig. 18-15a is uncharged at $t = 0^-$. What is the waveform of v?

SOLUTION.
At $t = 0^-$ there's no voltage anywhere in the circuit; therefore, the voltage across the 20-kΩ resistor is

$$v_p = 0$$

Just after the step at $t = 0^+$ the circuit looks like Fig. 18-15b, because the capacitor appears shorted. Voltage-divider action results in an output voltage of

$$v_i = \frac{R_2}{R} V = \frac{20}{30} 30 = 20 \text{ V}$$

Long after the switching instant, the capacitor appears open (Fig. 18-15c); therefore, the output voltage is

$$v_f = 0$$

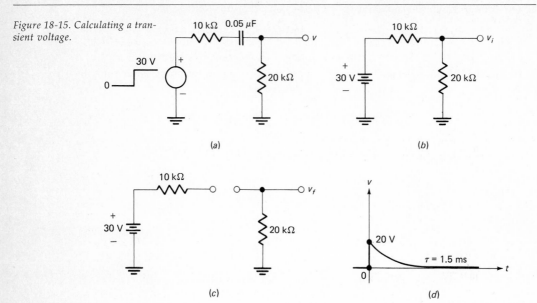

Figure 18-15. Calculating a transient voltage.

The time constant of Fig. 18-15a is

$$\tau = RC = 30(10^3) \times 0.05(10^{-6}) = 1.5 \text{ ms}$$

Figure 18-15d shows the waveform of v. Notice it steps from 0 V to 20 V at $t = 0$. Then it decays exponentially to 0 V. After 1.5 ms the voltage has dropped to 37 percent of the initial voltage; after 7.5 ms the voltage is down to 1 percent of the initial voltage.

EXAMPLE 18-16.
Before the current step in Fig. 18-16a, the capacitor is fully charged. What does the waveform of v look like?

SOLUTION.
At $t = 0^-$ the source appears open. The capacitor is fully charged so that the voltage across it equals 20 V. At $t = 0^+$ the voltage across the capacitor is still 20 V as shown in the equivalent circuit of Fig. 18-16b. Because of this,

$$v_i = v_p = 20 \text{ V}$$

Apply Thevenin's theorem left of the battery in Fig. 18-16b and you get the new circuit shown in Fig. 18-16c. In this single-loop circuit it's clear that

$$\tau = RC = 10^3 \times 100(10^{-12}) = 100 \text{ ns}$$

(Read this as 100 nanoseconds.) Figure 18-16c also makes it clear that the final voltage

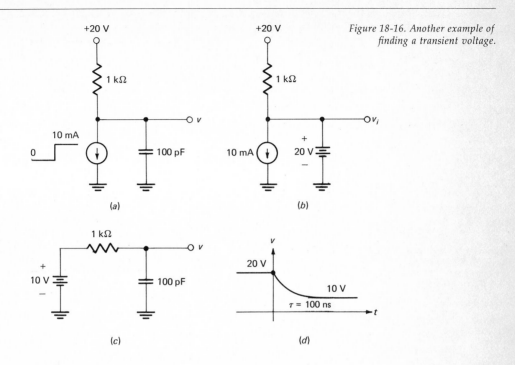

Figure 18-16. *Another example of finding a transient voltage.*

across the capacitor is

$$v_f = 10 \text{ V}$$

Figure 18-16*d* shows the transient voltage, an exponential from 20 V to 10 V with a time constant of 100 ns.

Test 18-5

1. A single time-constant circuit
    (*a*) has only one loop   (*b*) has only one resistor
    (*c*) can be reduced to a single loop   (*d*) has only one capacitor .... ( )
2. What do all exponential waves in a single time-constant circuit have in common?
    (*a*) peak value   (*b*) initial value   (*c*) final value   (*d*) time constant   ( )
3. In a single time-constant circuit each exponential voltage wave starts at
    (*a*) v_p (*b*) v_i (*c*) v_f (*d*) τ .... ( )
4. Period is as basic for sine waves as time constant is for
    (*a*) sine waves   (*b*) exponential waves   (*c*) sawtooth waves
    (*d*) square waves .... ( )
5. The following words may be rearranged to get a sentence about the exponential

waves in a single time-constant circuit: WAVES SAME ALL CONSTANT TIME HAVE THE. The sentence is

(*a*) true   (*b*) false   .... ( )

18-6.   THE SWITCHING FORMULA

In pulse and digital electronics, transistors or other devices sometimes terminate the exponential wave before it finishes. In the following discussion, get the answers to

How is the unit exponential function defined?
What is the 63 percent point?
What is the switching formula?

Unit exponential function

The base of natural logarithms is a number designated *e*. This number has a value of

$$e = 2.71828 \ . \ . \ .$$

followed by more digits. For most electronics work, the approximation

$$e \cong 2.72$$

is satisfactory.

In advanced mathematical analysis of single time-constant circuits, this function keeps turning up:

$$v = 1 - e^{-t/\tau} \qquad (18\text{-}8)$$

It's an easy function to work with. You substitute the value of t at any instant you're interested in, and then calculate v. For example,

When $t = 0$,

$$v = 1 - e^0 = 1 - 2.72^0$$

$$= 1 - 1 = 0$$

When $t = \tau$,

$$v = 1 - e^{-1} = 1 - 2.72^{-1} = 1 - \frac{1}{2.72}$$

$$1 - 0.367 \cong 0.63$$

When $t = 2\tau$,

$$v = 1 - e^{-2} = 1 - 2.72^{-2} = 1 - \frac{1}{2.72^2}$$

$$= 1 - 0.135 \cong 0.86$$

Continuing in this way,

When $t = 3\tau$,

$$v = 1 - e^{-3} = 1 - 0.049 \cong 0.95$$

When $t = 4\tau$,

$$v = 1 - e^{-4} = 1 - 0.018 \cong 0.98$$

When $t = 5\tau$,

$$v = 1 - e^{-5} = 1 - 0.007 \cong 0.99$$

When t approaches infinity, v approaches 1.

Table 18-1 summarizes these values. As shown, the progression is 0, 0.63, 0.86, 0.95, 0.98, and 0.99. By plotting the values in this table, the graph of Fig. 18-17 results.

TABLE 18-1.   UNIT EXPONENTIAL FUNCTION

t	$1 - e^{-t/\tau}$
0	0
τ	0.63
2τ	0.86
3τ	0.95
4τ	0.98
5τ	0.99

Notice how the wave starts at zero and finishes at unity. Because the shape is exponential and because the final value is unity, the graph is called the *unit exponential wave*.

The values of Fig. 18-17 are accurate to two digits, adequate for most work in electronics. The appendix lists more values for the unit exponential wave.

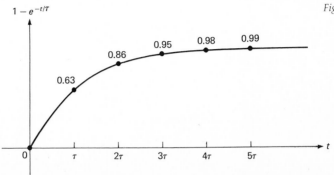

Figure 18-17. Unit exponential wave.

The 63 percent point

With sine waves the most natural amount of time to talk about is period T, because this is the duration of one cycle.

An exponential wave like Fig. 18-17 does not repeat; however, a natural amount of time to talk about is time constant τ. As shown in Fig. 18-17, after one time constant the wave has passed through 63 percent of its total change. In other words, the exponential wave is 63 percent finished after one time constant. Likewise, the wave is 86 percent finished after two time constants, 95 percent finished after three time constants, 98 percent after four time constants, and 99 percent finished after five time constants.

Because time constant τ plays a central role in the behavior of an exponential wave, it's as important in switching circuits as period T is in sine-wave circuits.

On an oscilloscope, an exponential wave like Fig. 18-17 appears so close to unity at $t = 5\tau$ that you can't notice the difference. This is why most people treat the exponential wave as finished after five time constants.

The switching formula

With advanced mathematics, it's possible to derive the formula for any exponential wave in a single time-constant circuit. This formula, referred to hereafter as the *switching formula,* says that after $t = 0$,

$$v = v_i + (v_f - v_i)(1 - e^{-t/\tau}) \tag{18-9}$$

where v = instantaneous voltage
v_i = initial voltage
v_f = final voltage
τ = time constant
t = elapsed time after switching instant

(The same formula applies to current if you replace the v's by i's.)

When graphed, Eq. (18-9) results in the exponential wave of Fig. 18-18. In other words, Eq. (18-9) is the formula for any exponential voltage in a single time-constant circuit. As shown, the exponential wave starts at v_i and finishes at v_f. To calculate the

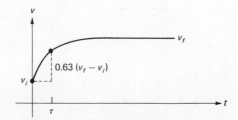

Figure 18-18. Exponential wave for any single time-constant circuit.

instantaneous voltage, substitute the values of v_i, v_f, and τ into Eq. (18-9); then, plug in whatever value of time you're interested in.

Equation (18-9) is useful in later studies of pulse and digital circuits, where you need to know the instantaneous voltage at certain points in time. In this book, however, it's enough to know the exponential wave starts at v_i and finishes at v_f. After one time constant, the wave has passed through 63 percent of the difference between v_f and v_i (see Fig. 18-18). After five time constants, the wave has passed through 99 percent of the difference between v_f and v_i.

EXAMPLE 18-17.
Write the switching formula for these conditions: $v_i = 0$, $v_f = 10$ V, and $\tau = 1$ ms. Calculate v for $t = 0$, 1 ms, 2 ms, and 3 ms.

SOLUTION.
Substitute the given values into Eq. (18-9):

$$v = v_i + (v_f - v_i)(1 - e^{-t/\tau})$$
$$v = 0 + (10 - 0)(1 - e^{-t/0.001})$$

or
$$v = 10(1 - e^{-t/0.001})$$

The quantity in the parentheses is the unit exponential function, whose values are given in Table 18-1. Therefore, you can calculate instantaneous voltages as follows:

When $t = 0$,
$$v = 10(1 - e^0) = 0$$

When $t = 1$ ms,
$$v = 10(1 - e^{-0.001/0.001})$$
$$= 10(1 - e^{-1}) = 10(0.63) = 6.3 \text{ V}$$

When $t = 2$ ms,
$$v = 10(1 - e^{-0.002/0.001})$$
$$= 10(1 - e^{-2}) = 10(0.86) = 8.6 \text{ V}$$

When $t = 3$ ms,
$$v = 10(1 - e^{-0.003/0.001})$$
$$= 10(1 - e^{-3}) = 10(0.95) = 9.5 \text{ V}$$

and so on. (If you need intermediate values like $t = 1.5$ ms, look up $1 - e^{-1.5}$ in the appendix; then multiply by 10 to get an answer of approximately 7.8 V.)

Figure 18-19a shows the exponential wave. It has an initial value of zero and a final value of 10 V. The time constant is 1 ms.

Figure 18-19. Examples of exponential waves.

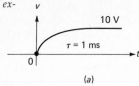

(a)

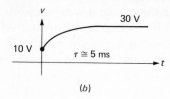

(b)

EXAMPLE 18-18.

Write the switching formula given these values: $v_i = 10$ V, $v_f = 30$ V, and $\tau = 5$ ms. What does the instantaneous voltage equal after one time constant?

SOLUTION.

Substitute the given values into Eq. (18-9):

$$v = v_i + (v_f - v_i)\,(1 - e^{-t/\tau})$$

$$v = 10 + (30 - 10)\,(1 - e^{-t/0.005})$$

or

$$v = 10 + 20(1 - e^{-t/0.005})$$

This is the switching formula for the given values. To get the instantaneous voltage after one time constant, substitute $t = 5$ ms:

$$v = 10 + 20(1 - e^{-0.005/0.005})$$
$$= 10 + 20(1 - e^{-1}) = 10 + 20(0.63)$$
$$= 22.6 \text{ V}$$

Figure 18-19b shows the exponential wave. It starts at 10 V and finishes at 30 V. The time constant is 5 ms.

Test 18-6

1. The quantity e most closely equals
    (a) 0.63   (b) 1   (c) 2.72   (d) 5 .... ( )
2. The value of the unit exponential function after one time constant is
    (a) 0   (b) 0.63   (c) 0.99   (d) 1 .... ( )
3. Period is to sine wave as time constant is to
    (a) square wave   (b) exponential wave   (c) sine wave
    (d) sawtooth wave .... ( )
4. Which does not belong?
    (a) five time constants   (b) τ (c) 63 percent finished
    (d) one time constant .... ( )
5. An exponential wave starts at 10 V and finishes at 0. After one time constant, the instantaneous voltage equals
    (a) 0   (b) 3.7 V   (c) 6.3 V   (d) 10 V .... ( )

6. If an exponential wave starts at -15 V and finishes at $+15$ V, what does the instantaneous voltage equal after one time constant?
   (*a*) -15 V   (*b*) 0   (*c*) 3.9 V   (*d*) 15 V .... (   )
7. The following words can be rearranged to form a sentence: HAS EXPONENTIAL MANY WAVE CYCLES. The sentence is
   (*a*) true   (*b*) false .... (   )

18-7. INDUCTOR CURRENT AND VOLTAGE

Resistive-inductive (*RL*) switching circuits are much less used than *RC* switching circuits, because inductors are more expensive and less convenient to work with than capacitors. Our discussion of *RL* switching circuits will be limited to a few essential ideas.

As a start, learn the answers to

What is the inductor formula?
What are the three load formulas?

Inductor voltage

Inductance is defined as

$$L = \frac{N\phi}{i}$$

which rearranges to

$$N\phi = Li$$

At $t = t_1$, this becomes

$$N\phi_1 = Li_1$$

and at $t = t_2$,

$$N\phi_2 = Li_2$$

The difference of the preceding equations is

$$N\phi_2 - N\phi_1 = Li_2 - Li_1$$

or

$$N(\phi_2 - \phi_1) = L(i_2 - i_1)$$

If t_1 and t_2 are extremely close points in time, we can use the special notation $d\phi$ and di for the flux change and the current change. The foregoing equation becomes

$$N \, d\phi = L \, di$$

Now we can divide both sides by dt to get

$$N\frac{d\phi}{dt} = L\frac{di}{dt}$$

With Faraday's law, Eq. (10-11), the foregoing equation becomes

$$v = L\frac{di}{dt} \tag{18-10}$$

This equation is useful, because it relates the voltage across the inductor to the current through it. It says induced voltage equals inductance times the rate of current change. If $L = 5$ H and $di/dt = 2$ A/s,

$$v = 5(2) = 10 \text{ V}$$

If at a later point in time the current changes at a slower rate like $di/dt = 0.1$ A/s, the induced voltage is only

$$v = 5(0.1) = 0.5 \text{ V}$$

Three load formulas

Resistance, capacitance, and inductance are the three basic properties of loads. Now we have all three load formulas:

$$v = Ri$$

$$i = C\frac{dv}{dt}$$

$$v = L\frac{di}{dt}$$

These three formulas, along with Kirchhoff's laws, are the backbone of circuit analysis. All derived formulas and theorems about currents and voltages in linear circuits ultimately come from these three formulas and Kirchhoff's laws.

Test 18-7

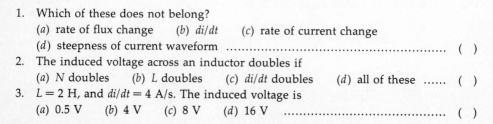

1. Which of these does not belong?
   (*a*) rate of flux change   (*b*) di/dt (*c*) rate of current change
   (*d*) steepness of current waveform .... ( )
2. The induced voltage across an inductor doubles if
   (*a*) N doubles   (*b*) L doubles   (*c*) di/dt doubles   (*d*) all of these ...... ( )
3. $L = 2$ H, and $di/dt = 4$ A/s. The induced voltage is
   (*a*) 0.5 V   (*b*) 4 V   (*c*) 8 V   (*d*) 16 V .... ( )

18-8. THE RL SWITCHING THEOREM

The main things to look for in this section are

> *What is a single time-constant RL circuit?*
> *How is its time constant defined?*
> *What is the RL switching theorem?*

Single time-constant circuit

Often, a complicated RL switching circuit can be reduced to a single loop like Fig. 18-20. If so, the original circuit is classified as a single time-constant circuit. The time constant is defined as the loop inductance divided by the loop resistance. In symbols,

$$\tau = \frac{L}{R} \tag{18-11}$$

where $L = L_1 + L_2 + L_3 + \cdots$
$R = R_1 + R_2 + R_3 + \cdots$

The theorem

Either experimentally or with advanced mathematics, it's possible to prove the RL *switching theorem* for a single time-constant circuit:

1. After $t = 0^+$ every voltage is an exponential wave starting at v_i and finishing at v_f.
2. After $t = 0^+$ every current is an exponential wave starting at i_i and finishing at i_f.
3. All exponential waves have the same constant. Its value is given by Eq. (18-11).

In applying this theorem, remember that the current through an inductor cannot change instantaneously; this was proved in Chap. 11. Because of this, inductors appear like current sources at $t = 0^+$; the value of current is the same as at $t = 0^-$.

Figure 18-20. Single time-constant RL circuit.

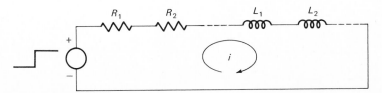

Given enough time, the magnetic field around the inductor becomes constant; neglecting winding resistance, an inductor appears like a short a long time after the switching instant.

The switching formula, Eq. (18-9), applies to a single time-constant RL switching circuit. The same formula can be used for current if the v's are replaced by i's.

EXAMPLE 18-19.

Before the switching instant, there's no current through the inductor of Fig. 18-21a. Sketch the waveform of voltage v.

SOLUTION.

First, Thevenize the circuit left of the inductor to get the single-loop circuit shown in Fig. 18-21b. In this circuit,

$$\tau = \frac{L}{R} = \frac{12}{3000} = 4 \text{ ms}$$

Figure 18-21.

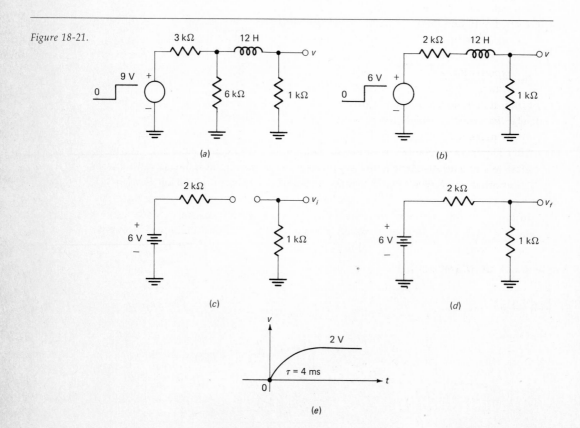

At $t = 0^+$ the inductor appears open (Fig. 18-21c), because there was no current through it at $t = 0^-$. Therefore, the initial voltage across the 1-kΩ resistor is

$$v_i = 0$$

Given enough time, the magnetic field around the inductor becomes constant and the induced voltage decreases to zero. Neglecting resistance, the inductor appears like a short (Fig. 18-21d). Voltage-divider action results in a final voltage across the 1-kΩ resistor of

$$v_f = 2 \text{ V}$$

Figure 18-21e shows the waveform of v, an exponential starting at 0 V and finishing at 2 V.

Test 18-8

1. A single time-constant RL circuit
   (a) has only one loop   (b) has only one resistor
   (c) can be reduced to a single loop   (d) has only one inductor .... ( )
2. Which does not apply to a single time-constant RL switching circuit?
   (a) exponential waves   (b) same time constant
   (c) inductor appears shorted long after switching instant
   (d) inductor always open at $t = 0^+$ .... ( )
3. After five time constants the transients in an RL switching circuit are
   (a) 63 percent finished   (b) 95 percent finished   (c) 99 percent finished
   (d) 100 percent finished .... ( )
4. If there's no current through an inductor at $t = 0^-$, it appears how at $t = 0^+$?
   (a) shorted   (b) open   (c) resistive .... ( )

SUMMARY OF FORMULAS

DEFINING

$$i = \frac{dq}{dt} \tag{18-3}$$

$$\tau = RC \tag{18-7}$$

$$\tau = \frac{L}{R} \tag{18-11}$$

DERIVED

$$i = C \frac{dv}{dt} \tag{18-4}$$

$$v = v_i + (v_f - v_i)\,(1 - e^{-t/\tau}) \tag{18-9}$$

$$v = L \frac{di}{dt} \tag{18-10}$$

Problems

18-1.   What is the waveform of node voltage v as in Fig. 18-22a?

18-2.   In Fig. 18-22b, what does node voltage v look like?

18-3.   Draw the waveform for node voltage v in Fig. 18-22c.

18-4.   Sketch the waveform of v in Fig. 18-22d.

10-5.   The capacitor of Fig. 18-23a is uncharged before the switch closes. Sketch the waveform of the voltage across the capacitor after the switch closes.

18-6.   If the capacitor of Fig. 18-23a has 10 V across it before $t = 0$, what does the charging waveform of voltage look like across the capacitor?

18-7.   The capacitor of Fig. 18-23b has 10 V across it. Describe the waveform of capacitor voltage after the switch closes. What is the time constant?

18-8.   Before the switching instant in Fig. 18-23c, the capacitor has no charge. What does the voltage waveform across the 2-kΩ resistor look like?

18-9.   The capacitor of Fig. 18-23d is uncharged before the switching instant. Draw the waveform of node voltage v.

18-10.   The capacitor is uncharged in Fig. 18-24a before the switch opens. What is the initial charging current into the capacitor at $t = 0^+$? The time constant? The final voltage across the capacitor?

18-11.   In Fig. 18-24a the capacitor is uncharged before the switch opens. Sketch the waveform of the current through the 3-kΩ resistor.

18-12.   At $t = 0$, the source of Fig. 18-24b steps from 0 to 16 V. If the capacitor is initially uncharged, what does the waveform of v look like?

18-13.   The capacitor of Fig. 18-24c is charged to 10 V before the switching instant. Sketch the voltage waveform across the capacitor.

18-14.   Before $t = 0$ in Fig. 18-24d the capacitor looks open. Draw the waveform of node voltage v.

18-15.   Figure 18-25a shows a transistor switching circuit. If Fig. 18-25b is the equivalent circuit for 18-25a, what does the waveform of node voltage v look like if the initial capacitor voltage is 10 V?

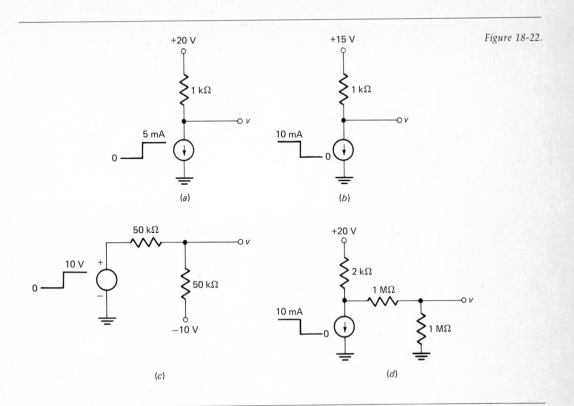

Figure 18-22.

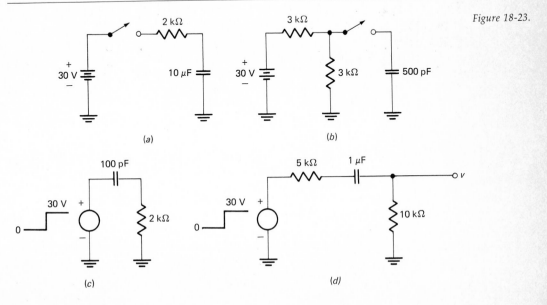

Figure 18-23.

Figure 18-24.

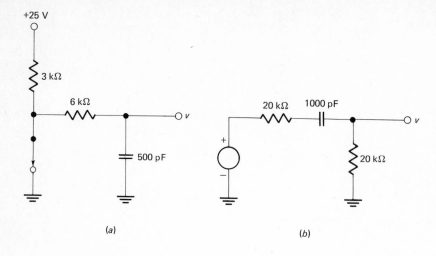

(a)   (b)

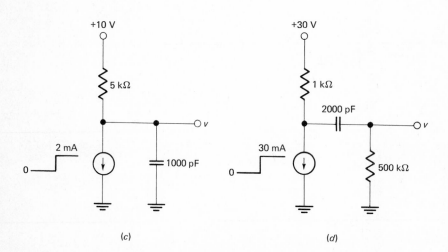

(c)   (d)

18-16.   Before the step in Fig. 18-26a there's no current through the inductor. Draw the waveform of current i.

18-17.   There's no current through the inductor of Fig. 18-26b before the switching instant. Sketch the waveform of current i. The waveform of voltage v across the inductor.

18-18.   The current through the inductor of Fig. 18-27a is constant before the switch opens. What is the time constant after the switch opens? The waveform of current i through the inductor? The waveform of node voltage v?

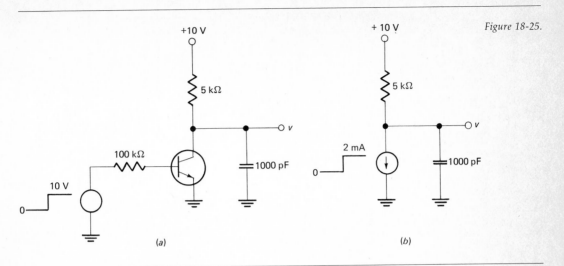

Figure 18-25.

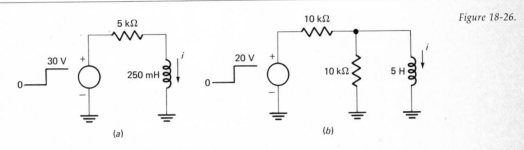

Figure 18-26.

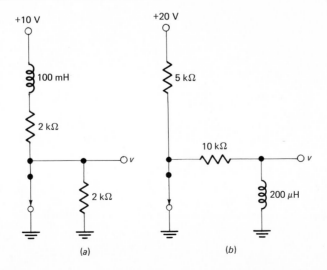

Figure 18-27.

18-19. There's no current through the inductor of Fig. 18-27*b* before the switch opens. Draw the waveform of current *i* through the inductor and the waveform of node voltage *v*.

ANSWERS TO TESTS

18-1. *a, b, d, d, c*
18-2. *d, b, b, a, a*
18-3. *c, c, d, b*
18-4. *a, c, b, a*
18-5. *c, d, b, b, a*
18-6. *c, b, b, a, b, c, b*
18-7. *a, d, c*
18-8. *c, d, c, b*

Appendix

UNIT SINE FUNCTION

θ	$\sin \theta$	θ	$\sin \theta$	θ	$\sin \theta$
0°	.0000	31°	.5150	61°	.8746
1°	.0175	32°	.5299	62°	.8829
2°	.0349	33°	.5446	63°	.8910
3°	.0523	34°	.5592	64°	.8988
4°	.0698				
		35°	.5736	65°	.9063
5°	.0872	36°	.5878	66°	.9135
6°	.1045	37°	.6018	67°	.9205
7°	.1219	38°	.6157	68°	.9272
8°	.1392	39°	.6293	69°	.9336
9°	.1564				
		40°	.6428	70°	.9397
10°	.1736	41°	.6561	71°	.9455
11°	.1908	42°	.6691	72°	.9511
12°	.2079	43°	.6820	73°	.9563
13°	.2250	44°	.6947	74°	.9613
14°	.2419				
		45°	.7071	75°	.9659
15°	.2588	46°	.7193	76°	.9703
16°	.2756	47°	.7314	77°	.9744
17°	.2924	48°	.7431	78°	.9781
18°	.3090	49°	.7547	79°	.9816
19°	.3256				
		50°	.7660	80°	.9848
20°	.3420	51°	.7771	81°	.9877
21°	.3584	52°	.7880	82°	.9903
22°	.3746	53°	.7986	83°	.9925
23°	.3907	54°	.8090	84°	.9945
24°	.4067				
		55°	.8192	85°	.9962
25°	.4226	56°	.8290	86°	.9976
26°	.4384	57°	.8387	87°	.9986
27°	.4540	58°	.8480	88°	.9994
28°	.4695	59°	.8572	89°	.9998
29°	.4848				
		60°	.8660	90°	1.0000
30°	.5000				

UNIT EXPONENTIAL FUNCTION

t/τ	$1 - e^{-t/\tau}$	t/τ	$1 - e^{-t/\tau}$
0	0.000	1.8	0.835
0.1	0.095	1.9	0.850
0.2	0.181	2.0	0.865
0.3	0.259	2.1	0.877
0.4	0.330	2.2	0.889
0.5	0.393	2.3	0.900
0.6	0.451	2.4	0.909
0.7	0.503	2.5	0.918
0.8	0.551	2.6	0.926
0.9	0.593	2.7	0.933
1.0	0.632	2.8	0.939
1.1	0.667	2.9	0.945
1.2	0.699	3.0	0.950
1.3	0.727	3.5	0.970
1.4	0.753	4.0	0.982
1.5	0.777	4.5	0.989
1.6	0.798	5.0	0.993
1.7	0.817	10.0	1.000

Answers To Odd-Numbered Problems

Chapter 1

1-1. Force, temperature, time, length, time. **1-3.** 2. **1-5.** *a.* 5000 μm. *b.* 8,000,000 ms. *c.* 5000 μs. *d.* 25(10^{-12}) s. **1-7.** *a.* 70 mV. *b.* 4000 μV. *c.* 0.02 mV. *d.* 5 V. **1-9.** *a.* 6,000,000 μA. *b.* 30 mA. *c.* 80 A. *d.* 30 nA. **1-11.** $f = W/d$, $d = W/f$. **1-13.** $P = RI^2$, $P = V^2/R$. **1-15.** 12 N. **1-17.** Outward.

Chapter 2

2-1. 10 A. **2-3.** 0.5 C. **2-5.** 3 J. **2-7.** 6 J, 3 J. **2-9.** 7 V. **2-11.** 15 J, 300 J. **2-13.** 250 Ω. **2-15.** 2.3 Ω. **2-17.** 300 kΩ. **2-19.** 40 mA. **2-21.** 20 Ω, 400 Ω. **2-23.** 35 V. **2-25.** 10 V, 20 V. **2-27.** 30 V, 60 V. **2-29.** 20 V, 40 V. **2-31.** 0.004 Ω. **2-33.** 1.89(10^{-5}) Ω, 4.72(10^{-6}) Ω. **2-35.** 825 Ω. **2-37.** 0.00201 Ω, 0.082 Ω. **2-39.** 4925 Ω, 5050 Ω. **2-41.** 1820°C.

Chapter 3

3-1. 0.00402 Ω, 4.02 mV. **3-3.** 7 A. **3-5.** 5 mA down. **3-7.** $I_1 = I_5 = 8$ mA, $I_2 = I_3 = I_4 = 7$ mA, $I_6 = 10$ mA. **3-9.** *a.* 19 V. *b.* 11 V. **3-11.** *a.* 12 V. *b.* 8 V. *c.* 5 V. **3-13.** $V_1 = 3$ V, $V_2 = 9$ V, $V = 18$ V. **3-15.** 2.5 mA, 5 V, 7.5 V, 2.5 V, 15 V. **3-17.** 43 kΩ. **3-19.** 1 kΩ. **3-21.** 1.34 Ω. **3-23.** 6 mA, 2 mA, 12 mA. **3-25.** 20.8 A; 10 A, 5 A, 3.33 A, 2.5 A. **3-27.** *a.* 3 to 6 kΩ. *b.* 1.5 to 3 kΩ. *c.* 50 to 100 kΩ. **3-29.** 5 MΩ. **3-31.** 4 kΩ. **3-33.** 75 kΩ. **3-35.** 480 Ω. **3-37.** 2 Ω. **3-39.** 40 kΩ. **3-41.** 3 kΩ.

Chapter 4

4-1. 6 V, 0, 3 V. **4-3.** 100, 30, 10, 3, 1, 0 V. **4-5.** 10 V, 40 kΩ. **4-7.** 5 V, 5 kΩ. **4-9.** 6 V. **4-11.** 3.33 V, 10 V. **4-13.** 0.75 Ω. **4-15.** 1 kΩ. **4-17.** 5 kΩ, 15 kΩ. **4-19.** 500 Ω. **4-21.** 32, 32, 0 V. **4-23.** 0.286 mA. **4-25.** 0.107 V. **4-27.** *a.* 2.18 mA. *b.* 0. *c.* −0.96 mA. **4-29.** Each battery in a separate circuit. **4-31.** 6 mA. **4-33.** 4 mA. **4-35.** 25 V. **4-37.** 1.25 mA, 15.6 V.

Chapter 5

5-1. 0.0025 mhos. **5-3.** 0.124 mhos. **5-5.** 1000 μmhos, 66.7 μmhos. **5-7.** 0, 50 V, 25 V. **5-9.** 40 V, 20 kΩ. **5-11.** 12 V, 9 kΩ. **5-13.** 11 V, 1 MΩ. **5-15.** 9 mA in parallel with 2 kΩ. **5-17.** 4 mA in parallel with 9 kΩ. **5-19.** ½ hp. **5-21.** 50 W. **5-23.** 100 W. **5-25.** 5 mW, 15 mW, 20 mW. **5-27.** 16 mW, 21.3 mW, 10.7 mW, 48 mW, 48 mW. **5-29.** 1 W, 9 W. **5-31.** 6 kΩ, 2.67 mW. **5-33.** 0, 10 mW. **5-35.** 41.7 mW. **5-37.** *a.* 0. *b.* 0.444 W. *c.* 0.5 W. *d.* 0.444 W. *e.* 0.

Chapter 6

6-1. $I_1 = 2$ A, $I_2 = 1$ A. **6-3.** *a.* Yes. *b.* No. *c.* Yes. *d.* No. **6-5.** 8, 5. **6-7.** 3, 4. **6-9.** $700I_1 - 500I_2 = 20$, $500I_1 - 800I_2 = 10$. **6-11.** 3, 3. **6-13.** 46.5 μA, 5 mA, 5 V. **6-15.** *a.* 3. *b.* 4. *c.* 6. *d.* 5. **6-17.** 2, 0. **6-19.** 5, 15, 3, 12, 5, 7, and 2 V. **6-21.** 6.67 kΩ. **6-23.** $I_1 = 36.4$ mA, $I_2 = 27.3$ mA, $I_3 = 9.09$ mA; $V_1 = 7.28$ V, $V_2 = V_3 = 2.73$ V. **6-25.** $I_1 = I_4 = 1.61$ mA, $I_2 = 0.968$ mA, $I_3 = 0.645$ mA; $V_A = 8.39$ V, $V_B = 6.44$ V.

Chapter 7

7-1. 100 μA, 95.2 μA. **7-3.** Approximately 5 mA, 50 mA, and 500 mA. **7-5.** 2.35 mA, 2.65 mA. **7-7.** 1 mA, 0.938 mA. **7-9.** 150 kΩ, 500 kΩ, approximately 1.5 MΩ. **7-11.** 50 kΩ/V. **7-13.** 15 kΩ, 50 kΩ, 150 kΩ, 500 kΩ. **7-15.** 6 V, 5.66 V. **7-17.** 7.5 V, 6.98 V. **7-19.** 150 kΩ, 450 kΩ. **7-21.** *a.* 15 Ω. *b.* 450 Ω. *c.* 13.5 kΩ.

Chapter 8

8-1. 4 V, 8 V. **8-3.** *a.* $v = 4t$. *b.* $v = 3000t$. **8-5.** 27 V, 3 V, 6 V, 13.5 V. **8-7.** 0.5 mA at t = 1 ms, 1 mA at t = 2 ms, 1.5 mA at t = 3 ms. **8-9.** Triangular current with a peak of 2 mA; triangular voltage with a peak of 8 V. **8-11.** Square-wave current with a peak of 0.667 mA; square-wave voltage with a peak of 1.33 V. **8-13.** A: triangular wave with a peak of 7 V. B: triangular wave with a peak of 0.7 V. C: triangular wave with a peak of 0.07 V. **8-15.** Triangular current with a peak of 5 μA; triangular current with a peak of 250 μA; triangular voltage with a peak of 2.5 V. **8-17.** Triangular wave; sine wave. **8-19.** Sawtooth current with a peak of 1 mA; average = 0.5 mA. **8-21.** 0, 24 V, 0 V, −24 V, 0. **8-23.** 25 Hz, 0.04 s. **8-25.** 4 μs, 250 kHz. **8-27.** 0.0145 μs or 14.5 ns.

Chapter 9

9-1. 5000 μF. **9-3.** −40 pC, 5 pF. **9-5.** 5 V. **9-7.** 22.1 pF. **9-9.** 750 pF, 1875 pF. **9-11.** 2600 pF, 3000 pF, 4000 pF, 4200 pF, 7500 pF, 10,000 pF. **9-13.** 5 to 1, 15 pF. **9-15.** 600 pF. **9-17.** 75 pF. **9-19.** 10 V, 10,000 pC. **9-21.** 3.64 V; 728 pC, 1092 pC, 2184 pC. **9-23.** 1.8 pF, 5.4 pF, 0.6 pF.

Chapter 10

10-1. 2.5 H. **10-3.** 0.0005 Wb. **10-5.** 806 μH. **10-7.** 1 H, 1 H. **10-9.** 0.5, 0.024 Wb. **10-11.** 0.05 H, 0.2. **10-13.** 60 μH. **10-15.** 0.671. **10-17.** 20 μH, 260 μH. **10-19.** 0.4 Wb/s. **10-21.** 5000 Wb/s, 1 Wb/s, −500 Wb/s. **10-23.** 4 μWb. **10-25.** Sine wave with a peak of 40 V, sine wave with a peak of 20 mA, sine wave with a peak of 5 mA, primary resistance of 32 kΩ.

Chapter 11

11-1. 5 mA. **11-3.** 20 ms, nothing. **11-5.** 750 ns. **11-7.** 20 ms, exponential from 0 to 30 V. **11-9.** 1000 μJ. **11-11.** 300 μs, 200 nJ. **11-13.** 20 ms, 100 ms. **11-15.** 10 V, 2 μs, 10 μs. **11-17.** 50 μs, 16.7 μs. **11-19.** 3.2 μJ. **11-21.** 0.4 ms, 0.2 ms. **11-23.** 0.36 μJ.

Chapter 12

12-1. 100 mV, 35.4 mV. **12-3.** 42.4 mV, 84.8 mV. **12-5.** 14.1 mA, 40 mA. **12-7.** 84.8 mV, 240 mV. **12-9.** 10.2 mA. **12-11.** 1.26 Ω, 796 kHz. **12-13.** Approximately 200 μA and 40 μA. **12-15.** 159 Hz. **12-17.** 5 Ω. **12-19.** 954 Ω. **12-21.** 220 kΩ. **12-23.** 796 Hz. **12-25.** 21.2 kHz. **12-27.** 2.83 μA. **12-29.** 10 kΩ, 13.3 H. **12-31.** 24.8 μA. **12-33.** 10 kΩ, 5 μA. **12-35.** 796 kHz. **12-37.** Approximately 1 μH. **12-39.** 7.12 kHz. **12-41.** 1.3 MHz, 10 V.

Chapter 13

13-1. a. 5.39. b. 6.71. c. 5.66. d. 5.39. **13-3.** a. 1.41 $\underline{/45°}$. b. 3.61 $\underline{/124°}$. c. 3.16 $\underline{/198°}$. d. 4.47 $\underline{/333°}$. **13-5.** a. 2 + j3.46. b. −4.49 + j5.36. c. −1.71 − j4.7. d. 6.93 − j4. **13-7.** The tip of each vector falls on each rectangular number given in Prob. 13-1. **13-9.** The tip of each vector falls on each polar number given in Prob. 13-5. **13-11.** a. −1 + j11. b. 1 − j6. c. −7 + j2. d. 8 + j6. **13-13.** a. 15$\underline{/75°}$. b. 28$\underline{/210°}$. c. 24$\underline{/29°}$. d. 30$\underline{/−65°}$. **13-15.** a. −18. b. −j7. c. j9. d. −10. **13-17.** a. 1.08 − j0.615. b. 0.1 + j0.7. c. 0.346 − j0.731. d. 2.15 − j0.231. **13-19.** a. Peak is 250 mA and phase is 60°; i = 0.25 sin (θ + 60°). b. Peak is 50 μA and phase is −50°; i = 0.00005 sin (θ − 50°). c. Peak is 10 V and phase is 75°; v = 10 sin (θ + 75°). d. Peak is 25 mV and phase is −35°; v = 0.025 sin (θ − 35°).

Chapter 14

14-1. 14.1 mA $\underline{/35°}$. **14-3.** *a.* 35.4 V $\underline{/35°}$. *b.* 1.41 V $\underline{/-120°}$. *c.* 0.283 V $\underline{/25°}$. *d.* 3.54 mV $\underline{/-45°}$. **14-5.** 17.7 mV $\underline{/30°}$. **14-7.** 500 Ω or 500 Ω $\underline{/0°}$. **14-9.** 2.5 kΩ $\underline{/90°}$ or $j2.5$ kΩ. **14-11.** 500 kΩ $\underline{/-90°}$ or $-j500$ kΩ. **14-13.** 79.6 Ω $\underline{/-90°}$ or $-j79.6$ Ω. **14-15.** 4 Ω $\underline{/75°}$. **14-17.** 20 V $\underline{/-15°}$. **14-19.** 25 μA. **14-21.** 24.7 V. **14-23.** Approximately 25 V and 5°. **14-25.** 0.0795 μF (0.1 μF is the next higher standard size). **14-27.** 3.54 V. **14-29.** $-17°$. **14-31.** 1.69 V. **14-33.** 0.111 mA. **14-35.** 0.314 V.

Chapter 15

15-1. 0.0002 mhos. **15-3.** *a.* 0.000125 mhos. *b.* 0.00159 mhos. *c.* 0.0000318 mhos. **15-5.** *a.* 0.001 mhos. *b.* $-j0.05$ mhos. *c.* $j0.0126$ mhos. *d.* $-j0.00142$ mhos. **15-7.** $0.001 + j0.00126$. **15-9.** 500 Ω, 1000 Ω. **15-11.** 80 Ω, 0.008. **15-13.** 20. **15-15.** 162 Ω in parallel with 65.6 mH. **15-17.** Approximately 200 μH in parallel with 126 kΩ. **15-19.** 250 kΩ in parallel with 500 μH. **15-21.** Approximately 27.4 Ω in series with 148 μH. **15-23.** 0.4 Ω. **15-25.** 159 kHz, 15.9 kHz. **15-27.** 21.2 μW. **15-29.** 0.125 W. **15-31.** *a.* 1.08 W. *b.* 1.25 W. *c.* 0.884 W.

Chapter 16

16-1. 1.55 mA. **16-3.** 200 mA, 2.67 mA. **16-5.** 223 V. **16-7.** 851 kHz. **16-9.** 7.96 MHz, 15.9 MHz, 8 V. **16-11.** 12.6, 25.1, 62.8, 126. **16-13.** 2.07 V. **16-15.** 58.6. **16-17.** 3.55 V, 100. **16-19.** 11.3 kHz. **16-21.** 80.

Chapter 17

17-1. 6.67 rad. **17-3.** 115°, 720°, 573°, 6300°. **17-5.** $160,000\pi$ rad/s. **17-7.** *a.* $40,000\pi$ rad/s. *b.* $2(10^6)\pi$ rad/s. *c.* $100,000\pi$ rad/s. **17-9.** *a.* $v = 14.1 \sin 4000\pi t$. *b.* $v = 0.0354 \sin (4000\pi t + 90°)$. *c.* $i = 0.00707 \sin (4000\pi t - 90°)$. *d.* $i = 0.0566 \sin (4000\pi t + 75°)$.

Chapter 18

18-1. Negative step from 20 to 15 V. **18-3.** Positive step from -5 to 0 V. **18-5.** Exponential from 0 to 30 V with a time constant of 20 ms. **18-7.** Exponential from 10 to 15 V; time constant is 0.75 μs. **18-9.** A positive step from 0 to 20 V, followed by an exponential from 20 to 0 V with a time constant of 15 ms. **18-11.** A negative step from 8.33 to 2.78 mA, followed by an exponential from 2.78 mA to 0 with a time constant of 4.5 μs. **18-13.** An exponential from 10 to 0 V with a time constant of 5 μs. **18-15.** Same answer as Prob. 18-13. **18-17.** Exponential current from 0 to 2 mA with a time constant of 1 ms. Voltage steps from 0 to 10 V, then decays exponentially from 10 to 0 V with a time constant of 1 ms. **18-19.** Exponential current from 0 to 1.33 mA with a time constant of 13.3 ns. Voltage steps from 0 to 20 V, then decays exponentially from 20 to 0 V with a time constant of 13.3 ns.

Index